CHEMISTRY RESEARCH AND APPLICATIONS

MOLECULAR AND SUPRAMOLECULAR BIOINORGANIC CHEMISTRY: APPLICATIONS IN MEDICAL AND ENVIRONMENTAL SCIENCES. VOLUME 2

CHEMISTRY RESEARCH AND APPLICATIONS

Additional books in this series can be found on Nova's website
under the Series tab.

Additional E-books in this series can be found on Nova's website
under the E-book tab.

CHEMISTRY RESEARCH AND APPLICATIONS

MOLECULAR AND SUPRAMOLECULAR BIOINORGANIC CHEMISTRY: APPLICATIONS IN MEDICAL AND ENVIRONMENTAL SCIENCES. VOLUME 2

ANA LUCIA RAMALHO MERCÉ
JUDITH FELCMAN
AND
MARIA ÁNGELES LOBO RECIO
EDITORS

Nova Science Publishers, Inc.
New York

For permission to use material from this book please contact us:
Telephone 631-231-7269; Fax 631-231-8175
Web Site: http://www.novapublishers.com

LIBRARY OF CONGRESS CATALOGING-IN-PUBLICATION DATA

Mercj, Ana Lucia Ramalho.
Molecular and supramolecular bioinorganic chemistry : applications in medical and environmental sciences / Ana Lucia Ramalho Mercj, Judith Felcman, Marma Angeles Lobo Recio.
p. cm.
Includes index.
ISBN 978-1-61728-785-5 (hardcover)
1. Bioinorganic chemistry. 2. Organometallic chemistry. 3. Supramolecular chemistry. I. Felcman, Judith. II. Recio, Marma Angeles Lobo III. Title.
QP531.M63 2010
612'.01524--dc22

2010044734

Published by Nova Science Publishers, Inc. ✝ New York

CONTENTS

Preface		**vii**
Chapter 1	Aluminium Neurotoxicosis. Should We Forget About *In Vitro* Models? *Paula P. Gonçalves*	**1**
Chapter 2	Bioinspired Synthetic Analogues for Purple Acid Phosphatases and Catalytic Promiscuity *Ademir Neves, Hernán Terenzi, Rosely A. Peralta and Patrícia Cardoso Severino*	**31**
Chapter 3	Phosphates in Biology: Some Interactions with Metal Ions and Other Biological Compounds *Judith Felcman and Ana Lucia Ramalho Mercê*	**73**
Chapter 4	Aluminium in Waters. Sources, Speciation and Removal Techniques *María Ángeles Lobo-Recio, Ana Lucia Ramalho Mercê, Maria Eliza Nagel-Hassemer and Flávio Rubens Lapolli*	**95**
Chapter 5	Advances in Metals Separations Technology from Liquid Effluents *F. J. Alguacil, M. Alonso, F. A. López, A. López-Delgado and I. Padilla*	**123**
Chapter 6	Review: Aging and Neurodegeneration – A Biochemical Approach *Luciana Dornnelas, Ana Lucia Ramalho Mercê and Judith Felcman*	**155**
Chapter 7	Radicals and Transition Metal Complexes in Aqueous Solutions *Alexandra Masarwa and Dan Meyerstein*	**187**
Index		**219**

PREFACE

We are proud to present to you the second volume of Molecular and Supramolecular Bioinorganic Chemistry: Applications in Medical Sciences. Volume 2.

This second volume intends to continue helping to fill the gap between research and practicing medical professionals as well as students in general. It will help to bring to anyone interested in the ultimate results and conclusions from research produced among interdisciplinary areas of knowledge, among other scientific informations, an understanding of the neurodegenerative diseases.

It is know today that metal ions participate in the enzymatic chemical reactions of the living bodies either acting as essential elements or being responsible for oxidative damaging processes.

The results, questions, doubts and partial conclusions of this discussion in the scientific literature up to now, are organized by all various areas of expertise contributions for this volume, in chapters.

The first chapter call us to remember to consider other aspects than *in vivo* and *in vitro* models from the scientific studies and results in understanding the role of aluminium as a neurotoxic agent as it is important to consider the deterioration of the antioxidant defence mechanism in the living beings in the presence of this metal ion, going from peroxidation of lipids till proteins, and consequently, the DNA.

In the second chapter, the syntheses and chemical behavior of some new promiscuous enzymatic compounds able to catalytically hydrolyze biomolecules such as proteins, phospholipids, ATP, DNA, RNA and model substrates of current interest, are discussed. They are important because studies on these synthetic analogues are a convenient strategy to gain insight into the structural and spectroscopic properties and the catalytic mechanism of the corresponding metalloenzymes,

The third chapter reviews another focus of the role of phosphates in the living bodies as well as their interaction with metal ions and their influence in promoting degenerative disorders.

In the fourth chapter, aluminium as an adjuvant in waste water treatment is discussed, considering the importance of the served waters of today being the clean waters of tomorrow. The traces of aluminium metal ion left by the treatments, often pose contamination problems. So, among other aspects discussed, alternative materials for remediating aluminium contaminated treated waste waters are considered.

In the fifth chapter, separation technologies for remediating metal ion contaminated water are presented, including organic extract compounds, liquid membrane technology, adsorption and ion exchange processing, where polymeric chelating ion exchange resins are discussed.

The sixth chapter reviews some basic and advanced aspects of some neurodegenerative diseases, old and new theories and results, emphasizing the discoveries on the role of metal ions in these kinds of neurological disorders.

The seventh chapter presents an understanding of electron transfer processes in biological systems as neurological diseases are known to be strongly related to oxidizing biochemical reactions.

Enjoy!

Ana Lucia Ramalho Mercê
Judith Felcman
María Ángeles Lobo Recio

In: Molecular and Supramolecular Bioinorganic Chemistry ISBN: 978-1-61728-785-5
Ed: A. L. R. Mercê, J. Felcman and M. A. L. Recio

Chapter 1

ALUMINIUM NEUROTOXICOSIS. SHOULD WE FORGET ABOUT *IN VITRO* MODELS?

Paula P. Gonçalves
Centre for Environmental and Marine Studies, Department of Biology,
University of Aveiro, Aveiro, Portugal

1. PREAMBLE

The consequences of damage to nervous system can be profound. Massive injury is well documented in the World History. Hippocrates recognized lead toxicity in mining industry (370 B.C.), and Pliny warned against inhalation of vapours from lead furnaces in Roman times! The "Minamata disease" (staggering mortality and morbidity due to mercury/methylmercury intoxication in Minamata Bay area of Japan from 1953 to 1960 and Niigata from 1964 to 1965) was a manifest illustration of the public health consequences of governmental disregard toward environmental pollution all around the world. Nowadays policy agencies and governments remain to have difficulty to deal with these issues. For instance, the "Camelford Incident" (U.K. in 1988), massive discharge of aluminium sulphate into the drinking water supplied to a large region of North Cornwall, remains to be clarified (it looks like that the most recent official report on the incident is the *Draft Lowermoor report - January 2005*).

The present scope to surpass "weaknesses" in precedent policies for managing global quality of life relies also on the right balance between different approaches to the risks of chemicals, and to the costs and benefits of their use, based on the judicious application of the precautionary principle. Increasingly, policymakers are moving towards an approach that is more supported on cooperation and incentives rather than on "command-and-control" regulations. Thereby, the current approaches to address the paradigm for chemical management include, among others, the precautionary principle, the cost-effectiveness and multiple benefit/costs approach, as well as voluntary and public right-to-know programmes. Bottlenecks in technological development towards higher performance and reduced negative impact on environment and health are often associated with economical aspects. So, economical production must require reduction of negative environmental and health impacts which is crucial for the intended transformation of society towards knowledge-based organizations, with better conformity to sustainability. Quality of life all over the world relies on durable social and economical development. Chemical management paradigms require global approaches, since they do not recognise frontiers between countries or continents. Life expectancy is raising and more elderly people exist in our societies. Nowadays features of neurodegeneration during aging are not compulsory, and are closely related to the catastrophic state of environmental quality to living organisms. Organophosphate pesticides, for instance, were reported by World Health Organisation to cause permanent cognitive and neurological damage in thousands and thousands people in developing countries. An upbringing challenge must be the achievement of life quality. The relatively fragility of the nervous system, that is, our most precious heritage, toward chemical aggression has been neglected. Specific neuronal coding, specialized subcellular systems, particular metabolic requirements and diversity of cell morphology, in addition to unusual anatomy and poor regenerative capacity, of the nervous system (central and peripheral), make it especially vulnerable to chemically induced damage. Besides, nervous system exhibits adaptive and compensatory mechanisms, as well as neurogenesis can occur in adult humans, and many kinds of injury to the nervous system are thought to be irreversible with consequences permanently loss of function. For that reason, the manifestation of a neurotoxic event during any time in the life cycle (from gestation through senescence) can occur immediately after the insult or become evident only as the nervous system matures and ages. Therefore, application of the precautionary principle is even more pertinent to address neurotoxicity paradigms. Actually, prevention of damage to the nervous system must be a main economic and/or political objective.

The conscience that slight nervous system damage represents a determinant factor for the present and future development of human societies must be diffused. Nervous system ensues safety and efficiency in the performance of daily activities, by coordinating motor function, attention, alertness, memory, communication and reasoning abilities, and this means behaviour. Slight impairment of these functions is difficult to detect in an individual and even more difficult is to identify its deficit in a population. Neurobiological functions are mostly expressed on a continuum and changes during individual development, as well as it seems like that many neurotoxicants fall on a continuum of severity, sometimes their deleterious effects could be evident after a prolonged time of exposure or even only a long period after exposure. It means that it is necessary to modelling effects of neurotoxicants within biological properties that themselves assume a spectrum of values in a population, taking into account the phenomenon of the continuum of both frequency and intensity of response related to dose. Due to this feature, neurotoxicologic risk assessment has been largely limited to the application of no-observed-effects levels and uncertain factors, so it has not generated specific risk for given magnitudes of exposure. This is why goodness-of-fit models with respect to variables such as age, sex, duration and routes of exposure, as well as in building non dichotomous risk-assessment models, which deal simultaneously with several endpoints produced by neurotoxicants, are required. From a safety standpoint, neurotoxic effects resulting from either a direct or indirect action of a chemical on the nervous system are important to define the toxicological profile of a substance. Even when neurotoxicity is secondary to some non-nervous system toxicity, neurotoxicity would typically represent one of the most sensitive endpoints. In fact, neuronal function simultaneously regulates and is regulated by the status of all organ systems in the body. The current neurotoxicity guidelines recommend that *in vitro* models should include cellular and subcellular biochemical endpoints, which address mechanism of neurotoxic action.

Critical mechanistic information on aluminium neurotoxicity is scattered across relevant scientific publications. The considerable progresses that have been made in understanding basic mechanisms involved in deleterious actions of aluminium revealed an almost numberless set of jeopardized cellular events under aluminium exposure. Nevertheless, the critical focus seems to remain in hazard identification that is the primary stage of risk assessment, given that major gaps in information require addressing.

If truth were told, the integration of neurotoxicologic screening of aluminium exposure in multidisciplinary programs of general toxicologic screening, which is mandatory to accomplish the cost-effectiveness and multiple benefit/costs approach, urges rapid development. Now please, take the following only as a suggestion to concern about aluminium toxicity and neurotoxicity in particular. For what is worth, it is merely how I was gathered by the aluminium paradigm.

2. Aluminium and Neurological Diseases

We can always argue whether, or to what extent, data from animal models are relevant to aluminium neurotoxicity in humans. Most current knowledge on aluminium neurotoxicity has been obtained from studies with small mammals and from the use of cell cultures and subcellular fractions. While this information is essential for understanding aluminium in human disease, there is sufficient lack of correspondence in the aluminium-induced effects among these species to justify cautious extrapolation to humans. At the moment Alzheimer's disease perhaps provides the most explicable of examples.

Alzheimer's disease is mainly characterized by the presence of intraneuronal neurofibrillary tangles, deposition of amyloid β peptide in neuritic plaques or senile plaques and neuronal loss in hippocampus and neocortex of the brain. The disruption of normal function of the amyloid β peptide and of the microtubule-associated protein tau is believed to lead to the accumulation of neurofibrillary tangles and neuritic plaques in a species-specific way. Rats and mice are recurrently used as animal models in aluminium neurotoxicity, which are among the species where these key processes for Alzheimer's disease pathogenesis seem not to occur. The microtubule-associated protein tau promotes assembly and stability of the microtubule network that is critical for many neuronal functions. As it can be observed in Fig. 1, in human, rat and mouse there are, respectively, at least 9, 8 and 6 isoforms of tau, which are generated by alternative mRNA splicing. The transcriptional splicing of mRNA and phosphorylation state of this primary neuronal phosphoprotein are developmentally regulated and of major importance for its physiological and pathological roles (Gong and Iqbal, 2008). The complete amino acid sequences of the tau isoforms used to infer the phylogenetic tree (Fig. 1) are presented in Fig. 2. All the 23 sequences were aligned unambiguously with gaps in well defined parts of the alignments of the amino acid chains. Besides the astonishing high degree of structural conservation during evolution amongst species, the variations in the amino acid sequence of tau proteins are thought to explain the differences among species in the propensity to develop neurofibrillary tangles. Tau function as a linker protein between axonal microtubules and the plasma membrane components is assured by the N- and C-terminal domains, respectively. The species-specific splicing in or out of exons 2 and 3 results on the presence of 0, 1 or 2 N-terminal inserts while the existence of 3 or 4 microtubule-binding domains results from the splicing in or out of exon 10 (Figs. 2 and 3). Neurofibrillary tangles are composed of paired helical filaments containing the six predominant tau isoforms in adult human brain (Tau-B, Tau-C Tau-D, Tau-E, Tau-F and Fetal-Tau), mostly in a hyperphosphorylated state. Tau function is predominantly regulated by phosphorylation. Tau proteins exhibited an enormous number of consensus sequences for several protein kinases, and, *in vivo*, it appears to be a substrate for glycogen synthase kinase 3 beta (GSK3β), cyclin-dependent kinase 5 (Cdk5), c-Jun N-terminal kinase (JNK), mitogen-activated protein kinase (p38), casein kinase-2 (CK2), calcium/calmodulin-dependent protein kinase II (CaMPKII), microtubule affinity regulating kinase (MARK), cyclic-AMP dependent kinase (PKA), protein kinase C (PKC) and tyrosine kinase (TK) (Fig. 2). Hyperphosphorylation of tau, particularly that mediated by MARK, CDKS and GSK3β decreases the ability of tau to bind microtubules, whereas tau phosphorylation at the proline-rich region and at the C-terminal tail region promotes its self-aggregation (Fig. 3). Hyperphosphorylated tau proteins aggregate into paired helical filaments that are deposited in neuronal cell bodies and in dystrophic neuritis. In addition to abnormal phosphorylation of tau, other structural features also contribute to develop neurofribrilary degeneration. The presence of a pair of cysteines in the microtubule binding repeats can form an intramolecular S-S bridge preventing the cross-link between two tau proteins. Moreover, the number of microtubule-binding domains and N-terminal inserts impacts the ability of tau proteins to form oligomeric proteins and to interact with microtubules. Briefly, isoforms with two N-terminal inserts are preferentially sequestered in aggregates and under oxidative conditions those with three microtubule-binding repeats easily dissociate from microtubules and self-aggregate. Just taking into consideration the above mentioned structural features of the rodent and human isoforms under analysis, it is possible to understand why exposure to aluminium of rodent brain has failed to cause accumulation of neurofibrillary tangles. Besides the mechanisms of developmental and species-specific differences in the regulation and distribution of tau isoforms are not well understood, experimental evidences suggest that in adult brain the rat and mouse genes produce predominantly tau with four microtubule-binding repeats, while the human gene produces tau with both three and four microtubule-binding repeats. Curiously, rodent isoforms exhibit an additional phosphorylation site by PKC at the beginning of the proline-rich region (Fig. 3). It has been shown that tau phosphorylation by PKC impairs its further phosphorylation by GSK3β, which, in turn, seems to contribute to the pathological hyperphosphorylation of tau in Alzheimer's disease (Isagawa et al, 2000). All tau isoforms with four microtubule-binding repeats also contain a putative cyclin recognition site (Fig. 3), suggesting that phosphorylation by Cdks could be facilitated in these cases. Interestingly, almost all putative sites for phosphorylation by GSK3β are included in consensus sequences for Cdk5. Moreover, isoforms with four microtubule-binding repeats also have a pair of cysteins in this region. As far as we know, tau proteins with one or two N-terminal inserts and three microtubule-binding repeats are not present in adult rodent brain, as illustrated in Fig. 3 (top panel), where the different tau isoforms are displayed in concentric circles according to these two structural features.

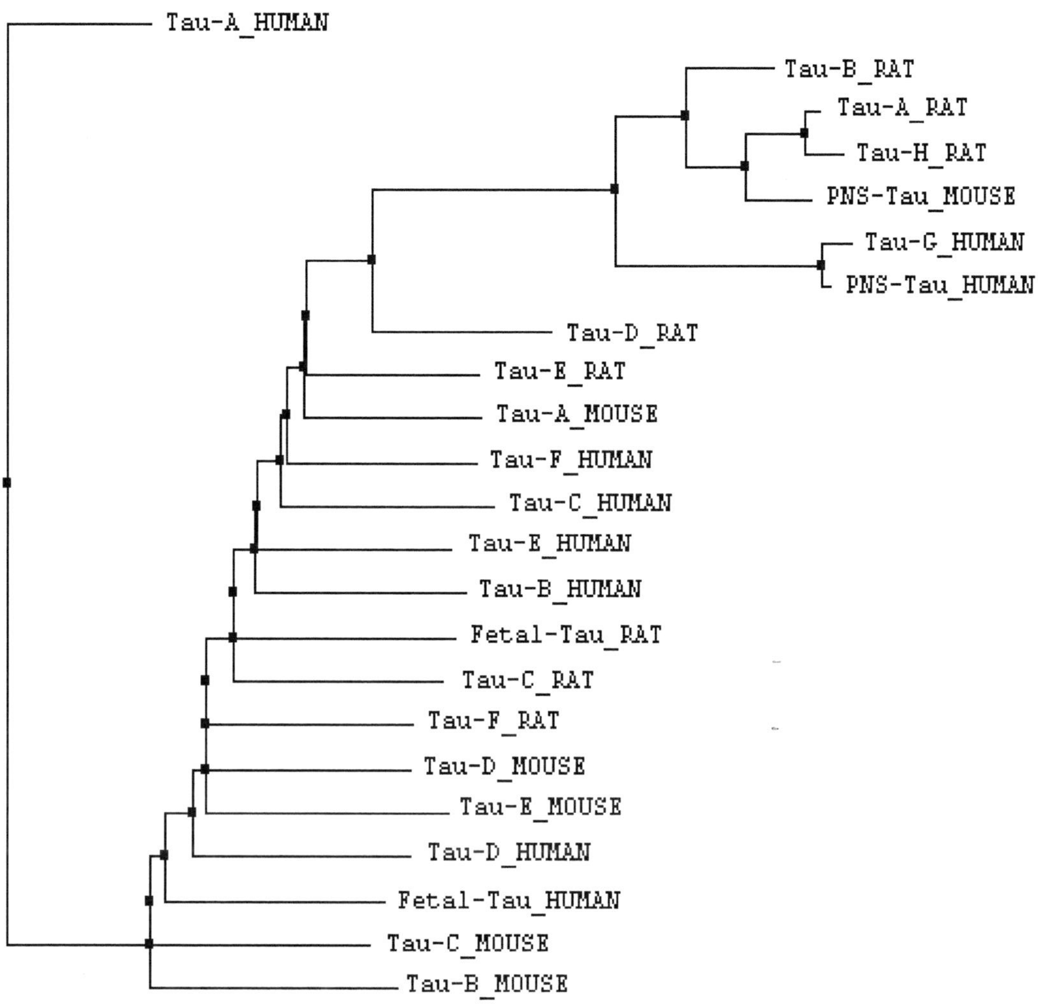

Figure 1. Phylogenetic relationships among the human, rat and mouse isoforms of the microtubule-associated protein tau.The phylogenetic tree was reconstructed by the neighbour joining as implemented in BLOSSUM62 using ClustalW2. Horizontal branch lengths are proportional to genetic distance. The accession numbers (AC) of the sequences are the following: P10636-1 (PNS-Tau_HUMAN); P10636-2 (Fetal-Tau_HUMAN); P10636-3 (Tau-A_HUMAN); P10636-4 (Tau-B_HUMAN); P10636-5 (Tau-C_HUMAN); P10636-6 (Tau-D_HUMAN); P10636-7 (Tau-E_HUMAN); P10636-8 (Tau-F_HUMAN); P10636-9 (Tau-G_HUMAN); P19332-1 (Tau-A_RAT); P19332-2 (Tau-B_RAT); P19332-3 (Tau-C_RAT); P19332-4 (Tau-D_RAT); P19332-5 (Tau-E_RAT); P19332-6 (Tau-F_RAT); P19332-7 (Feral-Tau_RAT); P19332-8 (Tau-H_RAT); P10637-1 (PNS-Tau_MOUSE); P10637-2 (Tau-A_MOUSE); P10637-3 (Tau-B_MOUSE); P10637-4 (Tau-C_MOUSE); P10637-5 (Tau-D_MOUSE); P10637-6 (Tau-E_MOUSE).

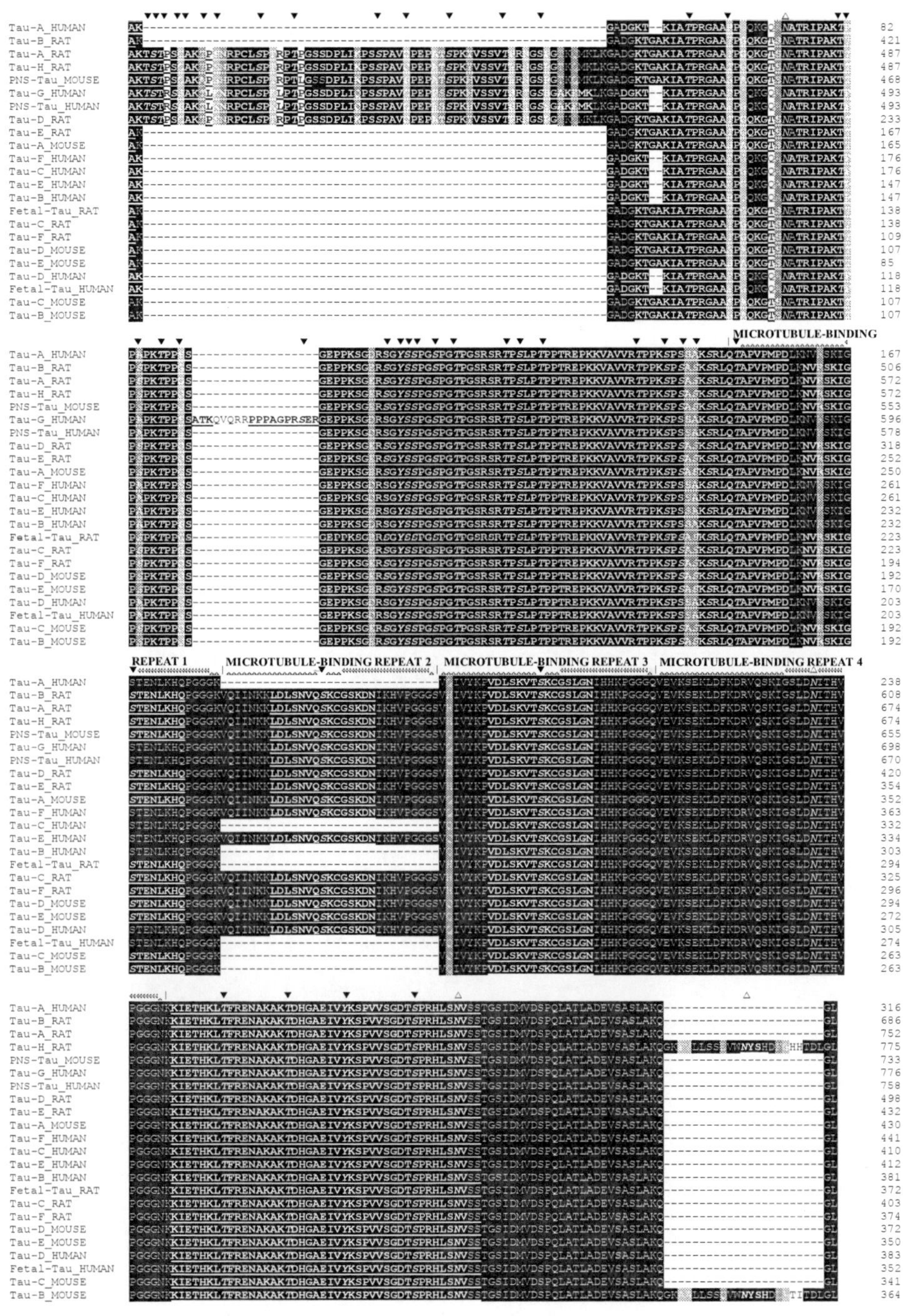

Figure 2. Amino acid sequence comparison of the human, rat and mouse isoforms of the microtubule-associated protein tau. The alignment of amino acid sequences was performed using ClustalW2. Identical amino acid residues are in black boxes and dark and light grey boxes represent conserved and semi-conserved substitutions, respectively. Italic caps represent phosphorylation (▼) and N-glycosylation (△) sites included in patterns with a high probability of occurrence that are underlined/bold (specific to the consensus sequences for GSK3β, Cdk5, JNK, p38, CK2, CaMPKII, MARK, PKA, PKC and TK) or underlined (specific to the consensus sequences for N-glycosylation). The conserved domains (<) that form the tandem repeats of amino acid residues which are implicated in

tubulin-binding are also shown. The accession numbers (AC) of the sequences are the following: P10636-1 (PNS-Tau_HUMAN); P10636-2 (Fetal-Tau_HUMAN); P10636-3 (Tau-A_HUMAN); P10636-4 (Tau-B_HUMAN); P10636-5 (Tau-C_HUMAN); P10636-6 (Tau-D_HUMAN); P10636-7 (Tau-E_HUMAN); P10636-8 (Tau-F_HUMAN); P10636-9 (Tau-G_HUMAN); P19332-1 (Tau-A_RAT); P19332-2 (Tau-B_RAT); P19332-3 (Tau-C_RAT); P19332-4 (Tau-D_RAT); P19332-5 (Tau-E_RAT); P19332-6 (Tau-F_RAT); P19332-7 (Feral-Tau_RAT); P19332-8 (Tau-H_RAT); P10637-1 (PNS-Tau_MOUSE); P10637-2 (Tau-A_MOUSE); P10637-3 (Tau-B_MOUSE); P10637-4 (Tau-C_MOUSE); P10637-5 (Tau-D_MOUSE); P10637-6 (Tau-E_MOUSE).

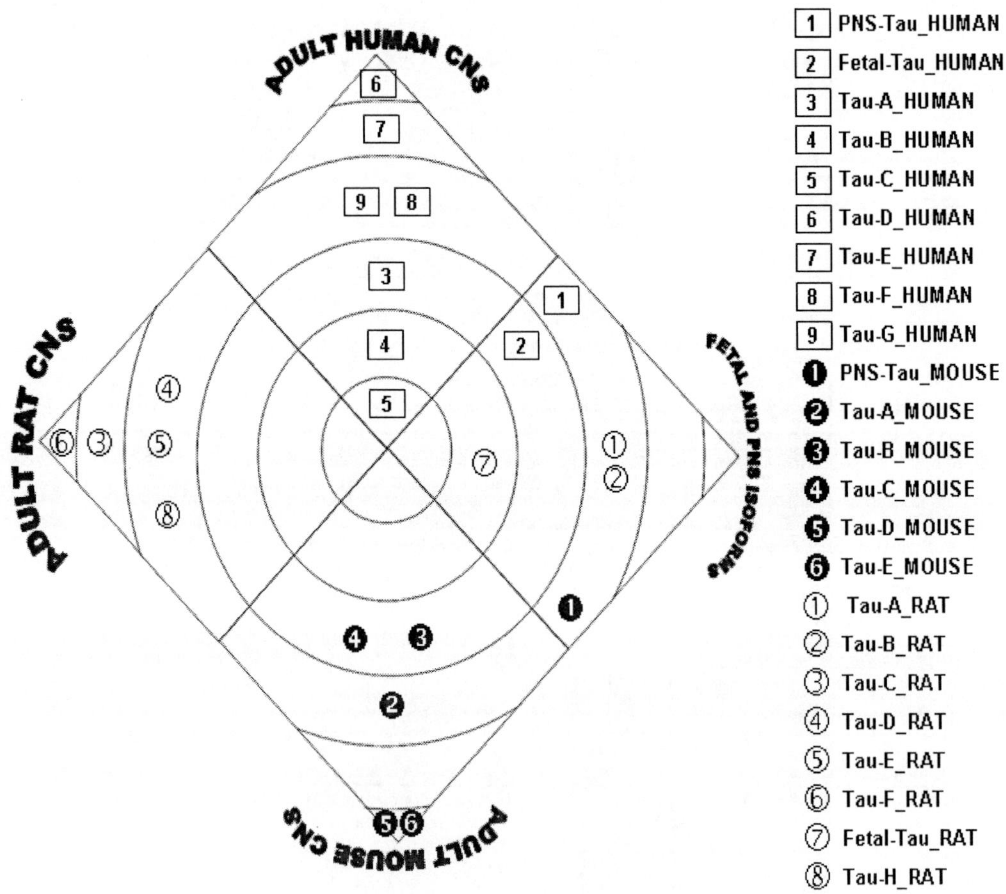

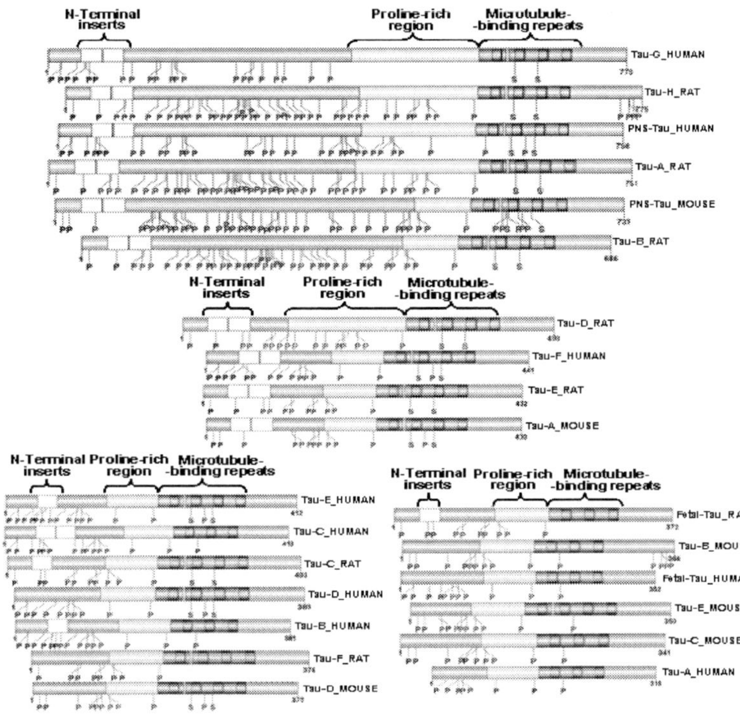

Figure 3. Structural features of the microtubule-associated protein tau that have been implicated in the assembling of the core of tau-paired helical filaments. The conserved domains t hat form N-terminal inserts (□), proline-rich regions (□), microtubule-binding repeats (■), tubulin-binding domain signatures (■) and cyclin recognition sites (□) are shown. P and S represent putative additional phosphorylation sites and disulfide bonds, respectively. Explanation in the text.

A major difference is also observed between man and rat and mouse with respect to the tendency to the deposition of β amyloid protein in senile plaques. Amyloid fibrils are formed by normally soluble proteins, which assemble to form insoluble fibers that are resistant to degradation. The bulk of amyloid plaques in Alzheimer's diseased brain specimens are Aβ peptides which are formed by enzymatic cleavage of the amyloid precursor protein. Briefly, the amyloid precursor protein is cleaved by β-secretase, shedding the amyloid β precursor proteins, a large ectodomain, and leaving the carboxy-terminal fragment in the membrane, which is cleaved by γ-secretase to produce Aβ. This protease produces mainly a peptide with 40 residues (Aβ40) but also the 42-residue peptide (Aβ42) and other minor species. Aβ42, which is the most prone to aggregation, is the major amyloid Aβ peptide found in cerebral plaques that characterize the Alzheimer's diseased brain (Hook et al, 2008). The normal physiological functions of the amyloid β precursor protein and its products of cleavage appear to be essential for protective and differentiation events during development and adulthood, including cell adhesion and migration (Wolfe and Guénette, 2007). Amino acid composition and sequence analysis of the predominant isoform of the amyloid precursor protein in human, guinea pig, mouse and rat neuronal tissue are presented in Fig. 4. The comparative analysis reveals that the four primary structures differ only on 24 amino acid residues, resulting from conserved or semi-conserved substitutions. Rat and mouse do not develop lesions identical to those observed in human brain with Alzheimer's disease, while amyloid Aβ self-aggregation into oligomers and formation of plaques occur in guinea pig aged brain. The comparative analysis of amino acid sequences from rat and mouse with human and guinea pig ones shows that nine of the amino acid substitutions occur only in the rodent isoforms. Remarkable is the fact that three out of these nine amino acid replacements are in the region that corresponds to the sequence of Aβ42 and Aβ40 (Fig. 4). Arginine 601, tyrosine 606 and histidine 609 are replaced by glycine, phenylalanine and arginine, respectively, in the rat and mouse sequences. The fibrillar assemblies are composed predominantly of β-sheet structure in a characteristic cross-β conformation. The two easily oxidized tyrosine and histidine residues are located precisely in a β-pleated sheet-forming region. The impact of the amino acid substitutions in rodent isoforms on the generation of oligomeric Aβ species has been experimentally addressed (Atwood et al, 2004). In contrast to human amyloid Aβ peptides, rat Aβ42 and Aβ40 neither form oligomeric species nor metal ion induced modifications were detected in these rat Aβ peptides. Actually, it is believed that the generation of ditytosine cross-links of Aβ peptides could form SDS-resistant oligomerization of human Aβ under oxidative conditions. Aluminium can account for exacerbating oxidative stress and other earliest known changes that occur in human brain with Alzheimer's disease, causing memory deterioration and behavioural changes associated with dementia in aged rats but not the classic neuropathological features of Alzheimer's disease (Walton, 2007). As a matter of fact, the expression of Alzheimer's tau as a transgene in the rat brain induced complete tau cascade of neuropathology characteristic of human Alzheimer's disease neurofibrillary degeneration (Kovacech et al, 2009).

The point is that we do not need animal models or *in vitro* assays to speculate about the deleterious effects of aluminium on human brain. We need them just to understand how core mechanisms and structures produce the insult. The society development has undertaken the demonstration of aluminium neurotoxicity in humans.

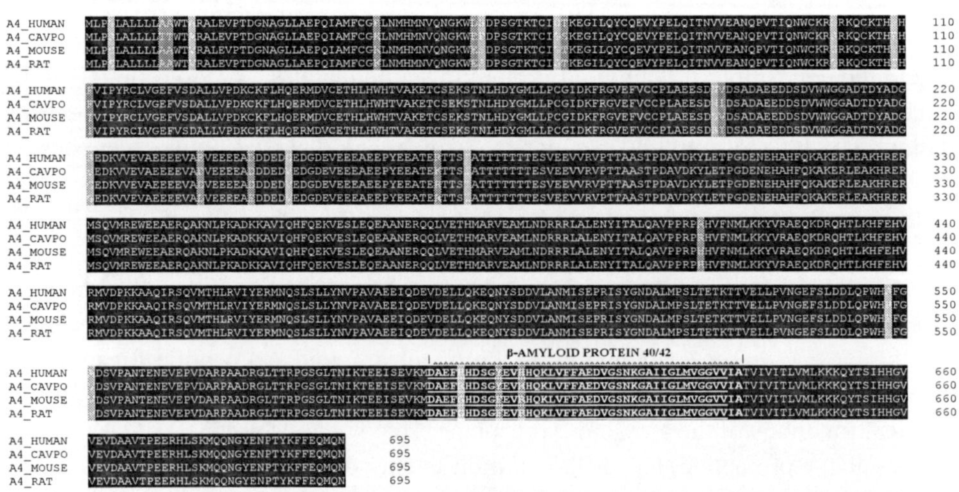

Figure 4.Amino acid sequence comparison of the human, guinea pig, mouse and rat amyloid β A4 protein. The alignment of amino acid sequences was performed using ClustalW2. Identical amino acid residues are in black boxes and dark and light grey boxes represent conserved and semi-conserved substitutions, respectively. The amino acid residues that form the chains of β-amyloid proteins 42 and 40 (^) are in bold and bold/ underlined caps, respectively. The accession numbers (AC) of the sequences are the following: P05067-4 (A4_HUMAN); Q60495-2 (A4_CAVPO); P12023-2 (A4_MOUSE) and P08592-2 (A4_RAT).

Starting from emerging concerns for aluminium safety, the Gulf War called attention again to the aluminium use in the production of vaccines. Aluminium salts are used to promote a stronger immune response to the antigen. The adjuvant effect is a benefit and vaccination is one of the most important prevention strategies. Do we have a justification for changing the current vaccination practices, like its schedule? Young, male colony CD-1 mice have been used to examine whether aluminium hydroxide doses equivalent to those given to US military service personnel as adjuvants in vaccines might contribute to neuronal deficits associated with Gulf War illness [Petrik et al, 2007]. Significant motor and cognitive deficits were shown in aluminium treated group when compared with the controls, which were concomitant to significant increase in the apoptotic effector, activated caspase-3, in lumbar spinal cord and primary motor cortex accompanied by significant motor neuron loss and increased numbers of astrocytes in the lumbar spinal cord. Outstanding is the fact that the receipt of multiple vaccines simultaneously during deployment has been linked to later ill health, including increased rates of amyotrophic lateral sclerosis, chronic fatigue and idiopathic pain among Gulf War veterans [Iversen et al, 2007]. Inquisitively, cognitive dysfunction, chronic fatigue and diffuse muscle pain are manifestations of macrophagic myofasciitis, a reaction to intramuscular injection of aluminium-containing vaccines first reported in France in 1998 [Authier et al, 2001]. Now a trade dispute is threatening up to motor delay and macrophagic myofasciitis in the pediatric population [Gruis et al, 2006; Lach and Cupler, 2008]. The list of case reports on putative aluminium-induced injury to human brain is so vast that I apologize for referring to so few studies.

Over 30-year of experience with the application of aluminium containing glass polyalkenoate cements in dentistry and orthopaedics surgery advises to consider other materials. During this period, the presence of aluminium in the glass phase of these cements has been hypothesized to elicit further degenerative brain diseases. In the particular case of a patient who died 6 weeks later because of septic complications after reconstructive otoneurosurgery, performed with a 2-component glass isomer cement consisting of a powder of calcium aluminium fluorosilicate glass and an aqueous solution of polycarboxylic acid, aluminium encephalopathy was confirmed [Reusche et al, 2001]. Direct aluminium access to brain parenchyma via a cerebrospinal fluid leakage caused disturbance of conciousness, subacute coma and grand mal seizures. In 1942, Kopeloff et al demonstrated that the application of discs containing hydrous oxides of aluminium to the cerebral cortex overlying the motor area induces acute, intermittent and recurrent convulsive seizures in monkeys [Kopeloff et al, 1942]. Since then the alumina cream technique became one of the most commonly employed experimental models for focal epilepsy, meaning that it is an effective and reproducible way to produce an effect. Interesting is to note that these studies suggested that aluminium creates focal epileptogenic lesions in experimental animals which differ on location, extension, frequency and latency period of the spontaneous seizure onset in a species-specific manner.

For the time being, the most documented type of aluminium neurological diseases with progressive and frequently fatal course is dialysis encephalopathy. For those who remember the beginnings of the second half of the twentieth century it is impossible to forget the impact of the industrial production of dialyzers as one of the advances in medical technology that improved the well being of many people on the planet. Finally, benefits of centuries of research have been offset. More and more renal patients have got access to hemodialysis or peritoneal dialysis in-center or at home. At almost the exact same time Alfrey et al [Alfrey et al, 1976] described dialysis encephalopathy as aluminium intoxication in patients undergoing long-term dialysis. Until the end of the last century, when the incidence of this fatal disease started to decline, there has been much speculation about its aetiology. However since the beginning and on present evidence, the most likely cause is accumulation of aluminium in the brain. Curiously, we are used to distinguish aluminium encephalopathy, sporadic and epidemic forms of dialysis encephalopathy and dialysis dementia. From a mechanicist point of view this is confusing. We are just emphasising different aspects of the intoxication of the human brain by aluminium. In this context, "epidemic" means intoxication by aluminium of a large percentage of patients in the same center or undergoing home dialysis in the same area occurring simultaneously. Unfortunately, such epidemic accidents have not been rare in the past and continue to happen all over the world. They are most related to engine failures that ultimately result on high levels of aluminium in dialysates. As a matter of fact, there is strong association between the incidence of epidemic dialysis encephalopathy and aluminium concentration in the brain and in the serum of who dies during these accidents and the high aluminium concentration in dialysates and often, in supplied water. So, two features deserve attention namely, overload of aluminium is lethal for humans and survey on aluminium content in supplied water from untreated sources and from the ones treated with aluminium compounds, should be mandatory. The above mentioned "Camelford Incident" [Exley and Esiri, 2006] is just one example of unknown consequences of aluminium overload for public health. When considering the sporadic form of dialysis encephalopathy, the chronic aspect of in aluminium neurotoxicity and its interindividual variability are highlighted. Conversely, the limit between encephalopathy and dementia, which is important in medical terms, appears to draw up the boundaries of different stage of aluminium encephalopathy or neurointoxication by aluminium. A gradual and progressive dysfunction of the Central Nervous System over the course of human exposure to aluminium seems to occur, which dependency on compound formulation, dosage, route, concentration and division over time, can be hypothesized in terms of most at risk populations. Briefly, aluminium neurotoxicosis in humans seems to be a progressive and fatal neurological disease, with a characteristic set of signs.

Among the first ones are loss of recent memory, impaired concentration and fatigue, as well as dysarthria, dysphasia, stammering and stuttering. Free radical generation, impairment of cerebrospinal fluid transport systems and damage of the blood-brain barrier are often suggested as factors involved in the pathogenesis of dialysis encephalopathy. The effects of aluminium on these processes have been addressed in experimental models.

Exacerbation of oxidative stress is among the most accepted hypotheses for the initiation of aluminium toxic insult to the brain. At a first glance it is not obvious, given that aluminium has no redox capacity in biological systems. Nevertheless, aluminium acts as a pro-oxidant agent and the brain is a particularly vulnerable environment for cell injury by oxidant agents. The occurrence of oxidative metabolism in neuronal tissues at high rates is associated with high levels of the redox-active metal iron in the brain. Actually, iron accumulation in the brain is a recurrent observation in patients suffering from various neurodegenerative disorders [Zecca et al, 2004]. Oxygen consumption is elevated when compared with other organs and the imminence of injury by oxygen-derived free radicals is accentuated by the relative low activities of protective antioxidant enzymes (superoxide dismutase, catalase and glutathione peroxidase) and the high ratio of the oxidizable polyunsaturated fatty acids enriched membrane surface area compared to cytoplasm. While the precise mechanism underlying the multiple actions of aluminium as a pro-oxidant agent remains to be demonstrated experimentally [Exley, 2004], immeasurable studies conducted with *in vivo* and *in vitro* models clearly showed that aluminium is able to enhance iron- and non-iron-supported oxidation, to deteriorate the antioxidant defence system, to facilitate the initiation and propagation of lipid peroxidation, causing oxidative damage to lipids, proteins and DNA [Verstraeten et al, 2008].

Injection of aluminium silicate into the *cisterna magna* in a wide range of species has been used to induce experimental hydrocephalus, which is usually caused by impaired flow or absorption of cerebrospinal fluid flow (da Silva Lopes et al, 2009). The blood-brain barrier and the blood-cerebrospinal fluid barrier are the two main membrane barriers that separate the blood and brain extracellular fluids. Aluminium distribution across brain barrier systems has been monitored using microdialysis probes implanted in the frontal cortex, lateral ventricle and jugular vein of rats (Yokel, 2002). Influx and efflux of aluminium across blood-brain barrier are likely more prominent than through the cerebrospinal fluid flow, with aluminium undergoing carrier-mediated uptake into brain.

Blood aluminium can reach the rat brain in few minutes and its half-life in the brain was estimated to be about 150 days. The complex issue of "exposure ↔ burden ↔ excretion" of aluminium in humans was addressed precisely by Christopher Exley (2008) in the previous volume of this book series. No one can dare to say that aluminium compounds can reach systemic circulation via ingestion, dermal absorption, injection or inhalation only in experimental models. Direct transport of aluminium to the brain via nasal olfactory pathways leading to excess accumulation in different rat brain regions has been demonstrated (Zatta et al, 1993). Actually, the incidence of neurological symptoms and subclinical neurotoxic effects among miners treated with the prophylactic McIntyre powder (aluminium oxide fine powder) and welding, potroom, smeltery and foundry workers chronically exposed to aluminium started to appear in the literature since 1962 (McLaughlin et al., 1962).

One of the most thrilling controversies about aluminium being an environmental factor in neurodegenerative diseases was burst fifty years ago by the fact that the incidence of amyotrophic lateral sclerosis among the Chamorros of Guam was estimated to be at least 50 times that of the rest of the world from which reports were available (Kurland and Mulder, 1954). Other studies led to conclude that the incidence of this fatal disease, characterized by progressive degeneration of motor neurons, was also high in other two regions of the Western Pacific, namely the Kii Peninsula of Honshu Island of Japan and the southern West New Guinea. These three geographically separated endemic foci of amyotrophic lateral sclerosis and parkinsonism dementia were shown to share high levels of bioavailable forms of aluminium, iron and manganese in association with very low levels of calcium and magnesium. The conclusions of the studies on the relative contributions of environmental factors and genetic predisposition, as well as on the mechanisms of disease among the tree genetically different homogenous populations were mastery reviewed by Meredith C. Hermosura and Ralph M. Garruto recently (Hermosura and Garruto, 2007). The Western Pacific amyotrophic lateral sclerosis, which etiology seems to rely on specific interactions between genes and the environment, likely shares similar molecular pathogenic mechanisms with the sporadic and familial variants of the disease. Studies were carried out in animal models of the disease. After exposure to mineral diets mimicking the environmentally observed Al^{3+}, Mn^{2+}, Ca^{2+} and Mg^{2+} levels, signs of neuronal damage, including neurofibrillary alterations, distorted Ca^{2+} homeostasis, mitochondrial degeneration and loss of dopaminergic neurons, were observed in conjugation with deposition of Al^{3+}, Mn^{2+} and Ca^{2+}

in the brain and spinal cord tissues. Briefly, oxidative stress, mitochondrial dysfunction, aberrant Ca^{2+} homeostasis and inflammatory responses have been considered the principal factors supposed of having involvement in these diseases. Curiously, the implication of these processes in aluminium-induced cell death has been confirmed experimentally: (1) oxidative stress and membrane lipid peroxidation boost (Abreo et al., 1991; Satoh et al., 2007); (2) protein phosphorylation and Ca^{2+} imbalance with subsequent disruption of the permeability barrier function of the cell membrane (Snyder et al., 1995); (3) cytoskeletal alterations (Müller and Bruinink, 1994); (4) phosphatidylserine exposure due to cell membrane scrambling (Niemoeller et al., 2006); and (5) activation of apoptotic signal transduction pathways and enzymes (Ghribi et al., 2001; Fu et al., 2003; Griffioen et al., 2004; Toimela and Tähti, 2004; Johnson et al., 2005; Lukiw et al., 2005; Platt et al., 2007). Exciting is the fact that two key etiological factors have been identified, namely members 2 (TRPM2) and 7 (TRPM7) of the transient receptor potential melastatin cation channel family (Hermosura and Garruto, 2007). The presence of genetic variants of these channels seems to be associated with the susceptibility to develop amyotrophic lateral sclerosis/parkinsonism dementia in response to exposure to specific environmental factors. These bifunctional proteins, with cation channel and enzymatic activities, are intrinsically linked to susceptibility to neuronal death following oxidative stress.

The mammalian transient receptor potential (TRP) ion channels, which contain six transmembrane spanning domains with a pore domain between the fifth and sixth segments and both C and N termini located intracellularly, are usually divided into three main families, the above mentioned TRPMx (melastatin), the TRPVx (vanilloid) and TRPCx (short) families, respectively. Besides the effect of aluminium on TRPM2 and TRPM7 has not been addressed experimentally. It has been shown that Al^{3+} causes a voltage dependent blockage of membrane currents through TRPV1 (Tousova et al., 2005) and AlF_4^--evokes inward currents in TRPC6-transfected cells (Kraft et al., 2006). In generally, signal transduction related to TRPs seems to be strongly associated with the phosphatidylinositol pathway, involving the phospholipase C beta (PLCβ) and phosphatidylinositol 4,5-bisphosphate (PIP_2). Interestingly, competitive inhibition of PIP_2 hydrolysis by aluminium has been suggested as one of the mechanism of aluminium-induced impairment of intracellular signal transduction pathways (Shafer and Mundy, 1995).

It is undeniable that aluminium might cause adverse health effects and *in vitro* models are superb tools to challenge the mechanisms of aluminium neurotoxic action. An illustration of how experimental models have contributed to clarify aluminium deleterious will be provided at the confined space of the synapse.

3. ALUMINIUM AT THE SYNAPSES

We can think about synapses as meeting points where bones of contention of past behaviour are dropped away and new ones come forward to reveal future behaviour. The term "synapse" has been coined by Sir Charles Scott Sherrington in 1897 to conceptualize the structural connection of one neuron with another cell. Its existence was confirmed by latter investigators, George Palade, Eduardo de Robertis and George Bennett, who observed for the first time a synapse in 1954. The original idea remains until now but the space and the partners were fashion afreshed (Araque et al., 1999). Nowadays, the synapse is tripartite, where neurons and astrocytes are dynamic cellular players in synaptic information processing by the nervous system (Perea and Araque, 2005). Recently, the most important effects of aluminium on the main steps of neurotransmission and on astrocyte function were reviewed by Gonçalves and Silva (2007) and Aremu and Meshitsuka (2006), respectively.

The nervous system can be imagined as a virtual communication network with two main streams, one relying on neurons and the other on astrocytes, which communication lines converge and diverge at approximately a quadrillion of nodes or synapses. Neuronal and glial excitabilities or communication protocols are displayed in distinct forms, mainly based on electrical changes in the plasma membrane and oscillations of cytosolic Ca^{2+} concentration, respectively. Many questions related to how synapses handle the integration and transfer of information flows, it means, the synaptic specific switching capabilities between the two communication protocols, remain to be worked out. Curiously, the first radical change that felt into synaptic communication in the middle of the 20th century, when the first intracellular recording of an action potential has been published (Hodgkin and Huxley, 1939) and its ionic nature described (Hodgkin et al., 1952) almost simultaneously came with the postulation of the calcium hypothesis of the stimulus-secretion coupling (Katz, 1969), the focus shifted on neurotransmission, with emphasis on the so called Ca^{2+} regulated exocytosis. Briefly, neuronal electrical excitability is believed to be unique due to the huge expression of voltage- and ligand-operated plasma membrane channels that assists on transient, large membrane potential variations due to ionic membrane currents coupled with oscillations of cytosolic Ca^{2+} concentration that activates synchronous transmitter exocytosis, contributing to synaptic transmission or information flow. Today, the pivotal and complex role of calcium in synaptic communication is even a more thoughtful idea than it was envisioned by the time of the '70s. Our present view must cover calcium spikes in neurons and astrocytes, multiple intracellular calcium pools, synchronous and asynchronous exocytosis of neurotransmitters and gliotransmitters and their nonexocytic release and even noncalcium-dependent mechanisms of neuro-glial transmission and secretion occurring via constitutive exocytosis.

Besides a clear picture of the machinery that controls the synaptic secretory apparatus still remains elusive, the synapse represents the basic functional unit of the nerve system. Simultaneous spatial and temporal integration of information flows through nerve and glial cells, generation of new information and its transmission to other cells of the nerve, muscular and endocrine systems takes place at synaptic level. This is, at subcellular level, the main feature of nerve system to accomplish their function in perceiving and responding to external stimuli, mediating communication with external environment and coordinating the activities of all other organ systems and individual behaviour. Over the past 20 years an extraordinary structural similarity for proteins that convey the same function across the evolution have been revealed by structure-function studies, which lean heavily on techniques for expressing the proteins of interest in foreign membranes.

Neurons and astrocytes developed complex secretion machineries to mediate intercellular communication that uses the same basic principles already presented in eukaryotic unicellular organisms for vesicle rejection giving poison and mineral protection. In generally, exocytic events require the participation of soluble NSF (N-ethylmaleimide-sensitive fusion protein) adaptor protein receptor (SNARE) proteins. These proteins are present on the cytoplasmic surfaces of secretory vesicles (v-SNAREs; i.e. vesicle associated membrane protein, VAMP) and of the plasma membrane (target or t-SNAREs; i.e. syntaxin and synaptosome-associated protein of 25 kDa, SNAP-25). According to the zipper hypothesis (Jahn et al, 2003) the coiling of the core complex (highly-stable quadruple α-helix complex of cognate v- and t-SNAREs) apposing the vesicle and target membranes drives the Ca^{2+}-dependent, tightly regulated fusion between the membranes, opening a fusion pore and releasing the vesicle contents into the synaptic cleft. Ultrahigh-resolution atomic force microscopy revealed fusion pores or porosome complexes in presynaptic nerve terminals and astrocytes as nanostructures, forming circular 'pits' containing a small number of cone-shaped depressions arranged in a circular array. Opposing vesicular and plasma membranes, which are brought closer to within a distance of few angstroms by SNARE-ring complexes, are supposed to fuse and form conducting channels due to calcium bound and subsequent loss of coordinated water from the hydrated calcium ion and oxygen of the phospholipid head groups in opposing membranes (Jena, 2009). So, it appears that the integration and flow of information through the nervous system is one of the fine tuning regulated physiological processes where structural and dynamical properties of the solvated calcium ion and its binding sites, play a role. Thereby, adaptation of the secretion machinery to the particular environmental conditions of each synapse at a given time should represent one of the most powerful mechanisms responsible for the multiplicity of patterns in synaptic communication. Aluminium is a trivalent metal cation of lower atomic mass (26.981539) than calcium (40.078). Hydrated calcium has more than one shell around the Ca^{2+}, with the first hydration shell having six water molecules in an octahedral arrangement. The positions of the first maxima in Ca-O and Ca-H radial distribution functions are located at 2.45 Å and 2.98 Å, respectively, with a second neighbour shell extending from 3.5 to 5 Å (Bakó et al, 2002). Interestingly, the third shells maximum of hydrated aluminium is located at *ca.* 6.1 Å, with the first and second shells maximum being located at *ca.* 1.91 and 4.2 Å, respectively (Hofer et al, 2005). Additionally, the chemical and physical properties of Al(III) ion, like its high polarizing effects on its surrounding, its amphoteric character and its high Lewis acidity, can contribute to disturb membrane fusion reactions at the synapses. In view of that, it seems profitable to summarize the already depicted effects of aluminium on the process designed to execute the specific, specialized function of the nervous function, since its impairment would certainly correspond to one of most early events and sensitive endpoints of global nerve system imbalance.

Besides the outstanding evolutionary importance of the enlargement of the SNARE family of proteins among multicellular organisms with respect to their unicellular ancestors, there are no studies on aluminium effects on this ubiquitous membrane fusion machinery, which mediate secretory vesicle trafficking in a multiplicity of physiological processes. However, synaptic configuration changes have been experimentally demonstrated following aluminium exposure (Jing et al, 2004). After 3 months of exposure to aluminium (500 mg/kg/day), the abnormal synaptic ultrastructure of Gray's type I synapses in hippocampus and frontal cortex of rats was characterized by decreased thickness of postsynaptic density, increased width of synaptic cleft, increased numbers of flat synapse, decreased numbers of positive curvature synapse and perforated synapse. These modifications of the synaptic configuration were accompanied by degradation in the ability to learn and memorize (Jing et al, 2004). Interestingly, a triple transgenic mouse model of Alzheimer′s disease showed a significant decrease in the fraction of perforated junctional areas in the CA1 pyramidal layer (Bertoni-Freddari et al, 2008). Perforated synapses, which are characterized by the appearance of discontinuities in postsynaptic densities, are thought to represent an index of synaptic plasticity, reflecting changes in synaptic activity and intermediate stages in synaptic remodelling and turnover. Thereby, aluminium-induced alterations on the synaptic ultrastructure might represent an untimely indication of synaptic dysfunction. Actually, it is believed that the potency and efficacy of vesicular cargo discharge through fusion pores depends on the perfect match of the curvature of the opposing membranes, since it was demonstrated that the size of the t-/v-SNARE complex is directly proportional to the diameter of the discharging vesicle (Cho et al, 2005). For instance, rapid and efficient fusion and release of neurotransmitters at the nerve terminal involves small synaptic vesicles (Ø, 40-50 nm), whereas larger secretory vesicles (Ø, 200-1200 nm) partner slow and prolonged release of digestive enzymes in the exocrine pancreas. Probably, one of the most studied factors affecting membrane curvature is cholesterol, a major component of mammalian biomembranes. According to the stalk-pore model of membrane fusion, which emphasizes the role of hydrophobic attractions and curvature stress in lipid packing/mixing and the transient formation of intermediate membrane structures, the intrinsic negative curvature of cholesterol contributes to a critical local negative curvature that supports formation of lipidic fusion intermediates (Furber et al, 2009). So, redistribution of cholesterol and alteration of its content in membranes should alter the synaptic microenvironment, causing modifications on the structure and functioning of synaptic communication. Additionally, other phenomena, like hydration, electrostatic and entropic interactions, also assist membrane fusion processes involving fusogenic proteins, metal ions and charged phospholipids. Besides the well-known properties of aluminium as a cross-linking agent, exposure to dietary aluminium appears to lead to a significant reduction of the cholesterol/phospholipid molar ratio in rat brain synaptosomes, which are a subcellular fraction enriched in presynaptic nerve terminals (Silva et al., 2002). Accordingly, several groups of researchers have shown aluminium-induced impairment of synaptic secretion using distinct experimental models and recording techniques.

Extracellular recording of electrical signals at the frog cutaneous pectoris neuromuscular junction has been used to monitor the effect of aluminium in the nerve terminal and on the millisecond synaptic response to acetylcholine discharge in the muscle fiber (Banin and Meiri, 1987; Meiri et al, 1993). A concentration dependent increase of acetylcholine release, mainly due to the increment of the average number of quanta released in response to a single stimulus (synchronous discharge) and in spontaneous secretion, was observed. However, irreversible cessation of spontaneous and evoked secretion in prolonged (> 30 minute) incubation occurs, suggesting that the life span of the synapse was significantly shortened by aluminium, while the excitable activity of the nerve terminal was maintained for a longer period of time. These data emphasizes the sometimes biphasic, even antagonic-like, effects of aluminium in living cells and the putative high sensitivity of synaptic secretion to aluminium insult, rather than the absence of its effects on electrical responsiveness and properties of electrically excitable cell membranes.

In early 1968, a slight elevation of Al^{3+} concentration (0.03 mEq/l) in seawater was shown to significantly reduce action potential evoked by a depolarizing current in the giant axon of the lobster (Blaustein and Goldman, 1968). A number of ensuing studies have confirmed the deleterious actions of aluminium on electrophysiological properties of neurons. For instance, membrane depolarization, increased firing activity and abnormal firing patterns were intracellular recorded in a molluscan (*Lymnaea stagnalis*) neuron (Campbell et al, 2000). *In vivo* and *in vitro* application of aluminium to rat hippocampal neurons have been showed to cause blockage of long-term potentiation at low concentrations (2.7 μg/ml), whereas a general reduction in cell excitability and in posttetanic potentiation were observed at higher concentrations (5.4 μg/ml) (Platt et al, 1995). It is important to note that in these studies and in many other studies of long-term potentiation, widely considered one of the major cellular mechanisms that underlies learning and memory, aluminium acted to impair synaptic transmission at levels which do not cause neuronal cell death. The passive and active electrical properties of isolated neurons have been examined to determine the source of the aluminium effects. In particular, voltage-operated and transmitter-gated ionic currents deserved attention, given that they assure the most relevant ionic currents across the synaptic plasma membranes.

The mechanism of aluminium action on voltage-activated calcium channel currents was extensively analyzed in mammalian brain neurons (Platt and Büsselberg, 1994). Briefly, N-, L- and T calcium currents were similarly reduced, in a concentration-dependent manner, by aluminium treatment, which shifts the extracellular Ca^{2+} concentrations required to achieve maximal Ca^{2+} uptake toward higher values. The action of aluminium was found to be use dependent and highly pH dependent, due to a putative interaction with protonable aluminium-sensitive binding sites of the calcium channel and/or by screening of fixed charges on the membrane surface. However, high aluminium concentrations appear to increase the Ca^{2+} current activated by high voltage stimulation, via activation of the cyclic AMP/protein kinase A pathway.

Paul Q. Trombley (1998), using whole cell recording techniques to examine the actions of aluminium on basic membrane properties and amino acid receptors on rat olfactory bulb mitral/tufted neurons in culture, showed a selective, biphasic modulation of $GABA_A$ receptors. The $GABA_A$ receptor is an ionotropic receptor, commonly a hetero-pentamer that belongs to the *Cys-loop* superfamily of receptor-coupled ion channels. The binding of two molecules of γ-aminobutyric acid (GABA), the major inhibitory neurotransmitter in adult human brain, to the recognition sites triggers the opening of the pore, allowing the influx of chloride ions and the initiation of an inhibitory input at the post-synaptic neuron. In the range of 10 to 100 μM aluminium, potentiation of GABA evoked current was monitored, while its blockage occurred in the range of 300 to 1000 μM. Remarkable, no alterations of macroscopic voltage-gated currents and of the membrane currents evoked by glutamate and glycine were observed. It has been suggested that $GABA_A$ receptors express two allosteric binding sites for aluminium: a high-affinity site that is potentiating and a low-affinity site that is inhibiting.

Little is known about aluminium action on regulated exocytosis at the synapse. GABAergic synapses and many other synapses express sodium-dependent transmitter transporters that are located on the plasma membrane of neurons and glial cells, where they primary remove neurotransmitters from the extracellular space following triggered secretion. These high-affinity transporters can reverse not only during pathological conditions but also in response to physiological changes in substrate gradients and membrane potential [Richerson and Wu, 2003]. Aluminium has been shown to interfere with the uptake of GABA, glycine, glutamate, 5-hydroxytryptamine, noradrenaline, dopamine, and choline. Synaptosomes isolated from distinct animal species show reduced capability to drive sodium-dependent transport of neurotransmitters when incubated with micromolar concentrations of aluminium (Lai et al, 1980; 1981; 1982; Wong et al, 1981; Sturman et al, 1983; Johnson and Jope, 1986; Cordeiro et al, 2003; Silva et al, 2007). This inhibitory action was also confirmed when applying *in vivo* exposure. Interestingly, during aluminium exposure, inhibition of (Na^+/K^+)ATPase activity occurs, preceding possible alterations of expression of catalytic subunits, cellular energy depletion, and disturbances in cellular membrane integrity (Silva et al, 2005). The (Na^+/K^+)ATPase electrogenic activity and its crucial role in maintaining sodium and potassium gradients have a major influence on neuronal functions, including excitability and sodium-dependent neurotransmitter uptake. The ubiquitously expressed (Na^+/K^+)ATPase is plasma membrane bound, oligomeric enzyme. It seems that micromolar aluminium can induce impairment of the interprotomeric interaction within the oligomeric ensemble of membrane-bound (Na^+/K^+)ATPase, reducing the rate-limiting step velocity of the reaction cycle (Silva and Gonçalves, 2003). Stimulation of GABA release mediated by reversal of its high-affinity uptake system was reported, which has been interpreted as a consequence of the abolishment of the transporter downregulation by Ca^{2+}-calmodulin-dependent calcineurin activity (Cordeiro et al, 2003). The GABA transport system, operating either in forward or backward directions, requires phosphorylation of internally localized calcineurin-sensitive sites to achieve maximal net translocation velocity. Conversely, changes in calcium channels and/or in intracellular calcium signalling pathways appear to be the underlying mechanisms of aluminium inhibition of glutamate secretion trigger by K^+ depolarization in rat hippocampal slices (Provan and Yokel, 1992).

As already aforesaid, cholinergic synapses were the first to be challenged by aluminium and remain the most studied ones until now. The half-century of progress on the machinery for acetylcholine synthesis, sequestration, secretion, signal transduction and inactivation revealed that this "rapid neurotransmission is like lightning: a spark of calcium in the nerve terminal, a spark of transmitter in the cleft, and the signal is over" (Dunant, 2006). Besides the *quickness* of cholinergic transmission being supported by particular features (i.e. mediatophore assisted secretion of acetylcholine and its enzymatic inactivation) continues to render cholinergic synapse unique, the common run of deleterious effects of aluminium at the synapse can be illustrated. Aluminium is actively accumulated in cultured neurons and astrocytes up to millimolar concentrations (Suarez-Fernandez et al., 1996). The neurotransmitter synthesizing enzyme, choline acetyltransferase, is inhibited either by *in vivo* or *in vitro* exposure to aluminium (Bilkei-Gorzó, 1994). In contrast to data obtained with non-mammalian models, in mammalian brain aluminium appears to increase basal release and to reduce its stimulated component when analysed in synaptosomes (Bielarczyk et al, 1998; Amador et al, 2001; Silva et al, 2007) and also in hippocampus using *in vivo* microdialysis (Yokel et al, 1994; Meyer et al, 1996). Acetylcholine inactivation of cholinergic signalling by acetylcholinesterase has been extensively analysed by many investigators and both, stimulatory and inhibitory effects have been reported. Phasic cholinergic transduction should be also impaired since aluminium administration seems to decrease nicotine binding to ionotropic acetylcholine receptors (Gulya et al, 1990), whereas it appears to reduce and to increase the number of acetylcholine binding sites of the metabotropic receptors in a subtype specific way (Gotti et al, 1987; Gulya et al, 1990; Grammas and Caspers, 1991; Harkány et al, 1996; Brus et al, 1997). Summing up, the so called cholinergic hypothesis of aluminium-induced neurodegeneration (Szutowicz, 2001) is mainly based on the fact that aluminium affects acetyl-CoA levels (Jankowska et al, 2000), which are central to energy and acetylcholine metabolism, rendering cholinergic neurons particularly vulnerable to aluminium insult. Thus, intense cholinergic activity should cause an energy deficit and an acetyl-CoA deficit should depress cholinergic activity more than other neurotransmitter systems (Szutowicz et al, 1998).

CONCLUSION

Many studies have been conducted on *in vivo* and *in vitro* models of aluminium neurotoxicosis. Too few studies have been screened here. After a forty year odyssey and a myriad of comprehensive hypotheses that try to explain the mechanism behind almost all known deleterious actions of aluminium in the human brain, just let finally referred to the 'Red Queen Theory', an analogy borrowed from Lewis Carroll's book *Through the Looking Glass* by Luigi F. Agnati: "The essence of… [Red Queen Theory] … is that a competition for the available plasticity exists between the compensatory responses to ageing-induced degeneration and the processes necessary for memory trace formation" (Agnati et al, 1992). The capability of living things to reproduce themselves and the outstanding adaptability/plasticity of living organisms are among the most magnificent properties of life. Synaptic plasticity, the ability of the synapse to change in structural and functional effective properties as a result of its own liveliness, is likely the height of life plasticity. Synaptic plasticity is believed to be the mechanism underlying learning and memory, which imply long-term changes in synaptic strength. Synapses appear to exhibit activity-dependent scaling of synaptic strength through coordinated actions at multiple levels in order to optimize the integrative functions of the Central Nervous System (Turrigiano, 1999). Decline of cognitive abilities like memory arise with senescence, but the development of neurodegenerative disorders and/or diseases seems only to occur when cells fail to respond adaptively to age-related increases in oxidative, metabolic and ionic stress, thereby resulting in the accumulation of damaged proteins, DNA and membranes (Mattson and Magnus, 2006).

Recently, learning and memory as well as ageing-related diseases were analysed from the perspective of Central Nervous System plasticity (Agnati et al, 2009). It has been postulated that neurons, astrocytes, microglia, exchange vessels and extracellular matrix form plastic multi-element structures that drive communication networks to adapt brain tissue to upcoming tasks and conditions in addition to guarantee cell survival. In this context, neurodegeneration may be contributed by a maladapted boosting of the plasticity mechanisms triggered by signals attempting to maximize neuronal plasticity. Presently, dopamine, glutamate, amiloyd β peptide and α-synuclein are thought to be the best candidates for the role of the so called "push-key-signal", given that they can switch from promoters of plasticity and cell survival in the central nervous system to triggers of neurodegenerative processes, especially during periods of considerable rearrangement of communication networks and insufficient repair processes. According to the 'Red Queen Theory', "…in ageing, processes responsible for plasticity must be forced to run at the highest possible rate to maintain the morphofunctional substrate of the existing networks as well as to allow the formation of new memory traces…". Like senescence, aluminium might diminish neuroprotective processes and synaptic plasticity while its accumulation in brain is ultimately a demanding task requiring increased synaptic plasticity. The studies that have been conducted on *in vivo* and *in vitro* models of aluminium neurotoxicosis can pretty much all support that aluminium can at least act as stressor or even as a ‘‘push-key-signal’’.

Detailed reviews of molecular mechanisms of aluminium toxicity argue that aluminium jeopardizes multiple cellular processes that affect the balance between physiological and pathological roles of "push-key-signals" and displace the delicate boundary between benefit and hazardous of the precious synaptic plasticity. Exacerbation of oxidative stress, disruption of calcium homeostasis, impairment of signal transduction cascades and induction of specific stress-related gene expression pattern, which are often associated with the modification of the structure and function of biomolecules, are among the well-documented effects of aluminium as revealed by studies carried out in experimental models of aluminium neurotoxicosis (for reviews see Deleers, 1985; Siegel, 1985; Martin, 1986; Trapp, 1986; McLachlan et al, 1989; Meiri et al, 1993; Haug et al, 1994; Shafer and Mundy, 1995; Julka and Gill, 1996; Strong et al, 1996; Yokel, 2000; Zatta et al, 2002; Savory et al, 2003; Exley, 2004; Kawahara, 2005; Lukiw et al., 2005; Verstraeten et al, 2008; and references therein). In addition to the examples pointed out all along the text, some further important features will be introduced to support the relevance of *in vitro* models to an understanding of aluminium in human brain.

As mentioned above dopamine, glutamate, amiloyd β peptide and α-synuclein are claimed to be linked to neuronal plasticity and survival as well as to neurodegenerative processes (Agnati et al, 2009). Recently, it was clearly demonstrated for the first time that aluminium promotes specific amyloid$_{1\text{-}42}$ aggregation in human neuroblastoma cells and consequent increase of membrane fluidity, alteration of the cellular morphology and cell death, when compared to the other metal (Cu, Zn, Fe) ions (Drago et al, 2008). It should be noted that abnormal conformation and aggregation of amiloyd β peptide and α-synuclein support the formation of toxic oligomers that impair memory and learning and thereafter cause progressive and massive neurodegeneration. Aluminium ion was demonstrated to be an approximately 2-fold more potent than zinc ion promoter of α-synuclein misfolding and aggregation at neutral pH (Spiegell et al, 2009). A strong and long-lasting impairment of neuronal signal transduction pathways associated with *N*-methyl-*D*-aspartate (NMDA) receptors (a preeminent ionotropic glutamate receptor on synaptic plasticity, memory function and excitotoxic cell death) as a deleterious effect of aluminium was also depicted in the past (Canales et al, 2001). Curiously, aluminium oxide has become the selected compound to adsorb catecholamines in rapid and routine analysis of these compounds in biological fluids (Bergquist et al, 2002). The 3,4-dihydroxyphenyl structure of the catecholamines form cyclic complexes with aluminium oxide and this property allows the selective purificiation of catecholamines. Recently, direct voltammetric recognition and determination of dopamine by using Al(III)-dopamine complexes at the hanging mercury drop electrode has been suggested as a selective electrochemical method to directly determine dopamine (Zhang et al, 2007). Aluminium enhances the ability of 6-hydroxydopamine to cause oxidative stress and neurodegeneration in the dopaminergic system and the disturbance of monoamine metabolism in the brain has been repetitively shown. Imbalance of brain biopterins, inhibition of dihydropteridine reductase, dopamine β-monooxygenase and β-hydroxylase, tyrosine hydroxylase and inhibition of dopamine-stimulated phospholipid methylation are among the deleterious effects of aluminium on the dopaminergic system that have been reported so far (Wenk and Stemmer, 1981; Flora et al, 1991; Tsunoda and Sharma, 1999; Amstrong et al, 1992; Milanese et al, 2001; Nishida, 2001; Altindag et al, 2003; Waly et al, 2004). Finally, Koudinov and Koudinova (2005) claim that cholesterol homeostasis failure is a unifying cause of synaptic degeneration. Aluminium exposure and cholesterol imbalance have been linked not only in experimental models but also in humans [i.e. Ramos et al, 1983; Sarin et al, 1997; Silva et al., 2002; Contini Mdel et al, 2006].

POST SCRIPTUM

Let's stop writing and start dreaming with the moment when aluminium behaviour in the human body will be completely understood.

During the second half of the eighteenth century enthusiasm was running on aluminium.

Charles Dickens referred to aluminium and cryolite as treasures.

"... a treasure [aluminium] has been divined, unearthed, and brought to light, which it is as hard to believe in as a prosaic fact, as it is to feel assured that by descending through a trap-door in a ruined vault, you will enter an Arabian Nights' garden, wherein the leaves are emeralds and the fruits on every tree are rubies, amethysts, topazes, and carbuncles."
Charles Dickens. Aluminium. In "Household Words", December 13, 1856.

"The probable future value of this mineral [cryolite] which, long known, has, until now, been considered worthless, might furnish the text of a discourse upon the hidden treasures of nature, and the possible sources of material prosperity to our descendants; a discourse in which no flight of imagination could be called visionary by sober men, and which might exhaust the dreams of an enthusiast without approaching to the events that will be seen."
Charles Dickens. Crystals. In "Household Words", May 2, 1857.

In 1872, Jules Verne used aluminium to arrive at the Moon.

"— Quoi donc? dit le major.
— De l'aluminium, répondit Barbicane.
— De l'aluminium! s'écrièrent les trois collègues du président.
— Sans doute, mes amis. Vous savez qu'un illustre chimiste français, Henry Sainte-Claire-Deville, est parvenu, en 1854, à obtenir l'aluminium en masse compacte. Or ce précieux métal a la blancheur de l'argent, l'inaltérabilité de l'or, la ténacité du fer, la fusibilité du cuivre et la légèreté du verre; il se travaille facilement, il est extrêmement répandu dans la nature, puisque l'alumine forme la base de la plupart des roches, il est trois fois plus léger que le fer, et il semble avoir été créé tout exprès pour nous fournir la matière de notre projectile!
— Hurrah pour l'aluminium! s'écria le secrétaire du Comité, toujours très-bruyant dans ses moments d'enthousiasme.
— Mais, mon cher président, dit le major, est-ce que le prix de revient de l'aluminium n'est pas extrêmement élevé?
— Il l'était, répondit Barbicane; aux premiers temps de sa découverte, la livre d'aluminium coûtait deux cent soixante à deux cent quatre-vingts dollars (— environ 1,500 francs); puis elle est tombée à vingt-sept dollars (— 150 fr.), et aujourd'hui enfin, elle vaut neuf dollars (— 48 fr. 75 c.).
— Mais neuf dollars la livre, répliqua le major, qui ne se rendait pas facilement, c'est encore un prix énorme!
— Sans doute, mon cher major, mais non pas inabordable.
— Que pèsera donc le projectile? demanda Morgan.

— Voici ce qui résulte de mes calculs, répondit Barbicane; un boulet de cent huit pouces de diamètre et de douze pouces d'épaisseur pèserait, s'il était en fonte de fer, soixante-sept mille quatre cent quarante livres; en fonte d'aluminium, son poids sera réduit à dix-neuf mille deux cent cinquante livres.
— Parfait! s'écria Maston, voilà qui rentre dans notre programme.
— Parfait! parfait! répliqua le major, mais ne savez-vous pas qu'à dixhuit dollars la livre, ce projectile coûtera...
— Cent soixante-treize mille deux cent cinquante dollars (— 928, 437 fr. 50 c.), je le sais parfaitement; mais ne craignez rien, mes amis, l'argent ne fera pas défaut à notre entreprise, je vous en réponds.
— Il pleuvra dans nos caisses, répliqua J.-T. Maston.
— Eh bien! que pensez- vous de l'aluminium! demanda le président.
— Adopté, répondirent les trois membres du Comité.
— Quant à la forme du boulet, reprit Barbicane, elle importe peu, puisque, l'atmosphère une fois dépassée, le projectile se trouvera dans le vide; je propose donc le boulet rond, qui tournera sur lui-même, si cela lui plaît, et se comportera à sa fantaisie.»
Ainsi se termina la première séance du Comité; la question du projectile était définitivement résolue, et J.-T. Maston se réjouit fort à la pensée d'envoyer un boulet d'aluminium aux Sélénites, «ce qui leur donnerait une crâne idée des habitants de la Terre!»"
Jules Verne. In "De la Terre à la Lune, Trajet direct en 97 heures 20 minutes", Chapitre VII "L'hymne du boulet".

And now we are taking advantage of aged dreams or of the fast cultural development. Biological development is a much more slow process. We learned how to profit from the multifaceted properties of aluminium in industrial terms. We do not recognize highly specific aluminium effects in our body other than the deleterious one. We do not control the elasticity and plasticity or whichever "biological" properties of aluminium that could be demonstrated in the future. We are not confident about the aluminium "minimum effective dose" that affects physiological processes or when their declines become a pathological condition. In the case of aluminium, Life cannot use the "natural delightful transformation" of metal ions into a less toxic form by modifying its oxidation state, since aluminium is a non-transition metal. Let's think that our relative advantage consists on the multiorgan barrier system reducing aluminium flow to our organism, which is high but not unlimited. In the meantime aluminium safety remains in the politic domain, where the precautionary principle is one of the achieves of our cultural development.

References

Abreo K., J. Jangula, S.K. Jain, M. Sella, J. Glass. Aluminum uptake and toxicity in cultured mouse hepatocytes. *J. Am. Soc. Nephrol.* 1 (1991) 1299-1304.

Agnati L.F., G. Leo, S. Genedani, L. Piron, A. Rivera, D. Guidolin, K. Fuxe. Common key-signals in learning and neurodegeneration: focus on excito-amino acids, β-amyloid peptides and α-synuclein. *J. Neural Transm.* 116 (2009) 953-974.

Agnati L.F., M. Zoli, G. Biagini, K. Fuxe. Neuronal plasticity and ageing processes in the frame of the 'Red Queen Theory'. *Acta Physiol. Scand.* 145 (1992) 301-309.

Altindag Z.Z., T. Baydar, A.B. Engin, G. Sahin. Effects of the metals on dihydropteridine reductase activity. *Toxicol. In Vitro* 17 (2003) 533–537.

Amador F.C., M.S. Santos, C.R. Oliveira. Lipid peroxidation and aluminium effects on the cholinergic system in nerve terminals. *Neurotox. Res.* 3 (2001) 223-233.

Amstrong R.A., J. Anderson, J.D. Cowburn, J. Cox, J.A. Blair. Aluminium administered in drinking water but not in the diet influences biopterin metabolism in the rodent. *Biol. Chem. Hoppe-Seyler* 373 (1992) 1075-1078.

Araque A., V. Parpura, R.P. Sanzgiri, P.G. Haydon. Tripartite synapses: glia, the unacknowledged partner. *Trends Neurosci.* 22 (1999) 208-215.

Aremu D.A., S. Meshitsuka. Some aspects of astroglial functions and aluminum implications for neurodegeneration. *Brain Res. Rev.* 52 (2006) 193-200.

Atwood C.S., G. Perry, H. Zeng, Y. Kato, W.D. Jones, K.Q. Ling, X. Huang, R.D. Moir, D. Wang, L.M. Sayre, M.A. Smith, S.G. Chen, A.I. Bush. Copper mediates dityrosine cross-linking of Alzheimer's amyloid-β. *Biochemistry* 43 (2004) 560-568.

Authier F.J., P. Cherin, A. Creange, B. Bonnotte, X. Ferrer, A. Abdelmoumni, D. Ranoux, J. Pelletier, D. Figarella-Branger, B. Granel, T. Maisonobe, M. Coquet, J.D. Degos, R.K. Gherardi. Central nervous system disease in patients with macrophagic myofasciitis. *Brain* 124 (2001) 974-983.

Bakó I., J. Hutter, G. Pálinkás. Car–Parrinello molecular dynamics simulation of the hydrated calcium ion. *J. Chem. Phys.* 117 (2002) 9838-9843.

Banin E., H. Meiri. Impaired control of information transfer at an isolated synapse treated by aluminum: is it related to dementia? *Brain Res.* 423 (1987) 359-363.

Bergquist J., A. Sciubisz, A. Kaczor, J. Silberring. Catecholamines and methods for their identification and quantitation in biological tissues and fluids. *J. Neurosci. Methods* 113 (2002) 1-13.

Bertoni-Freddari C., S.L. Sensi, B. Giorgetti, M. Balietti, G. Di Stefano, L.M. Canzoniero, T. Casoli, P. Fattoretti. Decreased presence of perforated synapses in a triple-transgenic mouse model of Alzheimer's disease. *Rejuvenation Res.* 11 (2008) 309-313.

Bielarczyk H., M. Tomaszewicz, A. Szutowicz. Effect of aluminum on acetyl-CoA and acetylcholine metabolism in nerve terminals. *J. Neurochem.* 70 (1998) 1175-1181.

Bilkei-Gorzó A. Effect of chronic treatment with aluminum compound on rat brain choline acetyltransferase activity. *Pharmacol. Toxicol.* 74 (1994) 359-360.

Blaustein M.P., D.E. Goldman. The action of certain polyvalent cations on the voltage-clamped lobster axon. *J. Gen. Physiol.* 51 (1968) 279-291.

Brus R., R. Szkilnik, I. Popieluch, R.M. Kostrzewa, K. Mengel. Effect of aluminium exposure on central serotonin and muscarine receptor reactivity in rats. *Med. Sci. Monit.* 3 (1997) 631-636.

Campbell M.M., R. Jugdaohsingh, K.N. White, J.J. Powell, C.R. McCrohan. Aluminum toxicity in a molluscan neuron: effects of counterions. *J. Toxicol. Environ. Health A.* 59 (2000) 253-270.

Canales J.J., R. Corbalán, C. Montoliu, M. Llansola, P. Monfort, S. Erceg, M. Hernandez-Viadel, V. Felipo. Aluminium impairs the glutamate-nitric oxide-cGMP pathway in cultured neurons and in rat brain in vivo: molecular mechanisms and implications for neuropathology. *J. Inorg. Biochem.* 87 (2001) 63-69.

Cho W.J., A. Jeremic, B.P. Jena. Size of supramolecular SNARE complex: membrane-directed self-assembly. *J. Am. Chem. Soc.* 127 (2005) 10156-10157.

Contini Mdel C., S. Mahieu, G. Bazzoni, C.A. Bernal, C.E. Carnovale. Study of hemorheological parameters following partial hepatectomy in rats with chronic aluminium intoxication. *Clin. Hemorheol. Microcirc.* 35 (2006) 431-439.

Cordeiro J.M., V.S. Silva, C.R. Oliveira, P.P. Gonçalves. Aluminium-induced impairment of Ca^{2+} modulatory action on GABA transport in brain cortex nerve terminals. *J. Inorg. Biochem.* 97 (2003) 132-142.

da Silva Lopes L., I. Slobodian, M.R. Del Bigio. Characterization of juvenile and young adult mice following induction of hydrocephalus with kaolin. *Exp. Neurol.* (2009) doi:10.1016/j.expneurol.2009.05.015.

Deleers M. Cationic atmosphere and cation competition binding at negatively charged membranes: pathological implications of aluminium. *Res. Commun. Chem. Pathol. Pharmacol.* 49 (1985) 277-294.

Drago D., M. Bettella, S. Bolognin, L. Cendron, J. Scancar, R. Milacic, F. Ricchelli, A. Casini, L. Messori, G. Tognon, P. Zatta. Potential pathogenic role of β-amyloid$_{1\text{-}42}$-aluminum complex in Alzheimer's disease. *Int. J. Biochem. Cell Biol.* 40 (2008) 731-746.

Dunant Y. Acetylcholine release in rapid synapses: two fast partners-mediatophore and vesicular Ca^{2+}/H^{+} antiport. *J. Mol. Neurosci.*30 (2006) 209-214.

Exley C. Aluminium and medicine. In: A.L.R. Mercê, J. Felcman, M.A.L. Recio (eds), *Molecular and Supramolecular Bioinorganic Chemistry: Applications in Medical Sciences*, (2008). Nova Science Publishers. pp. 45-68.

Exley C. The pro-oxidant activity of aluminum. *Free Radic. Biol. Med.* 36 (2004) 380-387.

Exley C., M.M. Esiri. Severe cerebral congophilic angiopathy coincident with increased brain aluminium in a resident of Camelford, Cornwall, UK. *J. Neurol. Neurosurg. Psychiatry.* 77 (2006) 877-879.

Flora S.J., M. Dhawan, S.K. Tandon. Effects of combined exposure to aluminium and ethanol on aluminium body burden and some neuronal, hepatic and haematopoietic biochemical variables in the rat. *Hum. Exp. Toxicol.* 10 (1991) 45-48.

Fu H., Q. Hu, Z. Lin, T. Ren, H. Song, C. Cai, S. Dong. Aluminum-induced apoptosis in cultured cortical neurons and its effect on SAPK/JNK signal transduction pathway. *Brain Res.* 980 (2003) 11-23.

Furber K.L., M.A. Churchward, T.P. Rogasevskaia, J.R. Coorssen. Identifying critical components of native Ca^{2+}-triggered membrane fusion. Integrating studies of proteins and lipids. *Ann. N. Y. Acad. Sci.* 1152 (2009) 121-134.

Ghribi O., D.A. DeWitt, M.S. Forbes, M.M. Herman, J. Savory. Co-involvement of mitochondria and endoplasmic reticulum in regulation of apoptosis: changes in cytochrome c, Bcl-2 and Bax in the hippocampus of aluminum-treated rabbits. *Brain Res.* 903 (2001) 66-73.

Gonçalves P.P., V.S. Silva. Does neurotransmission impairment accompany aluminium neurotoxicity? *J. Inorg. Biochem.* 101 (2007) 1291-1338.

Gong C.-X., K. Iqbal. Hyperphosphorylation of microtubule-associated protein tau: a promising therapeutic target for Alzheimer disease. *Curr. Med. Chem.* 15 (2008) 2321-2328.

Gotti C., D. Cabrini, E. Sher, F. Clementi. Exposure to aluminium, cadmium or lead on differentiation and cholinergic receptor expressin in a human neuroblastoma cell line. *Cell Biol. Toxicol.* 3 (1987) 431-440.

Grammas P., M.L. Caspers. The effect of aluminium on muscarinic receptors in isolated cerebral microvessels. *Res. Commun. Chem. Pathol. Pharmacol.* 72 (1991) 69-79.

Griffioen K.J., O. Ghribi, N. Fox, J. Savory, D.A. DeWitt. Aluminum maltolate-induced toxicity in NT2 cells occurs through apoptosis and includes cytochrome c release. *Neurotoxicology* 25 (2004) 859-867.

Gruis K.L., J.W. Teener, M. Blaivas. Pediatric macrophagic myofasciitis associated with motor delay. *Clin. Neuropathol.* 25 (2006) 172-179.

Gulya K., Z. Rakonczay, P. Kása. Cholinotoxic effects of aluminium in rat brain. *J. Neurochem.* 54 (1990) 1020-1026.

Harkány T., Z. Lengyel, P. Kása, K. Gulya. Chronic aluminium treatment results in aluminium deposits and affects M_I muscarinic receptors in rat brain. *Neurobiology* 4 (1996) 35-43.

Haug A., B. Shi, V. Vitorello. Aluminum interaction with phosphoinositide-associated signal transduction *Arch. Toxicol.* 68 (1994) 1-7.

Hermosura M.C., R.M. Garruto. *TRPM7* and *TRPM2* – Candidate susceptibility genes for Western Pacific ALS and PD? *Biochim. Biophys. Acta* 1772 (2007) 822-835.

Hodgkin A.L., A.F. Huxley, B. Katz. Measurement of current-voltage relations in the membrane of the giant axon of *Loligo*. *J. Physiol.* 116(1952) 424-448.

Hodgkin A.L., A.F. Huxley. Action potentials from inside a nerve fibre. *Nature* 144 (1939) 710-711.

Hofer T.S., B.R. Randolf, B.M. Rode. Influence of polarization and many body quantum effects on the solvation shell of Al(III) in dilute aqueous solution—extended *ab initio* QM/MM MD simulations. *Phys. Chem. Chem. Phys.* 7 (2005) 1382-1387.

Hook V., I. Schechter, H.U. Demuth, G. Hook. Alternative pathways for production of beta-amyloid peptides of Alzheimer's disease. *Biol. Chem.* 389 (2008) 993-1006.

Isagawa T., H. Mukai, K. Oishi, T. Taniguchi, H. Hasegawa, T. Kawamata, C. Tanaka, Y. Ono. Dual effects of PKNα and protein kinase C on phosphorylation of tau protein by glycogen synthase kinase-3β. *Biochem. Biophys. Res Commun.* 273 (2000) 209-212.

Iversen A., T. Chalder, S. Wessely, Gulf War Illness: lessons from medically unexplained symptoms. *Clin. Psychol. Rev.* 27 (2007) 842-854.

Jahn R., T. Lang, T.C. Südhof. Membrane fusion. *Cell* 112(2003) 519-533.

Jankowska A., B. Madziar, M. Tomaszewicz, A. Szutowicz. Acute and chronic effects of aluminum on acetyl-CoA and acetylcholine metabolism in differentiated and nondifferentiated SN56 cholinergic cells. *J. Neurosci. Res.* 62 (2000) 615-622.

Jena B.P. Porosome: the secretory portal in cells. *Biochemistry* 48 (2009) 4009-4018.

Jing Y., Z. Wang, Y. Song. Quantitative study of aluminum-induced changes in synaptic ultrastructure in rats. *Synapse* 52 (2004) 292-298.

Johnson G.V., R.S. Jope. Aluminum impairs glucose utilization and cholinergic activity in rat brain in vitro. *Toxicology* 40 (1986) 93-102.

Johnson V.J., S.H. Kim, R.P. Sharma. Aluminum-maltolate induces apoptosis and necrosis in neuro-2a cells: Potential role for p53 signaling. *Toxicol. Sci.* 83 (2005) 329-339.

Julka D., K.D. Gill. Altered calcium homeostasis: a possible mechanism of aluminium-induced neurotoxicity. *Biochim. Biophys. Acta* 1315 (1996) 47-54.

Katz B., The Release of Neural Transmitter Substances, Liverpool University Press, 1969, 60p.

Kawahara M. Effects of aluminium on the nervous system and its possible link with neurodegenerative diseases. *J. Alzheimers Dis.* 8 (2005) 171-182.

Kopeloff L.M., S.E. Barrera, N. Kopeloff. Recurrent convulsive seizures in animals produced immunologic and chemical means. *Am. J. Psychiatry* 98 (1942) 881-902.

Koudinov A.R., N.V. Koudinova. Cholesterol homeostasis failure as a unifying cause of synaptic degeneration. *J. Neurol. Sci.* 229-230 (2005) 233-240.

Kovacech B., N. Zilka, M. Novak. New age of neuroproteomics in Alzheimer's disease research. *Cell. Mol. Neurobiol.* (2009) doi: 10.1007/s10571-009-9358-6.

Kraft R., C. Grimm, H. Frenzel, C. Harteneck. Inhibition of TRPM2 cation channels by N-(p-amylcinnamoyl)anthranilic acid. *Br. J. Pharmacol.* 148 (2006) 264-273.

Kurland L.T., D.W. Mulder. Epidemiologic investigations of amyotrophic lateral sclerosis. 1. Preliminary report on geographic distribution, with special reference to the Mariana Islands, including clinical and pathological observations. *Neurology* 4 (1954) 355–378.

Lach B., E.J. Cupler. Macrophagic myofasciitis in children is a localized reaction to vaccination. *J. Child Neurol.* 23 (2008) 614-619.

Lai J.C., J.F. Guest, T.K. Leung, L. Lim, A.N. Davison. The effects of cadmium, manganese and aluminium on sodium-potassium-activated and magnesium-activated adenosine triphosphatase activity and choline uptake in rat brain synaptosomes. *Biochem. Pharmacol.* 29 (1980) 141-146.

Lai J.C., L. Lim, A.N. Davison. Differences in the inhibitory effect of Cd^{2+}, Mn^{2+} and Al^{3+} on the uptake of dopamine by synaptosomes from forebrain and from striatum of the rat. *Biochem. Pharmacol.* 30 (1981) 3123-3125.

Lai J.C., L. Lim, A.N. Davison. Effects of Cd^{2+}, Mn^{2+}, and Al^{3+} on rat brain synaptosomal uptake of noradrenaline and serotonin. *J. Inorg. Biochem.* 17 (1982) 215-225.

Lukiw W.J., M.E. Percy, T.P. Kruck. Nanomolar aluminum induces pro-inflammatory and pro-apoptotic gene expression in human brain cells in primary culture. *J. Inorg. Biochem.* 99 (2005) 1895-1898.

Martin R.B. The chemistry of aluminium as related to biology and medicine. *Clin. Chem.* 32 (1986) 1797-1806.

Mattson M.P., T. Magnus. Ageing and neuronal vulnerability. *Nat. Rev. Neurosci.* 7 (2006) 278-294.

McLachlan D.R., W.J. Lukiw, T.P. Kruck. New evidence for an active role of aluminium in Alzheimer's disease. *Can. J. Neurol. Sci.* 16 (1989) 490-497.

McLaughlin A.I.G., G. Kazantzis, E. King, D. Teare, R.J. Porter, R. Owen. Pulmonary fibrosis and encephalopathy associated with the inhalation of aluminium dust. *Brit. J. Ind. Med.* 19 (1962) 253-263.

Meiri H., E. Banin, M. Roll, A. Rousseau. Toxic effects of aluminium on nerve cells and synaptic transmission. *Prog. Neurobiol.* 40 (1993) 89-121.

Meyer J.J., D.D. Allen, R.A. Yokel. Hippocampal acetylcholine increases during eyeblink conditioning in the rabbit. *Physiol. Behav.* 60 (1996) 1199-1203.

Milanese M., M.I. Lkhayat, P. Zatta. Inhibitory effect of aluminium on dopamine β-hydroxylase from bovine adrenal gland. *J. Trace Elem. Med. Biol.* 15 (2001) 139-141.

Müller J.P., A. Bruinink. Neurotoxic effects of aluminium on embryonic chick brain cultures. *Acta Neuropathol.* 88 (1994) 359-366.

Niemoeller O.M., V. Kiedaisch, P. Dreischer, T. Wieder, F. Lang. Stimulation of eryptosis by aluminium ions. *Toxicol. Appl. Pharmacol.* 217 (2006) 168-175.

Nishida Y. Negligible ability of oxygen and peroxide ion activation by Al(III) ion is essential for Al(III)-induced neurodegeneration. *Z. Naturforsch* 56c (2001) 865–871.

Perea G., A. Araque. Glial calcium signaling and neuron–glia communication. *Cell Calcium* 38 (2005) 375–382.

Petrik M.S., M.C. Wong, R.C. Tabata, R.F. Garry, C.A. Shaw, Aluminum adjuvant linked to Gulf War illness induces motor neuron death in mice, *Neuromolecular Med.* 9 (2007) 83-100.

Platt B., A.J. Drysdale, C. Nday, E.L. Roloff, B.D. Drever, A. Salifoglou. Differential toxicity of novel aluminium compounds in hippocampal culture. Neurotoxicology 28 (2007) 576-586.

Platt B., D. Büsselberg. Actions of aluminum on voltage-activated calcium channel currents. Cell Mol Neurobiol. 14 (1994) 819-829.

Platt B., D.O. Carpenter, D. Büsselberg, K.G. Reymann, G. Riedel. Aluminum impairs hippocampal long-term potentiation in rats *in vitro* and *in vivo*. *Exp. Neurol.* 134 (1995) 73-86.

Provan S.D., R.A. Yokel. Aluminum inhibits glutamate release from transverse rat hippocampal slices: role of G proteins, Ca channels and protein kinase C. *Neurotoxicology* 13 (1992) 413-420.

Ramos J.M., R. Gokal, K. Siamopolous, M.K. Ward, R. Wilkinson, D.N. Kerr. Continuous ambulatory peritoneal dialysis: three years' experience. *Q. J. Med.* 52 (1983) 165-186.

Reusche E., P. Pilz, G. Oberascher, B. Lindner, R. Egensperger, K. Gloeckner, E. Trinka, B. Iglseder. Subacute fatal aluminum encephalopathy after reconstructive otoneurosurgery: a case report. *Human Pathol.* 32 (2001) 1136-1140.

Richerson G.B., Y. Wu. Dynamic equilibrium of neurotransmitter transporters: not just for reuptake anymore. *J. Neurophysiol.* 90 (2003) 1363-1374.

Sarin S., V. Gupta, K.D. Gill. Alterations in lipid composition and neuronal injury in primates following chronic aluminium exposure. *Biol. Trace Elem. Res.* 59 (1997) 133-143.

Satoh E., I. Yasuda, T. Yamada, Y. Suzuki, T. Ohyashiki. Involvement of NO generation in aluminum-induced cell death. *Biol. Pharm. Bull.* 30 (2007) 1390-1394.

Savory J., M.M. Herman, O. Ghribi. Intracellular mechanisms underlying aluminium-induced apoptosis in rabbit brain. *J. Inorg. Biochem.* 97 (2003) 151-154.

Shafer T.J., W.R. Mundy. Effects of aluminum on neuronal signal transduction: mechanisms underlying disruption of phosphoinositide hydrolysis. *Gen. Pharmacol.* 26 (1995) 889-895.

Siegel N. Aluminum interaction with biomolecules: the molecular basis for aluminium toxicity. *Am. J. Kidney Dis.* 6 (1985) 353-357.

Silva V.S., A.I. Duarte, A.C. Rego, C.R. Oliveira, P.P. Gonçalves. Effect of chronic exposure to aluminium on isoform expression and activity of rat (Na^+/K^+)ATPase. *Toxicol. Sci.* 88 (2005) 485-494.

Silva V.S., J.M. Cordeiro, M.J. Matos, C.R. Oliveira, P.P. Gonçalves. Aluminum accumulation and membrane fluidity alteration in synaptosomes isolated from rat brain cortex following aluminum ingestion: effect of cholesterol. *Neurosci. Res.* 44 (2002) 181-193.

Silva V.S., M.A. Nunes, J.M. Cordeiro, A.I. Calejo, S. Santos, P. Neves, A. Sykes, F. Morgado, Y. Dunant, P.P. Gonçalves. Comparative effects of aluminum and ouabain on synaptosomal choline uptake, acetylcholine release and (Na^+/K^+)ATPase. *Toxicology* 236 (2007) 158-177.

Silva V.S., P.P. Gonçalves. The inhibitory effect of aluminium on the (Na^+/K^+)ATPase activity of rat brain cortex synaptosomes. *J. Inorg. Biochem.* 97 (2003) 143-150.

Snyder J.W., A. Serroni, J. Savory, J.L. Farber. The absence of extracellular calcium potentiates the killing of cultured hepatocytes by aluminum maltolate. *Arch. Biochem. Biophys.* 316 (1995) 434-442.

Spiegell E.T., P. Jones, P. McLean, B. Hyman, W.R. Zipfel. Probing the effect of Heat Shock Protein 70 on the aggregation of α-Synuclein. *Biophys. J.* 96 Suppl. 1 (2009) 92a.

Strong M.J., R.M. Garruto, J.G. Joshi, W.R. Mundy, T.J. Shafer. Can the mechanisms of aluminium neurotoxicity be integrated into a unified scheme? *J. Toxicol. Environ. Health* 48 (1996) 599-613.

Sturman J.A., H.M. Wisniewski, J.W. Shek. High affinity uptake of GABA and glycine by rabbits with aluminum-induced neurofibrillary changes. *Neurochem. Res.* 8 (1983) 1097-1109.

Suarez-Fernandez M.B., A. Torreblanca-Pacias, H. Goenaga-Infanta, A.B. Soldado, A. Sanz-Mendel, J.A. Vega, A. Novelli. 1996. Aluminum accumulation and toxicity in cultured neurons and astrocytes. In: P. Caller, J. Carbella, J.L. Domingo, J.C. Etienne, J.M. Llabet (eds), *Metal Ions in Biology and Medicine*, vol 4 (1996). John Libbey Eurotext. Pp. 284-286.

Szutowicz A. Aluminum, NO, and nerve growth factor neurotoxicity in cholinergic neurons. *J. Neurosci. Res.* 66 (2001) 1009-1018.

Szutowicz A., M. Tomaszewicz, H. Bielarczyk, A. Jankowska. Putative significance of shifts in acetyl-CoA compartmentalization in nerve terminals for disturbances if cholinergic transmission in brain. *Dev. Neurosci.* 20 (1998) 485-492.

Toimela T., H. Tähti. Mitochondrial viability and apoptosis induced by aluminum, mercuric mercury and methylmercury in cell lines of neural origin. *Arch. Toxicol.* 78 (2004) 565-574.

Tousova K., L. Vyklicky, K. Susankova, J. Benedikt, V. Vlachova. Gadolinium activates and sensitizes the vanilloid receptor TRPV1 through the external protonation sites. *Mol. Cell. Neurosci.* 30 (2005) 207-217.

Trapp G.A. Interactions of aluminium with cofactors, enzymes, and other proteins. *Kidney Int.* 29 Suppl. 18 (1986) S12-S16.

Trombley P.Q. Selective modulation of $GABA_A$ receptors by aluminium. *J. Neurophysiol.* (1998) 755-761.

Tsunoda M., R.P. Sharma. Altered dopamine turnover in murine hypothalamus after low-dose continuous oral administration of aluminium. *J. Trace Elem. Med. Biol.* 13 (1999) 224-231.

Turrigiano G.G. Homeostatic plasticity in neuronal networks: the more things change, the more they stay the same. *Trends Neurosci.* 22 (1999) 221-227.

Verstraeten S.V., L. Aimo, P.I. Oteiza. Aluminium and lead: molecular mechanisms of brain toxicity. *Arch. Toxicol.* 82 (2008) 789-802.

Walton J.R. An aluminum-based rat model for Alzheimer's disease exhibits oxidative damage, inhibition of PP2A activity, hyperphosphorylated tau, and granulovacuolar degeneration. *J. Inorg. Biochem.* 101 (2007) 1275-1284.

Waly M., H. Olteanu, R. Banerjee, S.W. Choi, J.B. Mason, B.S. Parker, S. Sukumar, S. Shim, A. Sharma, J.M. Benzecry, V.A. Power-Charnitsky, R.C. Deth. Activation of methionine synthase by insulin-like growth factor-1 and dopamine: a target for neurodevelopmental toxins and thimerosal. *Mol. Psychiatry* 9 (2004) 358-370.

Wenk G.L., K.L. Stemmer. The influence of ingested aluminium upon norepinephrine and dopamine levels in the rat brain. *Neurotoxicology* 2 (1981) 347-353.

Wolfe M.S., S.Y. Guénette. APP at a glance. *J. Cell Sci.* 120 (2007) 3157-3161.

Wong P.C., J.C. Lai, L. Lim, A.N. Davison. Selective inhibition of L-glutamate and gammaaminobutyrate transport in nerve ending particles by aluminium, manganese, and cadmium chloride. *J. Inorg. Biochem.* 14 (1981) 253-260.

Yokel R.A . Brain uptake, retention, and efflux of aluminum and manganese. *Environ. Health Perspect.* 110 (2002) 699-704.

Yokel R.A. The toxicology of aluminium in brain: a review. *Neurotoxicology* 21 (2000) 813-828.

Yokel R.A., D.D. Allen, J.J. Meyer. Studies of aluminum neurobehavioral toxicity in the intact mammal. *Cell Mol. Neurobiol.* 14 (1994) 791-808.

Zatta P., M. Favarato, M. Nicolini. Deposition of aluminum in brain tissues of rats exposed to inhalation of aluminum acetylacetonate. *Neuroreport* 4 (1993) 1119-1122.

Zatta P., T. Kiss, M. Suwalsky, G. Berthon. Aluminium(III) as a promoter of cellular oxidation. *Coord. Chem. Rev.* 228 (2002) 271-284.

Zecca L., M.B.H. Youdim, P. Riederer, J.R. Connor, R.R. Crichton. Iron, brain ageing and neurodegenerative disorders. *Nat. Rev. Neurosci.* 5 (2004) 863-873.

Zhang F., M. Zhang, J. Cheng, L. Yang, M. Ji, S. Bi. Direct voltammetric specific recognition of dopamine using Al^{III}-DA complexes at the hanging mercury drop electrode. *Anal. Sci.* 23 (2007) 1325-1329.

In: Molecular and Supramolecular Bioinorganic Chemistry ISBN: 978-1-61728-785-5
Ed: A. L. R. Mercê, J. Felcman and M. A. L. Recio

Chapter 2

BIOINSPIRED SYNTHETIC ANALOGUES FOR PURPLE ACID PHOSPHATASES AND CATALYTIC PROMISCUITY

Ademir Neves*[1] *,Hernán Terenzi*[2] *,
***Rosely A. Peralta*[1] *and* [1]*Patrícia Cardoso Severino*[2]**
[1] Laboratório de Bioinorgânica e Cristalografia, Departamento de Química, Universidade Federal de Santa Catarina, Florianópolis-SC, Brazil.
[2]Centro de Biologia Molecular Estrutural, Departamento de Bioquímica, Universidade Federal de Santa Catarina, Florianópolis-SC, Brazil.

ABSTRACT

In this chapter we aim to illustrate how bioinspired synthetic metallocomplexes that mimic at least some structural, spectroscopic and functional properties of purple acid phosphatase enzymes have become suitable benchmark systems to study and understand such aspects of biological molecules. In addition, we aim to explore how these model complexes are able to cleave nucleic acids in a non-degradative manner while also being efficient catalysts in the hydrolysis of proteins and model substrates – "catalytic promiscuity".

The development of suitable biomimetic catalytic systems capable of mimicking the catalytic promiscuity properties of such enzymes represents a new challenge for bioinorganic chemists and biochemists, providing a wide range of biotechnological applications, for example, in organic synthesis and medicine.

1 HYDROLASES

1.1 Introduction

In hydrolytic reactions molecular bonds are cleaved by the addition of water. Hydrolases are enzymes that act as catalysts in the hydrolysis of proteins, nucleic acids, starch, fats, phosphate esters, and other macromolecular substances. For example, carboxypeptidase A, a

* E-mail corresponding author: malobo@ens.ufsc.br

digestive enzyme, and the bacterial enzyme thermolysin are zinc metalloproteases that belong to the major class of peptide-cleaving enzymes. At neutral pH and 25°C the uncatalyzed hydrolysis of peptides is a slow process with an estimated half-life of between seven and six-hundred years [1],[2] while the reaction catalyzed by the zinc-containing metalloenzymes carboxypeptidase A (Figure 1A) or thermolysin (Figure 1B) can reach a half-life in the order of milliseconds [3], [4].

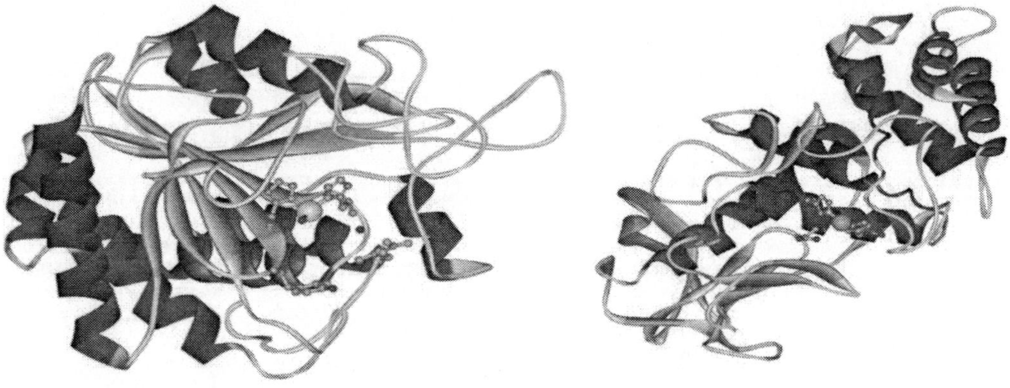

A. B

Figure 1. Structure of carboxypeptidase and of thermolysin.

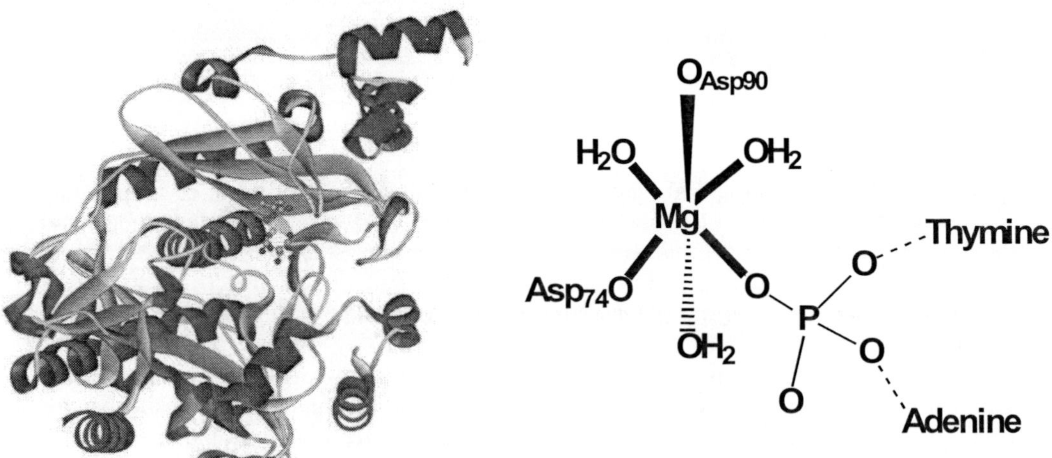

A B

Figure 2. (A) *EcoRV*, (B) *EcoRV* active site.

In fact, the active site of these enzymes contains a bound Zn^{II} ion that activates a water molecule which then acts as a nucleophile and attacks the peptide carbonyl group. Hydrolysis of phosphodiester bonds is also a slow process at neutral pH. For instance, the first-order rate constants for the uncatalyzed hydrolysis of the DNA model substrates bis(4-nitrophenyl) phosphate (BNPP) and bis(2,4-dinitrophenyl) phosphate (BDNPP) are respectively 1.1×10^{-11}

and 1.89 x 10^{-7} s^{-1}, at pH = 7.0 and 25°C [5],[6] and the half-life of phosphodiester bonds in DNA has been estimated as 130,000 years under the same conditions [7]. Indeed, the stability of these phosphodiester bonds indicates the need for the maintenance of the genetic code in order to use DNA as genetic material [7]-[9]. By analogy with peptide hydrolysis, metalloenzymes which catalyze the hydrolysis of phosphate ester bonds may utilize metal-bound H_2O/OH^- as the reaction-initiating nucleophile. For example, restriction endonucleases such as *Eco*RV require Mg^{2+} or some other similar divalent cation for activity. The M^{II} ion helps to activate a coordinated water molecule which is adequately positioned to attack the phosphorus atom on the phosphate group (Figure 2)) [10].

1.2 Purple Acid Phosphatases as Hydrolases

Purple acid phosphatases (PAPs) are dinuclear metallohydrolases, resistant to inhibition by L-tartarate, that catalyze the hydrolysis of a variety of phosphoester substrates *in vitro* at acid to neutral pH [11]. Indeed, they are the only dinuclear metallohydrolases where the necessity for a heterovalent active site (Fe^{III}-M^{II}, where M = Fe, Zn, or Mn) for catalysis has been clearly established. Heterodinuclear $Fe^{III}Zn^{II}$ and $Fe^{III}Mn^{II}$ PAPs have been isolated from plants while those containing a homodinuclear $Fe^{III}Fe^{II}$ center are found in animals. The enzymes isolated from mammals (pig, bovine, mouse, rat, and human [12]-[18]) are ~35 kDa monomeric proteins in which the amino acid sequences are highly conserved with at least 85% identity, whereas plant PAPs extracted from red kidney bean, soybean, sweet potato, duckweed, and tomato are dimeric with a subunit molecular weight of ~55 kDa and amino acid sequences sharing at least 65% identity [19]. However, sequence conservation between plant and mammalian PAPs is low (<20% sequence homology) and is limited predominantly to residues in the catalytic center [19]. The characteristic color of mammalian and plant PAPs (λ_{max} = 510-560 nm; ε= 3000-4000 $M^{-1}cm^{-1}$) is due to a LMCT transition in the active site from a conserved tyrosine ligand to the Fe^{III} center [20]-[22]. In fact, the catalytically active $Fe^{III}Fe^{II}$ form (λ_{max} = 515 nm) of mammalian PAPs can easily (and reversibly) be oxidized to its inactive $Fe^{III}Fe^{III}$ form (λ_{max} = 560 nm) due to the low redox potential (~340 mV) of the divalent iron center [11]. It has thus been suggested that PAP activity may be regulated *in vivo* by reversible oxidation/reduction of the active site [23],[24].

The biological function(s) of PAPs is still uncertain and is dependent on the organism. Mammalian PAPs have been implicated in iron transport in fetal pigs [25],[26], control of bone resorption [27],[28], and macrophage-specific generation of reactive oxygen species (ROS) through Fenton chemistry in response to the invasion of the host organism by pathogens [29]-[31]. It has been speculated that PAP participates in bone resorption by dephosphorylation of bone matrix proteins such as osteopontin [32]. Because of their high expression levels, PAPs have become the most commonly used histochemical marker for osteoporosis [33], making them a major target for the development of chemotherapeutic agents [34]. Recently, it has been demonstrated that active PAP is expressed in certain parts of the rat nervous system and this could be the starting point for understanding whether PAP may be involved in pathological processes such as Alzheimer's disease [35].

For plant PAPs it has been proposed that they may play a fundamental role in mobilizing organic phosphate esters in the soil during germination [36].

The crystal structures of several PAPs from mammals (rat [37],[38], pig [39] and human [40]) and from plants (red kidney bean–rkbPAP [41] and sweet potato [42]) have been reported. However, to date only structures of the inactive oxidized form in the presence of tetraoxoanions (phosphate, tungstate or sulfate) bound to the active site have been determined for mammalian PAPs [37]-[40]. It is interesting to note that despite low overall protein sequence homology, all PAPs contain highly conserved active sites with seven invariant amino acid residues coordinating the dinuclear metal sites. A general scheme for the common active site of PAPs is shown in Figure 3.

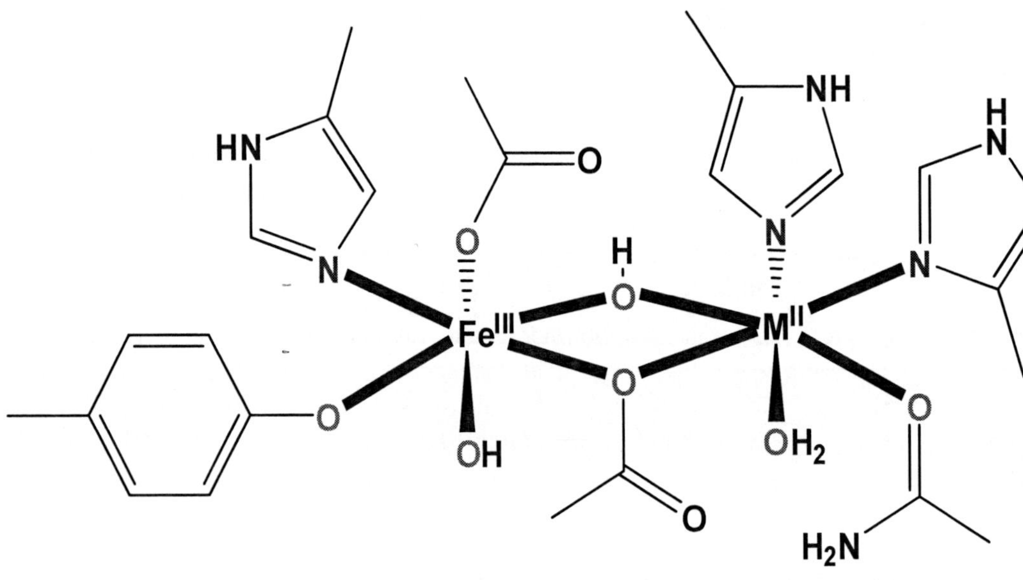

Figure 3. Common dinuclear active site for the animal and plant PAPs where M^{II} can be Fe, Zn or Mn.

In these $Fe^{III}M^{II}$ enzymes, the Fe^{III} and M^{II} metal centers show quite different coordination environments. In the mammalian and red kidney bean PAPs the Fe^{III} ion is coordinated by a tyrosine, a histidine, and a monodentate aspartate, while the divalent M^{II} has two histidines and an asparagine ligand. In order to complete the coordination spheres, exogenous ligands, such as a μ-hydroxo intermetallic bridge (a terminal aqua for M^{II} and a terminal hydroxo for Fe^{III}), were modeled [41] on the basis of spectroscopic [43], magnetic [21], and electrochemical [23],[24] measurements. A monodentate bridging carboxylate group of a 1,1-μ-aspartate residue completes the coordination spheres of the Fe^{III} and M^{II} centers. The equivalence of the coordination environment around the Fe^{III} and M^{II} centers has been demonstrated by M^{II} ion replacement studies which have shown that the Zn^{II}-derivatives of bovine, pig, rat, and human PAPs are kinetically indistinguishable from the native forms [44]-[50], while the di-iron derivative of rkbPAP has physicochemical properties very similar to those of the native $Fe^{III}Zn^{II}$ form [51],[52]. In addition, spectroscopic and magnetochemical studies further support similar active site geometries and properties. For instance, values for the antiferromagnetic exchange coupling in the range of J = -5 to -15 cm^{-1} have been determined for the $Fe^{III}Fe^{II}$ derivative of red kidney bean PAP and bovine spleen PAP, based on the Hamiltonian $H = -2JS_1S_2$ [53]. Also, electronic spectroscopy along with ^1H NMR studies of the di-iron derivative of the red kidney bean enzyme and native mammalian enzymes provide strong evidence that the coordination environments of the two chromophoric Fe^{III} centers are very similar [54]. Interestingly, in the active site of sweet potato PAP which was crystallized with phosphate, the dinuclear $Fe^{III}Mn^{II}$ center is bridged by the anion in a μ,η_2,η_2 mode with one of the phosphoryl oxygen atoms forming a 1,1-bridge while two other oxygen atoms bridge in a 1,3 manner, displacing the exogenous water molecules (Figure 4.) [11],[42].

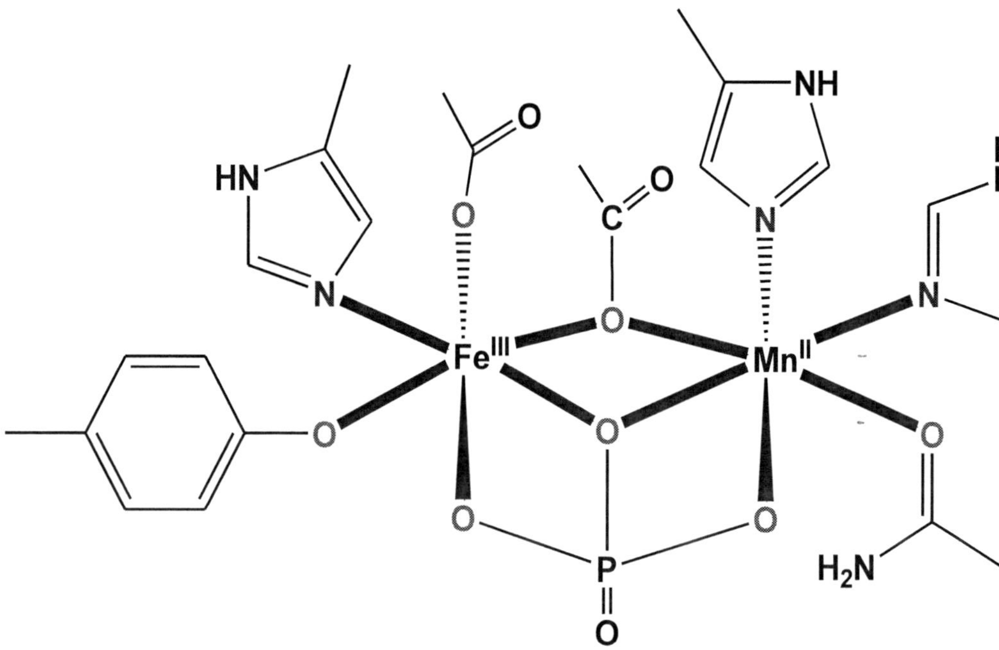

Figure 4. Dinuclear active site of $Fe^{III}Mn^{II}$ PAP.

In contrast to the $Fe^{III}(\mu\text{-}OH)Zn^{II}$ group observed in rkbPAP, based on the strong antiferromagnetic coupling ($-J > 70$ cm^{-1}, $H = -2JS_1S_2$) between the Fe^{III} and Mn^{II} ions and the absence of an EPR signal, it has been proposed that in the active site of sweet potato the metal ions are linked via a μ-oxo bridge [55]. Nevertheless, it should be noted that there are no previous examples of μ-oxo-bridged $Fe^{III}Mn^{II}$ enzymes or complexes described in the literature [56].

Despite the availability of detailed structural data, the catalytic mechanism of PAPs remains a matter of controversy. However, it is important to emphasize that the two chemically distinct environments around the Fe^{III} and M^{II} centers are very likely essential for the catalytic properties of this enzyme [57]. For rkbPAP, a mechanism in which, in the first step of the catalytic cycle, the substrate binds in a monodentate fashion to the Zn^{II} ion, facilitated by the rapid exchange of the bound water molecule, has been proposed (Figure 5a) [41].

The enzyme-substrate complex is oriented in such a way that a terminal Fe^{III}-bound hydroxo group can efficiently attack the phosphorus atom of the substrate, leading to the release of the alcohol product. The monodentate binding of the substrate to Zn^{II} is strongly corroborated by the fact that the addition of phosphate to the $Fe^{III}Zn^{II}$ derivative of bovine spleen PAP does not affect the spectroscopic properties of the Fe^{III} ion greatly at the pH of optimal activity (pH 6.5) [48].In contrast, at lower pH phosphate perturbs the Fe^{III} site, consistent with the formation of a μ-phosphate complex [48].

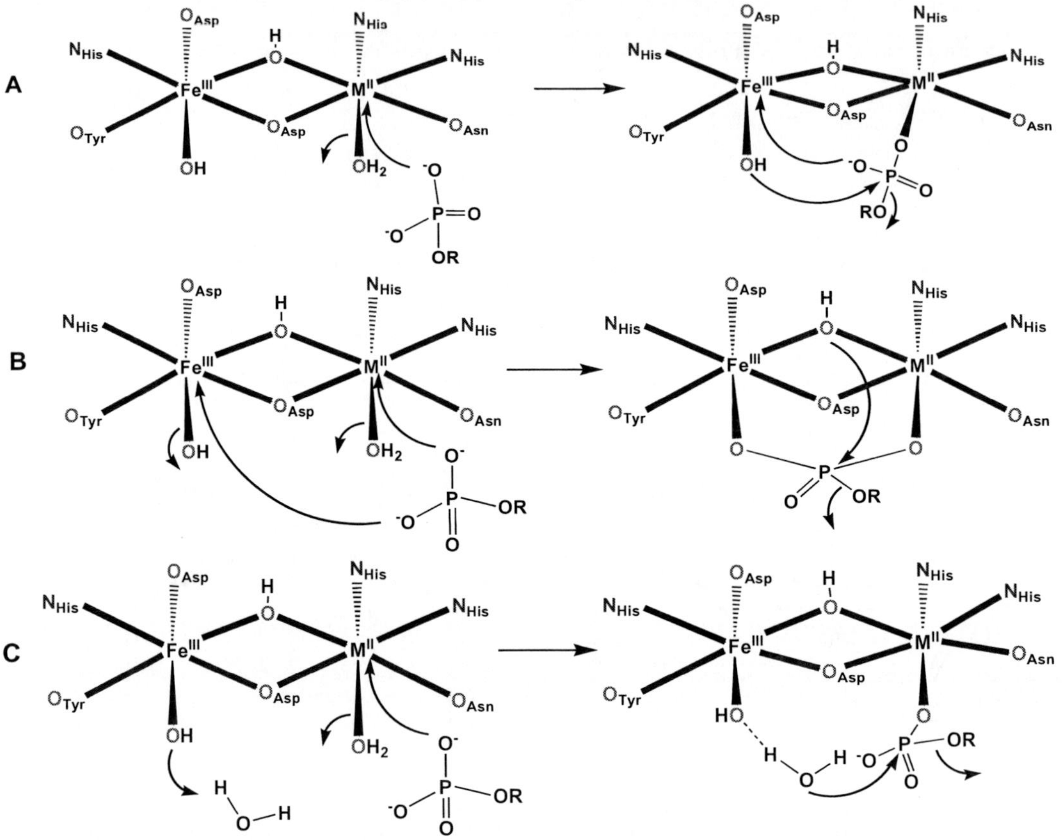

Figure 5. Proposed mechanisms for the hydrolytic PAP cleavage.

On the other hand, pig PAP has a lower pH optimum (pH ~ 5.0) [32],[58] and based on a combination of EXAFS, EPR, and resonance Raman spectroscopic data, it has been proposed that in this enzyme the substrate forms a μ-1,3-phosphate complex, thus placing the μ-hydroxo bridge in an ideal position to act as the reaction-initiating nucleophile (Figure 5b) [59]-[62]. In the case of sweet potato PAP, it has been proposed that a μ-oxo species rather than the μ-hydroxo bridge would be the attacking nucleophile under similar pH conditions [42]. This proposal is corroborated by the X-ray structure of the enzyme which reveals that the dinuclear $Fe^{III}Mn^{II}$ center is bridged by the phosphate in a μ,η_2,η_2 mode with one of the phosphoryl oxygen atoms forming a 1,1-bridge, while two other oxygen atoms bridge in a 1,3 mode (Figure 4) [42].

Alternatively, a mechanism in which the Fe^{III}–OH group can act as a general base to deprotonate a second water molecule, which then attacks the M^{II}-bound phosphate ester, has also been proposed (Figure 5c) [48].

1.3 Structural and Functional Biomimetics for PAPs

In recent years considerable attention has been directed towards the synthesis, X-ray structures and physicochemical characterization of dinuclear $Fe^{III}M^{II}$ complexes to mimic the active site of purple acid phosphatases [63]. Studies on synthetic analogues (i.e. small molecules that resemble the structural and functional sites of PAPs) are a convenient strategy to gain insight into the structural and spectroscopic properties and the mechanisms of action of the corresponding metalloenzymes. The construction of accurate synthetic unsymmetrical models is not a simple matter, since considerable attention must be given to the ligand design in order to achieve a coordination environment which is similar to that imposed by the unique topology of the protein. However, with some exceptions, most reported synthetic models are dinuclear complexes containing symmetrical phenoxo and alkoxo ligands, since the preparation of the ligands is greatly simplified [64]-[67]. Multidentate phenolate dinucleating ligands and their complexes have been studied extensively, and several reviews summarize some of this chemistry [63]-[70]. We will focus in this section mainly on the structural and physicochemical properties, as well as the hydrolase-like activity, of homo- and heterodinuclear $M^{III}M^{II}$ complexes containing the unsymmetrical ligands given in

Figure 6. It is important to emphasize that the availability of dinucleating ligands able to generate systems containing a single terminal Fe^{III}-phenolate bond is crucial in mimicking the structural, spectroscopic, electrochemical, and catalytic properties of PAPs [11],[70].

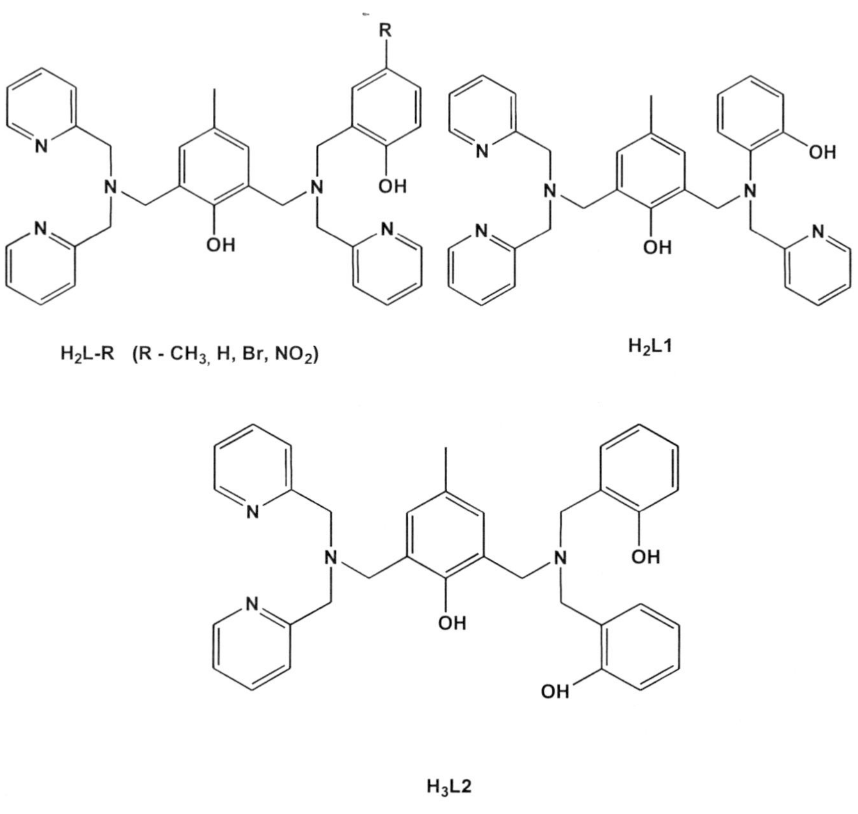

Figure 6. Ligands H_2L-R (R = H, CH_3, Br, NO_2), H_2L1, H_3L2 referred to in the text.

Accordingly to Sargeson and co-workers [71],[72] the catalytically active site of a synthetic hydrolase (Figure 7) should be able to (1) provide two labile sites, *cis*-oriented in order to coordinate the substrate and provide an H_2O/OH^- molecule as the nucleophile; (2) decrease the pK_a of a coordinated water molecule and, as a consequence, provide a hydroxide nucleophile coordinated to the metal at a pH close to neutral; (3) activate the substrate through a nucleophilic attack and/or stabilize the transition state; and (4) release the products at a reasonable rate.

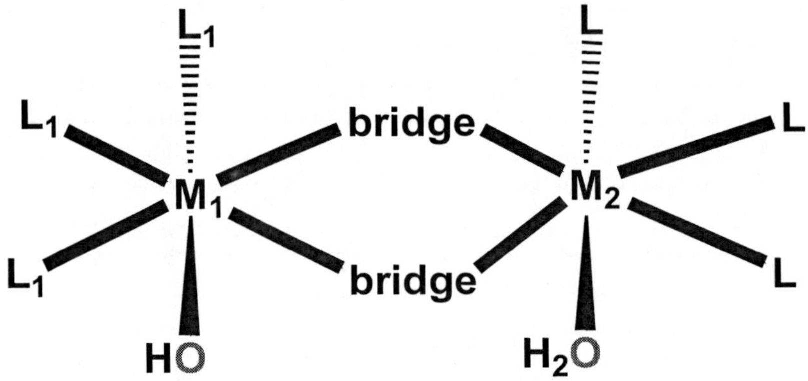

Figure 7. Proposed model for the catalytically active site of a dinuclear synthetic hydrolase.

1.3.1 Physicochemical Properties of [(L-R)FeIII(μ-OAc)2)(μ-PhO)MII]+ and Related Complexes

The ligand 2-{[bis(pyridin-2-ylmethyl)amino]methyl}-6-{[(2-hydroxybenzyl) (pyridin-2-ylmethyl)amino]methyl}-4-methylphenol (H_2L-H) has been widely employed in the preparation of unsymmetrical $M^{III}M^{II}$ mixed-valence complexes as biomimetic models for hydrolases/PAPs [11],[73]-[75]. A general synthetic route for the preparation of H_2L-H, according to Neves and co-workers [76],[77], and its *para*-phenolate substituted derivatives H_2L-R (R = CH_3, Br, NO_2), is given in Figure 8.

Indeed, our group has successfully developed a general method for the preparation of mixed-valence homo- and heterodinuclear $M^{III}M^{II}$ complexes, using the unsymmetrical ligand H_2L-H [57],[76]-[82]. The H_2L-H ligand is able to selectively generate homo- and heterodinuclear $[(L\text{-}H)Fe^{III}(\mu\text{-}OAc)_2)(\mu\text{-}PhO)\ M^{II}]^+$ complexes (PhO = phenoxo and M^{II} = Fe,Cu,Zn,Mn,Ni,Co) when the reactions are carried out with stoichiometric amounts of H_2L-H, $Fe(ClO_4)_3.9H_2O$ and $M(ClO_4)_2.6H_2O$ in CH_3OH in the presence of sodium acetate. The X-ray structures of these complexes have been determined [57],[76]-[80]. The structures of the $[(L\text{-}H)Fe^{III}(\mu\text{-}OAc)(\mu\text{-}PhO)(H_2O)_2Ni^{II}]^{2+}$, $[(L\text{-}H)Fe^{III}(\mu\text{-}PhO)(OH)(H_2O)_3Ni^{II}]^+$, $[(L\text{-}H)Ga^{III}(\mu\text{-}OAc)_2(\mu\text{-}PhO)Zn^{II}]^+$, and $[(L\text{-}H)Mn^{III}(\mu\text{-}OAc)_2(\mu\text{-}PhO)Mn^{II}]^+$ complexes have also been reported [57],[77],[81],[94].

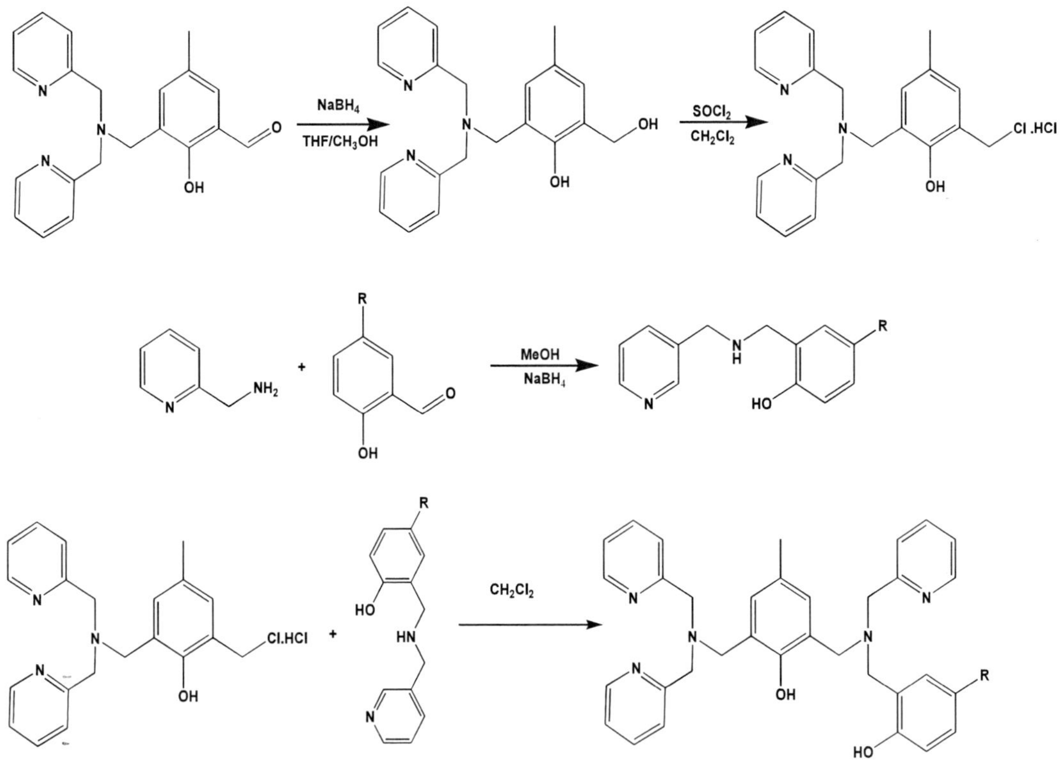

Figure 8. General route for the preparation of the H_2L-R ligands (R = NO_2, Br, H, CH_3).

The main physicochemical properties of the $\{M^{III}(\mu\text{-PhO})M^{II}\}$ complexes employed to model hydrolases/PAPs, which are discussed in this section and are most relevant to our work, are reported in Table 1.

Table 1. Physicochemical properties of complexes employed to mimic PAPs.[a]

Complex $[(L\text{-}H)M^{III}(\mu\text{-OAc})_2(\mu\text{-PhO})M^{II}]^+$	λ_{max}, nm (ε, $M^{-1}cm^{-1}$)	$E^1_{1/2}$, $E^2_{1/2}$(V)[b]	J (cm^{-1})	Ref.
$[Fe^{III}Fe^{II}]^+$	555 (4560), 1050 (60)	-0.49; 0.38	-7.4	[73]
$[Fe^{III}Zn^{II}]^+$	540 (3700)	-0.51		[76]
$[Fe^{III}Mn^{II}]^+$	544 (2680)	-0.47; 0.84	-7.4	[57]
$[Fe^{III}Cu^{II}]^+$	546 (3400)	-0.60; -0.90	-0.5	[79]
$[Fe^{III}Co^{II}]^+$	544 (3760)	-0.57; 0.24	-5.0	[80]
$[Fe^{III}Ni^{II}]^+$	522 (2600)	-0.54; 0.76	-13.3	[78]
$[Ga^{III}Co^{II}]^+$	423 (391)	0.30		[80]
Pig PAP - $Fe^{III}Fe^{II}$	515 (4000)	0.35		[23],[24]
Red kidney bean PAP - $Fe^{III}Zn^{II}$	544 (3200)		-5 to -15	[53]
Sweet potato PAP - $Fe^{III}Mn^{II}$	560 (3200)		-70.0	[55]

[a] All biomimetics have been generated using the ligand 2-{[bis(pyridin-2-ylmethyl)amino]methyl}-6-{[(2-hydroxybenzyl) (pyridin-2-ylmethyl)amino]methyl}-4-methylphenol (H_2L-H). For comparative purposes, corresponding data for PAPs are included. [b] Redox potentials are given *vs* NHE.

The geometric arrangement of the H_2L-H ligand around the metal centers in the mixed-valence $[(L\text{-}H)Fe^{III}(\mu\text{-}OAc)_2(\mu\text{-}PhO)Fe^{II}]^+$ was first proposed based on electrochemical studies and magnetochemical and Mössbauer data [73]. Recently, the X-ray structure of the perchlorate salt of the $Fe^{III}Fe^{II}$ complex has been determined and it shows an arrangement of the $L\text{-}H^{2-}$ ligand around the metal ions which is similar to the proposed structure based on the physicochemical properties of the complex [82]. In the structure of the cation $[(L\text{-}H)Fe^{III}(\mu\text{-}OAc)_2(\mu\text{-}PhO)Fe^{II}]^+$, and all of the other mixed-valence $Fe^{III}M^{II}$ complexes listed in Table 1, the Fe^{III} ion is coordinated by the tridentate N_2O hard site of $L\text{-}H^{2-}$ which contains one amine, one pyridine and the terminal phenolate, while the M^{II} ion lies in the other tridentate compartment containing the amine and soft pyridyl groups. In all of these structures, the metals are in distorted octahedral environments, separated by ~3.50 Å, and are bridged by two exogeneous carboxylate groups and the central phenolate donor group of $L\text{-}H^{2-}$. The $Fe^{III}\cdots M^{II}$ distances of ~3.50 Å are comparable to the 3.31 Å separation reported for uteroferrin [39] and the 3.26 Å reported for red kidney bean and sweet potato PAP [41,42]. The crystal structure of the $[(L\text{-}H)Fe^{III}(\mu\text{-}OAc)_2(\mu\text{-}PhO)Fe^{II}]^+$ cation is given in Figure 9 as an illustration [82].

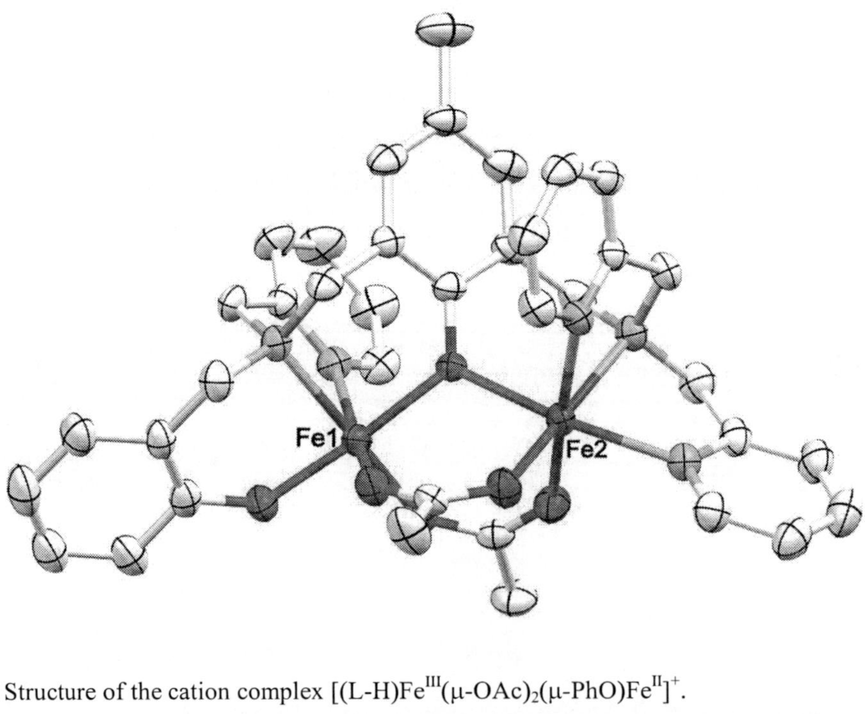

Figure 9. Structure of the cation complex $[(L\text{-}H)Fe^{III}(\mu\text{-}OAc)_2(\mu\text{-}PhO)Fe^{II}]^+$.

Cyclic voltammetry studies in CH_3CN have demonstrated that all $[(L\text{-}H)Fe^{III}(\mu\text{-}OAc)_2(\mu\text{-}PhO)M^{II}]^+$ (M^{II} = Fe,Cu,Zn,Mn,Ni,Co) complexes have a reversible half-wave potential in the -0.5 to -0.6 V range (Table 1), which is attributed to the $Fe^{III}M^{II}/Fe^{II}M^{II}$ redox couple and indicates that the coordination environment around the Fe^{III} ion is similar in these $[(L\text{-}H)Fe^{III}(\mu\text{-}OAc)_2(\mu\text{-}PhO)M^{II}]^+$ complexes when they are in solution. It has also been established that the $[(L\text{-}H)Fe^{III}(\mu\text{-}OAc)_2(\mu\text{-}PhO)Fe^{II}]^+$ complex is a good model for the electrochemical properties of uteroferrin, with a reduction potential for the $Fe^{III}Fe^{III}$ to

$Fe^{III}Fe^{II}$ couple of 380 mV (*versus* NHE), which compares well with the 367 mV measured for uteroferrin at pH 5.0, indicating that the two species have similar Fe^{II} Lewis acidity [73]. The antiferromagnetic coupling constant (J = - 7.4 cm^{-1}, $H = -2JS_1S_2$) and the Mössbauer data at 80 K showing two doublets with isomer shifts of 0.42 and 1.13, and quadrupolar splittings of 1.09 and 2.54 mms^{-1}, respectively, indicated a high spin valence localized $Fe^{III}Fe^{II}$ formulation with $S_1 = 2$ and $S_2 = 5/2$ and these parameters are also comparable to those of reduced uteroferrin, supporting a distorted structure for both the enzyme and the synthetic analogue [21].

Latour and co-workers reported a value of J = -3.3 cm^{-1} and isomer shifts of 0.49 and 1.12, with quadrupolar splittings of 0.59 and 2.37 mms^{-1}, for the $[(L1)Fe^{III}(\mu\text{-mpdp})(\mu\text{-PhO})Fe^{II}]^+$ complex containing mpdp = 1,3-benzenedipropionate as the bridging ligand [74], and this exchange coupling compares more favorably with that obtained (J = -3.0 cm^{-1}) for the phosphate derivative of reduced uteroferrin [83]. It has also been well documented that $[(L\text{-}H)Fe^{III}(\mu\text{-OAc})_2(\mu\text{-PhO})Fe^{II}](ClO_4)$ and its corresponding oxidized $[(L\text{-}H)Fe^{III}(\mu\text{-OAc})_2Fe^{III}](ClO_4)_2$ complex are good models for the resonance Raman properties of mammalian PAPs in their reduced and oxidized forms, respectively [84]. Interestingly, the mixed-valence complex has been used as an electrochemical biomimetic sensor for the determination of phenolic compounds with the sensor being accurate for six months and more than 800 determinations [85]. Importantly, the $[(L\text{-}H)Fe^{III}(\mu\text{-OAc})_2(\mu\text{-PhO})Fe^{II}]^+$ complex shows high catalytic activity in the hydrolysis of the model substrate BDNPP and DNA (See section 2).

Belle *et al.* have reported on the regioselectivity of the ligand H_3L2 to form $Fe^{III}Zn^{II}$ [86] and $Fe^{III}Cu^{II}$ [87] complexes containing μ-phenylphosphate and μ-ethoxo bridges, respectively. Particularly interesting in these studies is the illustration of regioselectivity, through crystallization of the monocopper species to which iron is added and the mixed-valence $Fe^{III}Cu^{II}$ complex is stabilized (Figure 10).

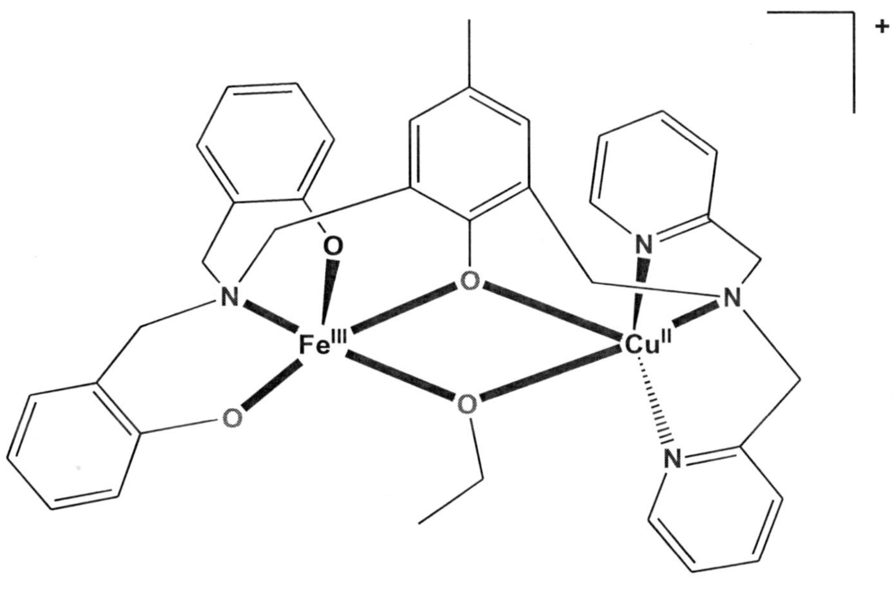

Figure 10. Representation of the structure of the complex $[(L2)Fe^{III}(\mu\text{-OEt})(\mu\text{-PhO})Cu^{II}]^+$.

From cyclic voltammetric measurements the authors have been able to demonstrate that the first reduction of the $Fe^{III}Cu^{II}$ complex is centered on the Cu^{II} ion, leading to the $Fe^{III}Cu^{I}$ species with a comproportionation constant of 1.5 x 10^{10} indicating that the complex has a stable $Fe^{III}Cu^{I}$ form. A μ-methoxo $Fe^{III}Fe^{III}$ complex with the same ligand has also been reported, and the mixed-valence form of this complex is stabilized. However, cyclic voltammetry studies reveal that the reversible $Fe^{III}Fe^{III}/Fe^{III}Fe^{II}$ electron transfer process ($E_{1/2}$ = -045 V *vs.* Ag/Ag^{+}) occurs on the metal center of the bis(pyridyl) moiety, a process which is significantly cathodic shifted when compared to the phenoxo-bridged $[(L\text{-}H)Fe^{III}(\mu\text{-}OAc)_2(\mu\text{-}PhO)Fe^{II}]^{+}$ [73] complex and uteroferrin [23],[24]. The results clearly demonstrate that to adequately mimetize the redox and UV-Vis properties of the $Fe^{III}Fe^{II}$ active site in uteroferrin, only one terminal Fe^{III}-bound phenolate, in conjunction with other softer ligands (amine, pyridyl, imidazol, carboxylate), must be present, in order to complete the hexa-coordination of the Fe^{III} and Fe^{II} ions in the model systems. Unfortunately, functional studies relating these complexes to the active site of PAPs have not yet been undertaken.

The electronic spectra of all $[(L\text{-}H)Fe^{III}(\mu\text{-}OAc)_2)(\mu\text{-}PhO)M^{II}]^{+}$ complexes in CH_3CN solution show absorption maxima in the range of 522-555 nm, which is attributed to the terminal phenolate-to-Fe^{III} charge-transfer transition. The $[(L\text{-}H)Fe^{III}(\mu\text{-}OAc)_2)(\mu\text{-}PhO)Fe^{II}]^{+}$ complex shows an additional intervalence band in the NIR region (1050 nm) and is classified as being a type II mixed-valence species in the scheme of Robin and Day [88]. Spectroelectrochemistry experiments have been employed in order to identify the electronic absorption and redox properties of the oxidized $[(L\text{-}H)Fe^{III}(\mu\text{-}OAc)_2)(\mu\text{-}PhO)Fe^{III}]^{2+}$ species (Figure 11) generated in CH_3CN solution from the reduced complex.

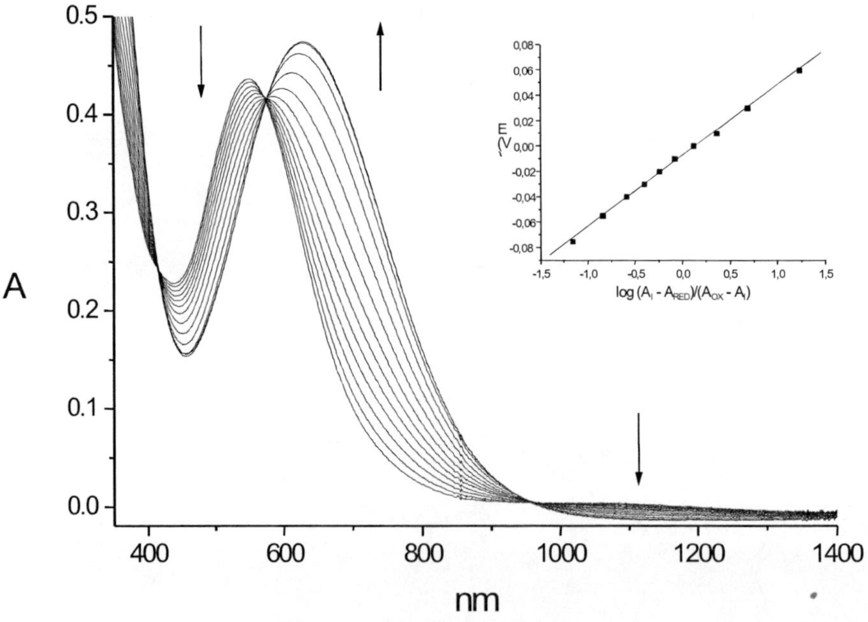

Figure 11. Spectra recorded during spectro-potentiostatic experiments on 1×10^{-5} M $[(L\text{-}H)Fe^{III}(\mu\text{-}OAc)_2)(\mu\text{-}PhO)Fe^{II}]^{+}$ (0.1 M $TBAPF_6$) in CH_3CN. Applied potentials in V *vs* SCE are as follows: -0.075; -0.055; -0.040; -0.030; -0.020; -0.010; 0.000; 0.010; 0.030; 0.060.

During the electrolysis the absorption maximum at 551 nm of the mixed-valence species changes to 633 nm with concomitant disappearance of the broad intervalence charge-transfer band centered at ~1050 nm. The $E^{0'}$ = +0.392 V *vs.* NHE and n = 1.0±0.1 electrons obtained from the Nernst plot are in very good agreement with the results obtained through cyclic voltammetry. Interestingly, the oxidation of the mixed-valence $Fe^{III}Fe^{II}$ complex (λ_{max} = 551 nm, ε = 4600 M^{-1} cm^{-1}) to the corresponding $Fe^{III}Fe^{III}$ species (λ_{max} = 633 nm, ε = 4900 M^{-1} cm^{-1}) is accompanied by essentially no change in the intensity of the charge-transfer absorption, suggesting that these bands in the $Fe^{III}Fe^{II}$ and $Fe^{III}Fe^{III}$ species originate mainly due to the terminal phenolate to Fe^{III} charge-transfer transition. In fact, this conclusion is supported by a theoretical calculation using molecular mechanics and an analysis of molecular orbitals which indicate that the contribution of the bridging phenolate to the charge-transfer transition is negligible [89]. The oxidized $[(L\text{-}H)Fe^{III}(\mu\text{-}OAc)_2(\mu\text{-}PhO)Fe^{III}]^{2+}$ complex has also been synthesized and characterized in the solid state through magnetochemistry and Mössbauer spectroscopy (Table 1) [90]. The isomer shifts and quadrupole splittings are consistent with the presence of distinct Fe^{III} centers (N_3O_3 and N_2O_4 coordination environments) as observed in the Mössbauer spectrum of ^{57}Fe-enriched oxidized purple form of uteroferrin [91]. The unusually large quadrupole splittings suggest strong distortions from octahedral symmetry and are comparable to those observed for uteroferrin.

As expected, the UV-Vis spectrum of the oxidized $[(L\text{-}H)Fe^{III}(\mu\text{-}OAc)_2(\mu\text{-}PhO)Fe^{III}]^{2+}$ complex in CH_3CN solution shows the phenolate to Fe^{III} charge transfer process at 633 nm, ε = 4900 M^{-1} cm^{-1} which is similar to the spectrum obtained from the spectro-potentiostatic experiments. The cyclic voltammogram of the oxidized $[(L\text{-}H)Fe^{III}(\mu\text{-}OAc)_2(\mu\text{-}PhO)Fe^{III}]^{2+}$ species in CH_3CN solution revealed two reversible redox couples at +0.38 and -0.49 V *vs* NHE which were ascribed to the successive redox reactions of $Fe^{III}Fe^{III}$ to $Fe^{III}Fe^{II}$ and $Fe^{III}Fe^{II}$ to $Fe^{II}Fe^{II}$, respectively. These values are identical to those obtained from the CV experiments using the reduced $[(L\text{-}H)Fe^{III}(\mu\text{-}OAc)_2(\mu\text{-}PhO)Fe^{II}]^{+}$ complex, thus indicating that the coordination ligands around the iron centers are identical in the two species. In summary, from the data presented above, of the three ligands shown in

Figure 6, $H_2L\text{-}H$ seems to be the most appropriate to mimetize the spectral, as well as the electrochemical, properties of mammalian PAPs.

Red kidney bean PAP (rkbPAP), the most intensively studied plant enzyme, has a dinuclear $Fe^{III}Zn^{II}$ catalytically active center (Figure 3) in which the metal centers are bridged by a monodentate aspartate residue and one exogeneous μ-hydroxo group with a $Fe^{III}\cdots Zn^{II}$ separation of 3.26 Å in its native form [41]. This distance is increased to 3.33 Å when the metal ions Fe^{III} and Zn^{II} are additionally bridged by a bidentate phosphate group which replaces the solvent molecules bound to the dimetal center of the phosphate-free enzyme. In 2002, Neves and coworkers [76] reported the first $Fe^{III}Zn^{II}$ complex containing a single terminal Fe^{III}-Ophenolate bond with the unsymmetrical ligand $H_2L\text{-}H$ as a structural and functional model for the active site of rkbPAP. The structure of the $[(L\text{-}H)Fe^{III}(\mu\text{-}OAc)_2(\mu\text{-}PhO)Zn^{II}]^{+}$ cation is similar to that presented in Figure 9 for the corresponding $Fe^{III}Fe^{II}$ complex. The crystal structure of the complex reveals that the Fe···Zn distance of 3.490 (9) Å is close to the 3.33 Å separation observed for rkbPAP in the presence of phosphate [41]. On the other hand, the terminally coordinated Fe^{III}-phenolate bond distance (1.890 Å) in the $Fe^{III}Zn^{II}$ model is significantly shorter compared to the $Fe^{III}O_{tyr}$ (2.05 Å) in the native rkbPAP enzyme. The electronic spectrum of the $[(L\text{-}H)Fe^{III}(\mu\text{-}OAc)_2(\mu\text{-}PhO)Zn^{II}]^{+}$ complex in

CH_3CN solution shows a terminal phenolate-to-Fe^{III} charge-transfer transition at 540 nm (ε = 3700 M^{-1} cm^{-1}) which is comparable to the tyrosine-to-Fe^{III} absorption observed at 544 nm (ε = 3200 M^{-1} cm^{-1}) for the rkbPAP enzyme [76].

Based on detailed spectroscopic and potentiometric studies, it has been proposed that upon dissolving the $[(L\text{-}H)Fe^{III}(\mu\text{-}OAc)_2(\mu\text{-}PhO)M^{II}]^+$ complexes in an aqueous solution the dissociation of the carboxylate groups leads to the formation of the catalytically active $[(OH)Fe^{III}(\mu\text{-}OH)M^{II}(OH_2)]$ species, similar to that proposed to be present in the active site of rkbPAP ([40],[41]). In fact, these authors were able to isolate and crystallographically characterize the dinuclear $[L\text{-}H(OH_2)Fe^{III}(\mu\text{-}OH)Zn^{II}]^{2++}$ complex containing the structural $[Fe^{III}(\mu\text{-}OH)Zn^{II}]$ unit and the terminal Fe-bound water molecule at a position equivalent to that of the proposed nucleophile in red kidney bean PAP (Figure 12) [92].

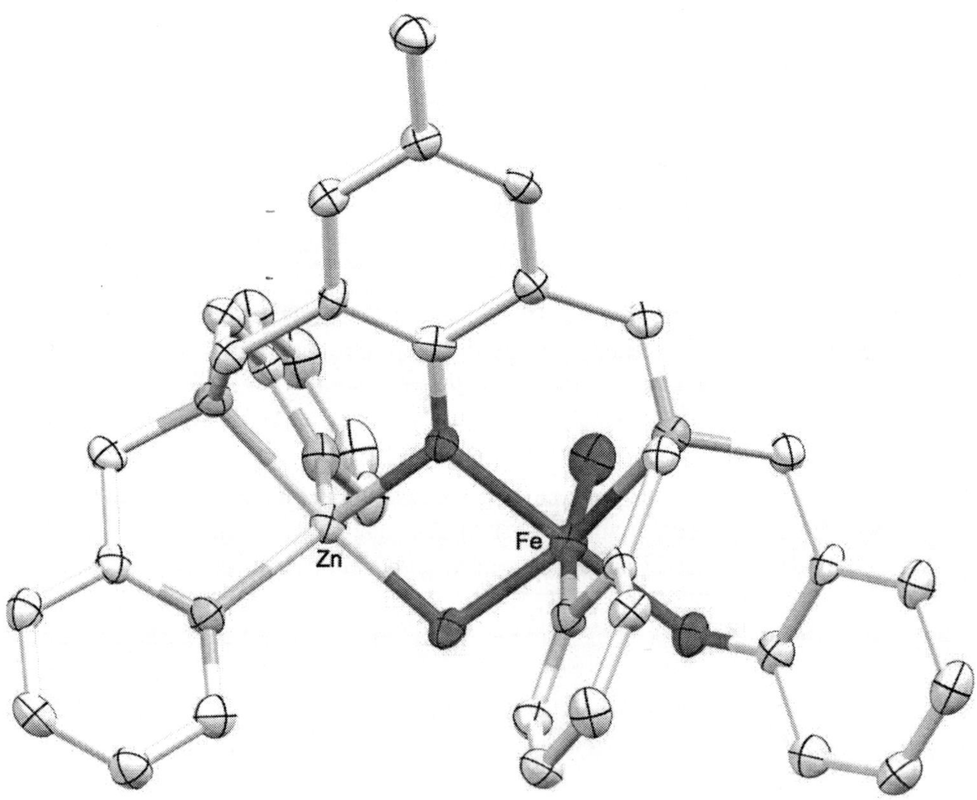

Figure 12. Structure of the complex $[(L\text{-}H)(OH_2)Fe^{III}(\mu\text{-}OH)Zn^{II}]^{2+}$.

In this structure, the Fe···Zn distance of 3.040(1) Å is shorter but comparable to the 3.26 Å found in rkbPAP [41]. In order to probe the role of the Fe - $O_{phenolate}$ bond in the Lewis acidity of the Fe^{III} center, and consequently in the spectroscopic, electrochemical, and kinetic properties of the $[Fe^{III}(\mu\text{-}OH)Zn^{II}]$ unit, the phenolate moiety was modified to contain electron-donating or -withdrawing groups ($-CH_3$, -H, -Br, $-NO_2$) in the 5-position [97]. The structures of the $-CH_3$ and $-NO_2$ derivatives have also been solved by X-ray crystallography [97]. Although the distinct substituents ($-CH_3$, -H, $-NO_2$) at the terminal Fe^{III}-bound phenolate do not have any significant effect on the arrangement of the donor atoms around the metal centers, or on bond lengths and angles within this series of complexes, the same is not true regarding the physicochemical properties of the complexes. For example, the Fe^{III}-to - phenolate charge transfer process is shifted to a lower wavelength (higher energy) as the electron-withdrawing effect of the substituent groups increases while the half-wave ($E_{1/2}$) potentials become more positive as the electron-withdrawing capacity of the *para*-substituent group on the terminal Fe^{III}-bound phenolate increases ($-NO_2 > -Br > -H > -CH_3$) (data not shown), a clear indication that the electron spin density around the Fe^{III} center is strongly dependent on the electronic nature of the *para*-substituent.

The heterodinuclear mixed-valence $[(L\text{-}H)Fe^{III}(\mu\text{-}OAc)_2(\mu\text{-}PhO)Mn^{II}]^+$ complex has been reported as a model for PAP in sweet potato [57]. Both, the Fe···Mn distance of 3.510(9) Å and the terminal phenolate-to-Fe^{III} charge-transfer transition at 544 nm ($\varepsilon = 2680\ M^{-1}\ cm^{-1}$) in the model are comparable to the corresponding parameters (3.26 Å and 560 nm/$\varepsilon = 3600\ M^{-1}\ cm^{-1}$) observed in sweet potato PAP [42]. It is important to note that the terminal Fe^{III}-phenolate bond distance of 1.882(2) Å in the $Fe^{III}Mn^{II}$ biomimetic is similar to the Fe^{III}-tyrosine oxygen Tyr166 distance (1.88 Å) observed in the enzyme [42]. Considering the redox potentials for the $Fe^{III}Mn^{III}/Fe^{III}Mn^{II}$ and $Fe^{III}Mn^{II}/Fe^{II}Mn^{II}$ couples, determined by cyclic voltammetry (Table 1) and the comproportionation reactions, it was concluded that the mixed-valence $[(L\text{-}H)Fe^{III}(\mu\text{-}OAc)_2(\mu\text{-}PhO)Mn^{II}]^+$ complex is the most stable when compared to the homodinuclear $Fe^{III}Fe^{II}$ and $Mn^{III}Mn^{II}$ analogues [77]. Magnetic susceptibility studies of $[(L\text{-}H)Fe^{III}(\mu\text{-}OAc)_2(\mu\text{-}PhO)Mn^{II}]^+$ revealed weak antiferromagnetic coupling of the Fe^{III} and Mn^{II} centers with an exchange coupling constant $J = -\ 6.8\ cm^{-1}$ ($H = -2JS_1S_2$, $S_1 = S_2 = 5/2$). The significantly lower value of J in $[(L\text{-}H)Fe^{III}(\mu\text{-}OAc)_2(\mu\text{-}PhO)Mn^{II}]^+$ when compared to the enzyme ($J > -\ 70\ cm^{-1}$) [55] was attributed to the presence of a μ-oxo bridge in the enzyme in contrast to a μ-phenolate bridge in the model complex.

1.3.2 Hydrolase-like Activity of the [(L-H)MIII(μ-OAc)2(μ-PhO)MII]+ and [(L-R)(OH2)FeIII(μ-OH)ZnII]2+ Complexes

The design and development of suitable biomimetic catalytic systems capable of mimicking the functional properties of enzymes continues to represent a challenge for bioinorganic chemists. Indeed, metalloenzymes such as purple acid phosphatases (PAPs), for which mechanistic studies are available, have been used as a successful starting point in the development of a class of synthetic metal complexes known as synthetic hydrolases and/or chemical nucleases [93]. In this regard, new compounds able to catalytically hydrolyze biomolecules such as proteins, phospholipids, ATP, DNA, RNA and model substrates are of current interest because studies on these synthetic analogues are a convenient strategy to gain insight into the structural and spectroscopic properties and the catalytic mechanism of the corresponding metalloenzymes.

As emphasized in section 1.3.1., unsymmetrical dinucleating ligands able to generate mixed-valence $[Fe^{III}(\mu\text{-}OH)M^{II}]$ systems containing a single terminal Fe^{III}-phenolate bond have been considered crucial to mimicking the structure of the active site and the physicochemical properties and hydrolase-like activity of the purple acid phosphatase metalloenzymes (Figure 3), [11],[69],[70]. Among the unsymmetrical phenolate-type ligands reported in the literature [74],[86],[87], H_2L-H (

Figure 6) has been assigned as being probably the best and most widely employed ligand for mimicking the mixed-valence $Fe^{III}M^{II}$ (M^{II} = Fe, Zn, Mn) coordination environment of the metals in the active site of PAP [11],[70]. In this section we will focus mainly on the catalytic activity of the [L-R(OH)M^{III}(μ-OH)M^{II}]$^+$ complexes (see H_2L-R in

Figure 6) in the hydrolysis of the DNA model substrate 2,4-bis(dinitrophenyl)phosphate (Table 2).

Solution potentiometric titrations studies of the [(L-H)M^{III}(μ-OAc)$_2$(μ-PhO)M^{II}]$^+$ complexes listed in Table 2 (with the exception of $Fe^{III}Fe^{II}$) in water/ethanol (30:70%) have revealed three protonation equilibria in the pH range 4-10 [57],[76],[78]-[80],[82],[92],[94],[97].

Table 2. Protonation equilibria for the [(L-H)M^{III}(μ-OAc)$_2$(μ-PhO)M^{II}]$^+$ complexes.

Complex^a	pK_{a1}	pK_{a2}	pK_{a3}	Reference
[(L-H)Fe^{III}(μ-OAc)$_2$(μ-PhO)Zn^{II}]$^+$	4.84	5.99	7.97	[76]
[(L-H)Fe^{III}(μ-OAc)$_2$(μ-PhO)Mn^{II}]$^+$	5.80		7.76	[57]
[(L-H)Fe^{III}(μ-OAc)$_2$(μ-PhO)Fe^{II}]$^+$	4.13	5.76	7.53	[82]
[(L-H)Fe^{III}(μ-OAc)$_2$(μ-PhO)Cu^{II}]$^+$	5.25	6.20	7.82	[79]
[(L-H)Fe^{III}(μ-OAc)$_2$(μ-PhO)Ni^{II}]$^+$	5.30	6.80	8.61	[78]
[(L-H)Fe^{III}(μ-OAc)$_2$(μ-PhO)Co^{II}]$^+$	5.00	6.58	8.31	[80]
[(L-H)Ga^{III}(μ-OAc)$_2$(μ-PhO)Co^{II}]$^+$	5.46	6.57	8.30	[80]
[(L-H)Ga^{III}(μ-OAc)$_2$(μ-PhO)Zn^{II}]$^+$	5.59	6.19	7.96	[94]

^{a} Data were obtained through potentiometric titration in water/ethanol (30:70%), 298 K, μ 0.1M KCl.

For the [(L-H)M^{III}(μ-OAc)$_2$(μ-PhO)M^{II}]$^+$ complexes it has been proposed that upon dissolving the complexes, the dissociation of the first carboxylate leads to a species described as [(OH)M^{III}(μ-OAc)M^{II}(OH_2)] (M^{III} = Fe, Ga and M^{II} = Zn, Mn, Cu, Ni, Co), where the pK_{a1} value is consistent with the dissociation of the Fe^{III}-OH_2 bound water molecule [39],[41],[42],[57],[80],[92],[94],[97]. Dissociation of the second carboxylate bridge results in the formation of a species in which a second water molecule is bound to the M^{III} center and upon its deprotonation (pK_{a2}) an M^{III}(μ-OH)M^{II} species is formed. This second pK_a apparently does not affect the rate of catalysis (*vide infra*). Finally, deprotonation of the M^{II}-OH_2 bond (pK_{a3}) leads to the [(OH)Fe^{III}(μ-OH)M^{II}(OH)] species. The distinct pK_{a3} values observed for the $Fe^{III}M^{II}$ (M^{II} = Zn, Mn, Fe, Cu, Ni) complexes are in agreement with the trend observed for the first deprotonation of the [M(OH_2)$_6$]$^{2+}$ complexes. Moreover, it is important to note that the pK_{a1} values related to the deprotonation of the Fe^{III}-OH_2 bond are comparable to the corresponding pK_a values observed for PAPs. The pK_{a1} value corresponding to deprotonation of a terminal M^{III}-bound water molecule is higher for Ga^{III} (5.59) when compared to the corresponding Fe^{III} (4.84) in the isostructural $Fe^{III}Zn^{II}$ complex [94]. A similar trend has been observed in the [(L-H)Fe^{III}(μ-OAc)$_2$(μ-PhO)Co^{II}]$^+$ and [(L-H)Ga^{III}(μ-OAc)$_2$(μ-PhO)Co^{II}]$^+$ complexes [80] and these values are in agreement with those determined for the first deprotonation of the hexaaquo complexes of Ga^{III} and Fe^{III}, which have pK_a values of 3.1 and 2.7, respectively. As expected, it has been observed that the values for pK_{a3}, which is related to the deprotonation of the M^{II}-bound water molecule, are similar in the $Fe^{III}Zn^{II}$/$Ga^{III}Zn^{II}$

(~7.96) and $Fe^{III}Co^{II}/Ga^{III}Co^{II}$ (8.30) complexes [76],[80],[94]. It has also been observed that the Mn^{II} to Ni^{II} substitution in the $[(L\text{-}H)Fe^{III}(\mu\text{-}OAc)_2(\mu\text{-}PhO)Mn^{II}]^+$ complex results in a decrease of 0.8 pH units for the M^{II}-bound water molecule, in agreement with the trend observed for the first deprotonation of the hexaaqua complexes of Mn^{II} and Ni^{II}, which are 10.9 and 10.2, respectively, at 0.1 M [57],[78].

For the homodinuclear $[(L\text{-}H)Fe^{III}(\mu\text{-}OAc)_2(\mu\text{-}PhO)Fe^{II}]^+$ complex, potentiometric as well as spectrophotometric titrations in 1:1 $CH_3CN:H_2O$ solution indicated the presence of four protonation equilibria [82],[95] although the fit of the potentiometric data did not allow the determination of the lowest pK_a value (~3.02). In this case, pK_{a2} = 5.76 was attributed to deprotonation of the bridging hydroxo to a μ-oxo bridge (Figure 13 and Table 2) and this process is presumably not catalytically relevant, while pK_{a1} = 4.13 is in good agreement with the pK_{es1} of uteroferrin of 4.15 [34]. Similarly, potentiometric titration of the oxidized complex $[(L\text{-}H)Fe^{III}(\mu\text{-}OAc)_2(\mu\text{-}PhO)Fe^{III}]^{2+}$ also showed the presence of four protonation equilibria and, in this case, four reliable pK_{as} values (3.14; 4.22; 6.45; 9.65) could be obtained in the pH range of 2.4-10.9. While the three acidic pK_as values have been attributed to the corresponding protonation/deprotonation equilibria proposed for the $Fe^{II}Fe^{III}$ complex, pK_a = 9.65, was assumed to be deprotonation of a Fe^{III}-bound water molecule after substitution of one pyridine group bound to one Fe^{III} center. It is important to note that replacement of the divalent Fe^{II} ion in the mixed-valence $[(L\text{-}H)Fe^{III}(\mu\text{-}OAc)_2(\mu\text{-}PhO)Fe^{II}]^+$ complex with a trivalent Fe^{III} ion decreases the pK_{a2} by 1.5 pH units, suggesting that the assignment of this pK_a to the deprotonation of a bridging hydroxo to a μ-oxo bridge is reasonable.

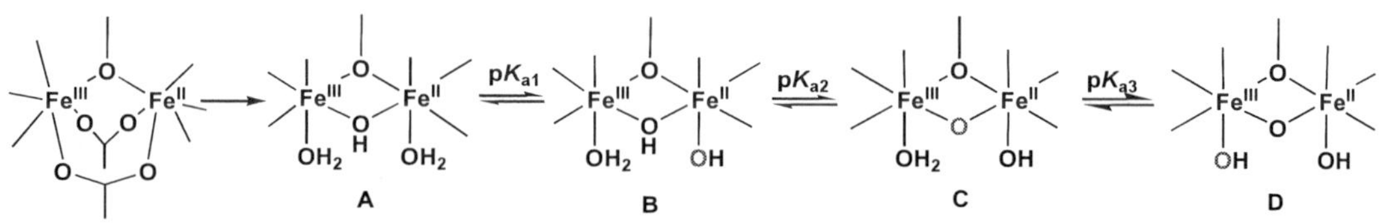

Figure 13. Proposed equilibrium for $[(L\text{-}H)Fe^{III}(\mu\text{-}OAc)_2(\mu\text{-}PhO)Fe^{II}]^+$.

Potentiometric titration studies of the $[L\text{-}R(OH_2)Fe^{III}(\mu\text{-}OH)Zn^{II}]^{2+}$ complexes (R= CH_3, H, Br, NO_2) in a water:ethanol (30:70) solution showed the neutralization of 3 mol of KOH/mol of complex in the pH range of 2–10, which resulted in the pK_{as} values given in Table 3 [92],[97].

Table 3. Protonation equilibria and Hammett constants for the $[L\text{-}R(OH_2)Fe^{III}(\mu\text{-}OH)Zn^{II}]^{2+}$ complexes.

Complex[a]	pK_{a1}	pK_{a2}	pK_{a3}	σ_p	Ref.
$[L\text{-}CH_3(OH_2)Fe^{III}(\mu\text{-}OH)Zn^{II}]^{2+}$	3.28	4.88	8.43	-0.14	[97]
$[L\text{-}H(OH_2)Fe^{III}(\mu\text{-}OH)Zn^{II}]^{2+}$	3.01	4.66	8.25	0	[92],[97]
$[L\text{-}Br(OH_2)Fe^{III}(\mu\text{-}OH)Zn^{II}]^{2+}$	2.73	4.64	7.80	0.26	[97]
$[L\text{-}NO_2(OH_2)Fe^{III}(\mu\text{-}OH)Zn^{II}]^{2+}$	2.55	4.25	8.25	0.81	[97]

[a] Data were obtained through potentiometric titration in water/ethanol (30:70%), 298 K, μ=0.1M KCl.

These pK_a values have been attributed to the following equilibria: [(OH$_2$)Fe(μ-OH$_2$)Zn(OH$_2$)] $\rightleftharpoons$ [(OH$_2$)Fe(μ-OH)Zn(OH$_2$)] $\rightleftharpoons$ [(OH)Fe(μ-OH)Zn(OH$_2$)] $\rightleftharpoons$ [(OH)Fe(μ-OH)Zn(OH)]. The [(OH$_2$)Fe(μ-OH)Zn(OH$_2$)] species is the one determined by X-ray crystallography, containing an additional Zn-coordinated water molecule. EXAFS spectroscopy studies of [L-H(OH$_2$)Fe^{III}(μ-OH)Zn^{II}]$^{2+}$ indicated that both metal centers are six-coordinated in solution. In this study [92], the authors demonstrated that the pK_{a2} values attributed to deprotonation of the terminal Fe^{III}-bound water for these complexes have a linear dependence on the Hammett constants, indicating the strong influence of the ligand on the electron spin density of the Fe^{III} center as the electron donor/acceptor nature of the ligand substituent is changed. Thus, as expected within the series of complexes, the [(L-CH$_3$)Fe^{III}Zn^{II}(μ-OH)(H$_2$O)]$^{2+}$ complex shows the highest pK_{a2} value, while the [(L-NO$_2$)Fe^{III}Zn^{II}(μ-OH)(H$_2$O)]$^{2+}$ derivative has the lowest.

Phosphatase-like activity of all the [(L-H)Fe^{III}(μ-OAc)$_2$(μ-PhO)M^{II}]$^+$ complexes has been investigated with the DNA model substrate 2,4-bis(dinitrophenyl) phosphate (BDNPP) and the kinetic parameters are summarized in Table 4.

Table 4. Kinetic parameters for the hydrolysis of BDNPP by the [(L-H)Fe^{III}(μ-OAc)$_2$(μ-PhO)M^{II}]$^+$ complexes, at 298 K, μ = 0.1 M.

[M^{III}M^{II}]$^+$	K_M, M	k_{cat}, s^{-1}	k_{cat}/K_M M^{-1} s^{-1}	k_{cat}/k_{unc}	**Ref.**
[Fe^{III}Mn^{II}]	2.1x10^{-3}	7.1x10^{-4}	3.3x10^{-1}	3.9x10^{3}	[57]
[Fe^{III}Zn^{II}]	8.1x10^{-3}	1.1x10^{-3}	1.3x10^{-1}	6.1x10^{3}	[76]
[Fe^{III}Cu^{II}]	1.1x10^{-2}	1.8x10^{-3}	1.6x10^{-1}	1.0x10^{4}	[79]
[Fe^{III}Ni^{II}]	3.8x10^{-3}	4.4x10^{-4}	1.2x10^{-1}	2.4x10^{3}	[78]
[Fe^{III}Co^{II}]	9.3x10^{-2}	1.4x10^{-3}	1.5x10^{-2}	7.6x10^{3}	[80]
[Ga^{III}Zn^{II}]	7.1x10^{-3}	1.4x10^{-3}	2.0x10^{-1}	7.6x10^{3}	[94]
[Ga^{III}Co^{II}]	8.8x10^{-2}	1.9x10^{-3}	2.1x10^{-2}	1.0x10^{4}	[80]

These hydrolysis reactions are strongly dependent on pH giving rise to bell-shaped pH:rate profiles and pH optimum in the range of 5.5-7.0, and they are approximately four orders of magnitude faster than the hydrolysis rate of the uncatalyzed reaction (1.8 x 10^{-7} s^{-1}) of the BDNPP substrate [6].

In each case, the two kinetically relevant pK_a values correspond to the lowest and highest pK_a values calculated from the potentiometric titration experiments (pK_{a1} and pK_{a3}), suggesting that the catalytically active species is of the type [(OH)M^{III}(μ-OH)M^{II}(OH$_2$)]. In addition, at higher pH values (>6.5) a decrease in reactivity occurs because of the presence of the fully deprotonated form [(OH)M^{III}(μ-OH)M^{II}(OH)] which reduces the leaving tendency of the OH$^-$ ion from the M^{II}-OH group, thus preventing coordination by the substrate. As a general mechanism, it has been postulated that hydrolysis of the diester BDNPP occurs by intramolecular nucleophilic attack of the terminally M^{III}-bound OH$^-$ group on the phosphorus atom after monodentate binding of the substrate to the divalent metal ion (Figure 14).

For the [(L-H)Fe^{III}(μ-OAc)$_2$(μ-PhO)Cu^{II}]$^+$ complex, an alternative mechanism in which a μ-hydroxo/oxo is the initiating nucleophile has been proposed based on the observation by EXAFS that upon dissolving the complex in ethanol:water (70:30%) the coordination number of Fe changes from 6 to 5 thus supporting the assignment of pK_{a1} (Table 2) to deprotonation of the [Fe^{III}(μ-OH)Cu^{II}] group [79].

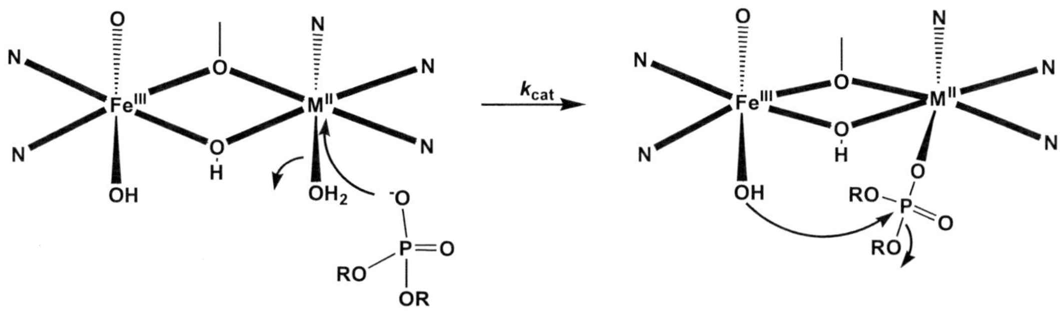

Figure 14. Proposed mechanism of the hydrolysis of BDNPP catalyzed [(L-H)Fe^{III}(μ-OAc)$_2$(μ-PhO) M^{II}]$^+$ complexes (M^{II} = Zn, Mn, Ni, Co).

In order to probe the role of the metal ions in enzymatic catalysis, the effect of metal ion substitutions on reactivity has been studied. The Fe^{III} center was replaced by Ga^{III} to produce $Ga^{III}Fe^{II}$ and $Ga^{III}Zn^{II}$ in pig PAP and bsPAP. These derivatives exhibit catalytic activities similar to those of the native $Fe^{III}Fe^{II}$ forms, providing evidence that diamagnetic Ga^{III} can act as a functional analog of paramagnetic Fe^{III} [81],[94],[95]. The [(L-H)Ga^{III}(μ-OAc)$_2$(μ-PhO)Zn^{II}]$^+$ and [(L-H)Ga^{III}(μ-OAc)$_2$(μ-PhO)Co^{II}]$^+$ biomimetics show higher phosphatase-like activity than that observed for the corresponding $Fe^{III}Zn^{II}$ and $Fe^{III}Co^{II}$ complexes at optimal pH, and this observation suggests an M^{III}-bound hydroxo as the nucleophile [76],[80],[94]. In fact, the higher pK_{a2} values obtained for the Ga^{III}-bound hydroxo when compared with the corresponding Fe^{III} complexes are in agreement with both the more alkaline pH optimum for the activity of the $Ga^{III}Zn^{II}$ and $Ga^{III}Co^{II}$ complexes when compared with the related $Fe^{III}Zn^{II}$ and $Fe^{III}Co^{II}$ complexes, and the relative pK_a values for the first deprotonation of the hexaaqua complexes of Ga^{III} and Fe^{III}. It is relevant to note that, similarly to the biomimetics described by Neves and co-workers, the Ga^{III} derivatives of bsPAP and Uf are more reactive than their Fe^{III} counterparts [81],[94], suggesting a common catalytic role of the M^{III} center in both the PAPs and the model systems.

Through an extensive study of the $Fe^{III}Zn^{II}$ complexes, Neves et al. [92] provided relevant information on the mechanism of the biomimetics and of the rkbPAP itself. The structures of the [(L-R)(OH_2)Fe^{III}(μ-OH)Zn^{II}]$^+$ (R = H, CH_3) complexes presented in Figure 12 are similar to the catalytically active species obtained upon dissolving [(L-R)Fe^{III}(μ-OAc)$_2$(μ-PhO)Zn^{II}]$^+$ in aqueous solution with an additional Zn-coordinated water molecule. This structure is also similar to that proposed to be present in the active site of rkbPAP [41]. In a further study using the [(L-R)(OH_2)Fe^{III}(μ-OH)Zn^{II}]$^{2+}$ (R = CH, H, Br, NO_2) as catalysts these authors were able to demonstrate that phosphatase activity increases in the order -NO_2 <-Br <-H <-CH_3 when 2,4-bis(dinitrophenyl)phosphate (BDNPP) was used as the substrate [97].

The resulting bell-shaped pH rate profiles (Figure 15) show pH optima in the range 6.5-7.0, and pK_as values in good agreement with those obtained from potentiometric titration, demonstrating that the catalytically active species is the monodeprotonated form [(OH)Fe^{III}(μ-OH)Zn^{II}(OH_2)]$^+$ within the series of complexes.

Table 5. Kinetic parameters for the hydrolysis of BDNPP, catalyzed by the $[(L\text{-}R)(OH_2)Fe^{III}(\mu\text{-}OH)Zn^{II}]^{2+}$ complexes, at 298 K, μ = 0.1 M.

Complex	K_M, M	k_{cat}, s^{-1}	k_{cat}/K_M M^{-1} s^{-1}	k_{cat}/k_{unc}	Ref.
$[L\text{-}CH_3(OH_2)Fe^{III}(\mu\text{-}OH)Zn^{II}]^+$	5.2×10^{-3}	9.74×10^{-4}	19×10^{-2}	5.2×10^3	[97]
$[L\text{-}H(OH_2)Fe^{III}(\mu\text{-}OH)Zn^{II}]^+$	4.2×10^{-3}	9.13×10^{-4}	20×10^{-2}	4.8×10^3	[97]
$[L\text{-}Br(OH_2)Fe^{III}(\mu\text{-}OH)Zn^{II}]^+$	6.0×10^{-3}	6.55×10^{-4}	11×10^{-2}	3.5×10^3	[97]
$[L\text{-}NO_2(OH_2)Fe^{III}(\mu\text{-}OH)Zn^{II}]^+$	3.5×10^{-3}	4.20×10^{-4}	12×10^{-2}	2.2×10^3	[97]

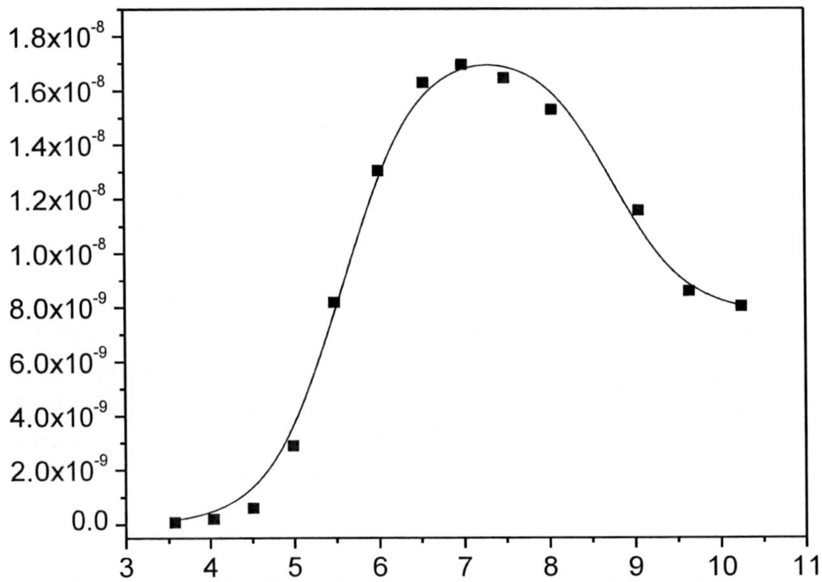

Figure 15. Kinetic profile of the hydrolysis of BDNPP catalyzed by $[(L\text{-}CH_3)Fe^{III}(\mu\text{-}OH)(\mu\text{-}PhO)Zn^{II}]^{2+}$, in CH_3CN:H_2O (50:50%), μ = 0.1M at 298K.

UV-Vis and EPR studies indicated that the diester BDNPP does not interact with the Fe^{III} center, and the measured kinetic isotope effect k_H/k_D of 0.93-1.34 suggests that no proton transfer is involved in the rate-determining step of the reaction, thus supporting an intramolecular nucleophilic attack by the terminal Fe^{III}-bound hydroxo group and the concomitant release of 2,4-dinitrophenolate. Finally, the authors were also able to obtain a Brønsted plot of log k_{cat} vs the pK_{a2} of the Fe^{III}-OH_2 group which exhibited a good linear correlation with a β_{nuc} = 0.60 [97]. This value is substantially higher than those observed for the uncatalyzed hydrolyses of phosphate diesters and triesters (β_{nuc} values around 0.30) [98][99], therefore indicating that the hydroxide-catalyzed hydrolysis of BDNPP involves more nucleophilic contribution in the transition state.

Based on the fact that the $[(L\text{-}R)(OH_2)Fe^{III}(\mu\text{-}OH)Zn^{II}]^{2+}$ (R = CH_3, H, Br, NO_2) complexes are structurally very similar and that the difference in reactivity results from varying the strength of the nucleophile (Fe-OH) due to the change of the *p*-substituent on the terminal Fe-bound phenolate, a common mechanism as shown in

Figure 16, has been proposed for the hydrolysis reaction of the phosphate diester BDNPP.

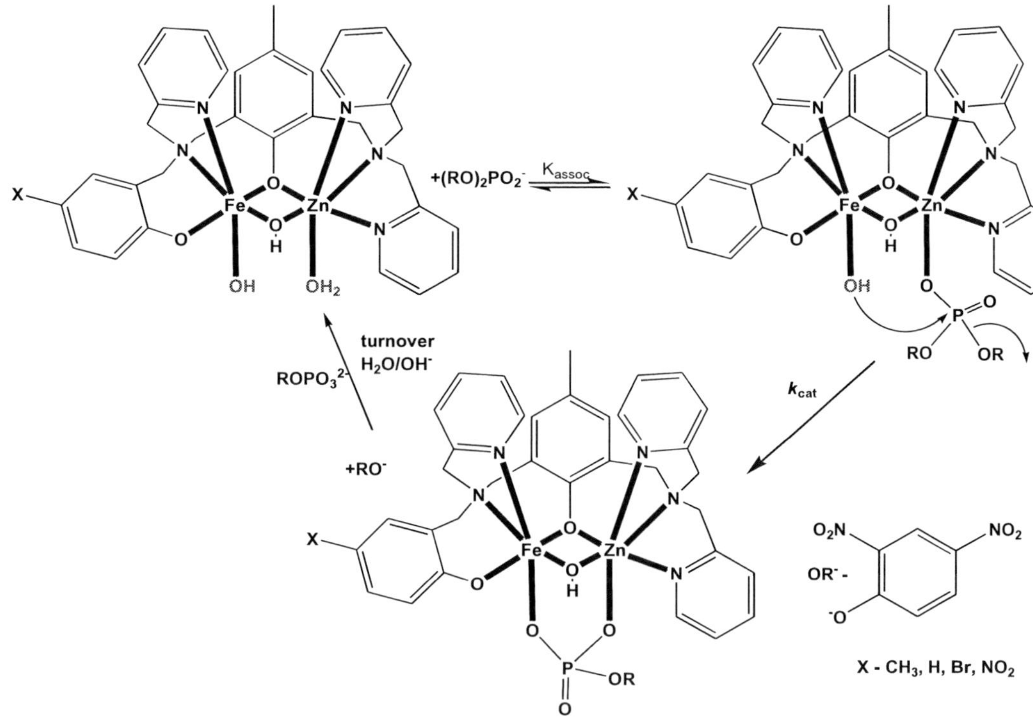

Figure 16. Proposed common mechanism for the hydrolysis of BDNPP by the $[(L\text{-}R)(OH_2)Fe^{III}(\mu\text{-}OH)Zn^{II}]^{2+}$ complexes.

In the first phase of the mechanism $[(OH)Fe^{III}(\mu\text{-}OH)Zn^{II}(OH_2)]^+$ is the catalytically active species and possesses a labile Zn^{II} site that permits the monodentate binding of the substrate. Next, an intramolecular nucleophilic attack by the *cis* oriented Fe^{III}-bound hydroxide leads to the release of 2,4-dinitrophenolate with the monoester 2,4-dinitrophenylphosphate remaining coordinated to the complex; and finally, the μ-1,3-coordinated monoester intermediate undergoes substitution by two water molecules and regenerates the active site for the next catalytic cycle. Therefore, it has been concluded that the μ-hydroxo is a significantly poorer nucleophile than the terminally Fe^{III}-bound hydroxo group, and that hydrolysis of the monoester to form inorganic phosphate and 2,4-dinitrophenolate occurs exclusively due to the background reaction.

2 Hydrolytic DNA Cleavage

Nucleic acids are important targets for biotechnology, medicine and basic molecular biology and biochemistry. Nucleases, the enzymes capable of DNA/RNA cleavage, are a group of metalloenzymes with a great variety of active site motifs linked to many different metal ions. DNA-metal complex interactions have been studied in order to gain insight into the mechanisms governing the cleavage of phosphodiester bonds. To this end, several mononuclear and binuclear metal complexes, as described earlier in this chapter, have been synthesized and carefully assayed toward BDNPP hydrolysis. Herein, we will examine in detail the interaction of these complexes with DNA, and of particular complexes obtained to mimic hydrolases/nucleases, which were also able to cleave proteins – catalytic promiscuity.

It is clear that, with a view to their use in DNA-targeted drugs, metal complexes may be capable of the hydrolytic cleavage of the phosphodiester bond, leading to reaction products containing 5'-phosphate and 3'-hydroxyl groups (Figure 17). This is important because these products are obtained after natural metallonucleases catalyze the cleavage reaction. This implies that after the chemical attack of the metal complexes on DNA, the reaction products may be bound by natural enzymes. The search for efficient synthetic metallonucleases, following a hydrolytic pathway for the cleavage of DNA, and also an understanding of the mechanisms governing the catalysis, will be important in the design of more effective and more specific synthetic nucleases, able to recognize DNA target sequences.

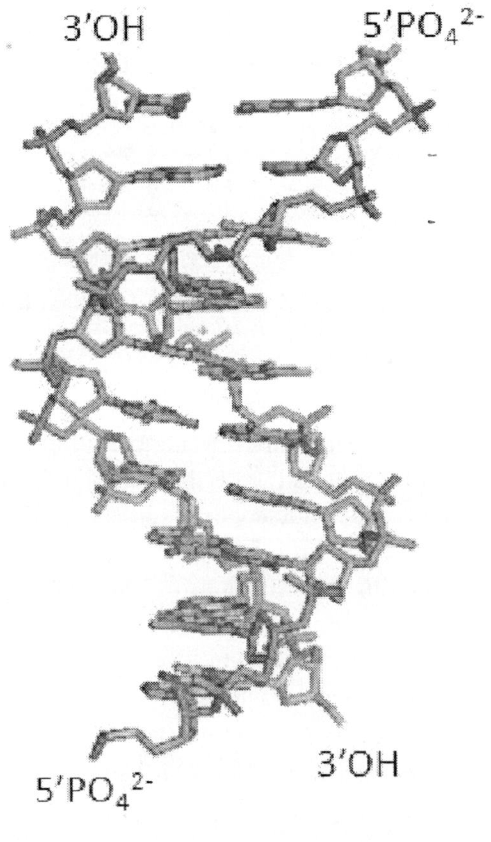

Figure 17. DNA molecule.

It is very interesting to note that until now the majority of DNA metallonucleases are active in the DNA hydrolysis at concentrations higher than those of the natural substrate, and turnover numbers are rarely reported. This indicates that the model systems mimicking natural hydrolysis may be further improved in terms of catalytic efficiency and selectivity.

Magnesium is the most common metal ion present in nuclease enzymes, in virtue of its abundance and redox inertness. However, other metal sites are frequently found, such as Zn^{II}, $Fe^{II/III}$ and Mn^{II}. The exact role of the metal ions in catalysis is still under investigation for many enzymes, however, their role in the reaction mechanism is evidenced from the following relevant indications: metal ions (i) define a geometrical scaffold around the active site; (ii) facilitate nucleophilic attack; and (iii) activate phosphodiester bonds.

As in the case of any enzyme, natural metallonucleases use a variety of strategies to achieve efficient chemical cleavage of phosphodiester bonds, such as correct three-dimensional positioning of the substrate and catalytic amino acid side chains, stabilization of the transition state, and high turnover numbers. Artificial or synthetic metallonucleases are generally incapable of achieving all these functions and, thus, the stronger the nucleophile of the model, the better the artificial metallonuclease. Model metallonucleases can further be improved by increasing their ability to interact with DNA, or by adding intercalators, groove-binders or positive charges to the metal complexes. These modifications may emulate the surface of the natural enzyme as well as its catalytic center.

2.1 Copper Complexes

The tridentate Schiff base [(2-(imidazol-4-yl)ethyl)(1-methylimidazol-2-yl)methyl]imine (HISMIMI) and its reduced form HISMIMA have been synthesized and characterized [93].

The mononuclear species described in Figure 18 (complex [HISMIMICuCl_2]) is able to catalyze the cleavage of linear DNA obtained from mussel genomic DNA, and also of supercoiled plasmid DNA.

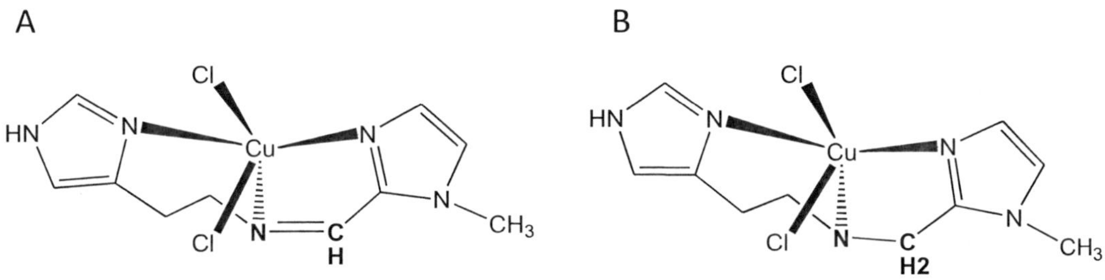

Figure 18. Schematic structural representation for the complexes: (A) [HISMIMICuCl_2]) and (B) [HISMIMACuCl_2]) [93].

The pH of the incubation medium has a great influence on the activity, with higher pH (9.3) being more effective than lower pH (8.0) in the cleavage reaction. This establishes a clear relationship between the presence of the hydroxo species, identified by titration experiments, and the extent of cleavage reaction. Also, the ability to induce double-strand cleavages on the DNA molecule was observed for this complex, since the linear form of plasmid DNA was detected after reaction at pH 9.3. However, the product of this reaction could not be religated by T4 DNA ligase, suggesting that the extremities generated after the cleavage were either still bound to the complex or did not present 5'-phosphate and 3'-OH groups, that is, the natural substrates for the ligase. To establish the possible participation of molecular oxygen in the cleavage of DNA by [HISMIMICuCl_2], assays have been performed under anaerobic conditions. These data showed that the DNA cleavage occurs in the absence of oxygen and, thus, the cleavage of DNA by [HISMIMICuCl$_2$] in anaerobic experiments is oxygen-independent, and also that this reaction most probably occurs through a hydrolytic mechanism as proposed in the reaction of [HISMIMICuCl$_2$] with the model BDNPP substrate. Although simple, this assay is highly conclusive in this respect.

The apparent pseudo-first-order rate constant for DNA hydrolysis is estimated to be 0.28 h^{-1} and the half-life 2.4 h. On comparing these parameters with others reported in the literature, it is evident that complex [HISMIMICuCl$_2$] is around five times more effective than the Cu^{II}-triamine complexes reported by Burstyn and co-workers (0.04-0.09 h^{-1} at 50ºC), and similar in potency to that of lanthanide cations and their complexes (e.g., [Eu^{3+}] 2.5 mM with k 0.25 h^{-1} at catalyst concentration used for Michaelis-Menten analysis). Higher efficiencies have been reported for a Co^{III}-polyamine system with k_{cat} 0.82 h^{-1} at [catalyst] of 1 mM and for a Cu^{II}-1,3,5-triaminocyclohexane system (k_{cat} 4.3 h^{-1}), under identical experimental conditions [93].

Complex [HISMIMICuCl$_2$] effectively promotes the hydrolytic cleavage of double-stranded plasmid DNA under anaerobic and aerobic conditions, with a rate constant of 0.28 h^{-1}, which represents a rate increase of approximately 10^7 compared with the estimated uncatalyzed hydrolysis rate.

2.2 [(L-H)FeIII(μ-OAc)2(μ-PhO)CoII]ClO4 and [(L-H)GaIII(μ-OAc)2(μ-PhO)CoII]ClO4 Complexes

Other interesting examples of hydrolytic DNA cleavage by binuclear complexes are the unsymmetrical [(L-H)Fe^{III}(μ-OAc)$_2$(μ-PhO)Co^{II}]ClO_4 and [(L-H)Ga^{III}(μ-OAc)$_2$(μ-PhO)Co^{II}]ClO_4 complexes with the unsymmetrical dinucleating ligand H_2L-H (
Figure 6) [80].

In studies carried out to evaluate the influence of ionic strength on the DNA cleavage by these two complexes it was observed that an increase in NaCl concentration greatly inhibited the hydrolytic cleavage of pBSK II DNA. This suggests that opposite charges are involved in the reaction mechanism for both complexes [100].

The effects of the metal complex concentration and of pH on the nucleic acid cleavage were also evaluated. Cleavage of DNA by [(L-H)Fe^{III}(μ-OAc)$_2$(μ-PhO)Co^{II}]$^+$ and [(L-H)Ga^{III}(μ-OAc)$_2$(μ-PhO)Co^{II}]$^+$ is higher at pH 6.0, and according to the potentiometric titration species distribution curves, there is a higher percentage of [(L-H)(HO)M^{III}(μ-OH)Co^{II}(H_2O)]$^+$ at this pH. Thus, the [(L-H)(HO)M^{III}(μ-OH)Co^{II}(H_2O)]$^+$ species is most probably associated with the catalytic activity in the DNA cleavage, as observed in the hydrolysis of the model substrate BDNPP. At pH 6.5 and 7.0, the catalytic activities of [(L-H)Fe^{III}(μ-OAc)$_2$(μ-PhO)Co^{II}]$^+$ and [(L-H)Ga^{III}(μ-OAc)$_2$(μ-PhO)Co^{II}]$^+$ diminished, which is most probably associated with deprotonation of a water molecule bound to the Co^{II} center. It should be noted that pH optima in the cleavage of DNA by [(L-H)Fe^{III}(μ-OAc)$_2$(μ-PhO)Co^{II}]$^+$ and [(L-H)Ga^{III}(μ-OAc)$_2$(μ-PhO)Co^{II}]$^+$ are approximately 0.5 pH units lower when compared to the values observed in the hydrolysis of BDNPP. A possible interpretation for these distinct results may be related to the interaction of the catalyst with the complex molecular surface of DNA.

Both complexes were able to cleave DNA at low micromolar concentrations under mild pH and temperature conditions, being complex $[(L\text{-}H)Ga^{III}(\mu\text{-}OAc)_2(\mu\text{-}PhO)Co^{II}]^+$ two times more efficient in DNA cleavage than complex $[(L\text{-}H)Fe^{III}(\mu\text{-}OAc)_2(\mu\text{-}PhO)Co^{II}]^+$. We can thus confirm that DNA cleavage catalyzed by $[(L\text{-}H)Fe^{III}(\mu\text{-}OAc)_2(\mu\text{-}PhO)Co^{II}]^+$ and $[(L\text{-}H)Ga^{III}(\mu\text{-}OAc)_2(\mu\text{-}PhO)Co^{II}]^+$ is in accordance with the proposed mechanism for the hydrolysis of the model substrate BDNPP in which the terminal hydroxo group bound to M^{III} (M^{III}=Fe, Ga) acts as the nucleophile during the catalysis. As mentioned previously, metal complexes interact (covalently or not) with DNA through its nitrogen bases and phosphate groups, establishing strong interactions.[101],[102],[103],[104]

As a preliminary observation of the type of interaction with DNA exhibited by these systems, a spectrophotometric titration of complex $[(L\text{-}H)Fe^{III}(\mu\text{-}OAc)_2(\mu\text{-}PhO)Co^{II}]^+$ was performed in the presence of increasing concentrations of *calf thymus DNA (*CT-DNA). Maximum absorption of the free complex was observed at 485 nm, and after DNA addition, a molar absorptivity decrease (hypochromism) with a red shift (λ_{max} from 485 to 503 nm) was observed, indicating a possibly intercalation of the complex in the major groove of DNA. The calculated intrinsic binding constant (K_b=1.35 x 10^5 M^{-1}) and red shift suggested a strong binding of $[(L\text{-}H)Fe^{III}(\mu\text{-}OAc)_2(\mu\text{-}PhO)Co^{II}]^+$ with DNA, as compared to other metal complexes described in the literature [105],[106],[107],[108].

2.3 [(L-H)FeIII(μ-OAc)2(μ-PhO)NiII]ClO4 Complex

The heterodinuclear mixed valence complex $[(L\text{-}H)Fe^{III}(\mu\text{-}OAc)_2(\mu\text{-}PhO)Ni^{II}]^+$ with the unsymmetrical N_5O_2 donor ligand H_2L-H (Figure 6) has been synthesized and characterized by Neves and co-workers [78]. It seems that Ni^{II} would be a good candidate as a substitute for M^{II} in purple acid phosphatases, and we thus analyzed its interaction with DNA. The binding of drugs to the DNA helix has been characterized through absorption or fluorescence spectral titrations, monitoring the change in absorbance (hypochromism) and shift in wavelength (red shift) due to DNA-drug interaction. The absorption maximum for free complex $[(L\text{-}H)Fe^{III}(\mu\text{-}OAc)_2(\mu\text{-}PhO)Ni^{II}]^+$ was observed at 486 nm in 25 mM Pipes buffer, pH 6.5 at 37 °C and ε = 2962 M^{-1} cm^{-1}. The absorption spectrum of $[(L\text{-}H)Fe^{III}(\mu\text{-}OAc)_2(\mu\text{-}PhO)Ni^{II}]^+$ is significantly affected by the presence of DNA resulting in a substantial red shift and an appreciable decrease in molar absorptivity. The resulting estimate of the intrinsic binding constant (K_b) of 3.0 x 10^4 M^{-1} and the significant red shift suggest a weak interaction of complex $[(L\text{-}H)Fe^{III}(\mu\text{-}OAc)_2(\mu\text{-}PhO)Ni^{II}]^+$ with DNA, possibly *via* an intercalation mechanism involving a stacking interaction between the aromatic phenolate chromophore and the DNA base pairs.

2.4 [(L-H)FeIII(μ-OAc)2(μ-PhO)CuII]ClO4 Complex

A heterobinuclear mixed valence complex $[(L\text{-}H)Fe^{III}(\mu\text{-}OAc)_2(\mu\text{-}PhO)Cu^{II}]^+$, with the unsymmetrical N_5O_2 donor ligand H_2L-H has been demonstrated to be an excellent model to mimic the function of nucleases [79].

As previously described, an $[(L\text{-}H)Fe^{III}(\mu\text{-}OAc)_2(\mu\text{-}PhO)Ni^{II}]^+$ complex, isostructural to $[(L\text{-}H)Fe^{III}(\mu\text{-}OAc)_2(\mu\text{-}PhO)Cu^{II}]^+$, can bind tightly to plasmid DNA, although no data supporting hydrolytic cleavage of this substrate have been reported. Thus, the cleavage of nucleic acids by the complex $[(L\text{-}H)Fe^{III}(\mu\text{-}OAc)_2(\mu\text{-}PhO)Cu^{II}]^+$ at pH 7.0 was investigated. The active $[(L\text{-}H)(HO)Fe^{III}(\mu\text{-}OH)Cu^{II}(H_2O)]^+$ species of $[(L\text{-}H)Fe^{III}(\mu\text{-}OAc)_2(\mu\text{-}PhO)Cu^{II}]^+$ should be present, as established by potentiometric titration, mass spectrometry and BDNPP hydrolysis. Complex $[(L\text{-}H)Fe^{III}(\mu\text{-}OAc)_2(\mu\text{-}PhO)Cu^{II}]^+$ is able to cleave plasmid DNA even at very low concentrations (2.5 mM), under mild pH and temperature conditions. The estimated EC_{50} (the effective concentration promoting 50% cleavage of the substrate) for complex $[(L\text{-}H)Fe^{III}(\mu\text{-}OAc)_2(\mu\text{-}PhO)Cu^{II}]^+$ in relation to DNA was 55 μM, one of the lowest values reported in the literature. [109],[110],[111]

In addition, it has been observed that the hydroxyl radical scavenging of glycerol and dimethyl sulfoxide (DMSO) did not interfere with the DNA cleavage activity of complex $[(L\text{-}H)Fe^{III}(\mu\text{-}OAc)_2(\mu\text{-}PhO)Cu^{II}]^+$, thus suggesting that diffusible hydroxyl radicals are not involved in the cleavage reaction. Even the absence of oxygen does not affect DNA cleavage by $[(L\text{-}H)Fe^{III}(\mu\text{-}OAc)_2(\mu\text{-}PhO)Cu^{II}]^+$, and we assume that complex $[(L\text{-}H)Fe^{III}(\mu\text{-}OAc)_2(\mu\text{-}PhO)Cu^{II}]^+$ is likely to cleave DNA via a hydrolytic mechanism.

Kinetics studies of the DNA cleavage by $[(L\text{-}H)Fe^{III}(\mu\text{-}OAc)_2(\mu\text{-}PhO)Cu^{II}]^+$ revealed the following parameters: k_{cat}=0.98 h^{-1}, K_M=3.97·10^{-5} M and k_{cat}/K_M=2.46·10^4 M^{-1} h^{-1}, providing a rate enhancement of 2.71·10^7 over the uncatalyzed double-stranded DNA cleavage. This highlights $[(L\text{-}H)Fe^{III}(\mu\text{-}OAc)_2(\mu\text{-}PhO)Cu^{II}]^+$ as one of the most active complexes described to date, comparable to the Cu neamine complex described by Sreedhara and coworkers [109].

The heterobinuclear complex $[(L\text{-}H)Fe^{III}(\mu\text{-}OAc)_2(\mu\text{-}PhO)Cu^{II}]^+$ displays high efficiency towards the activated substrate BDNPP, as previously shown and, interestingly, it is also very proficient in the cleavage of DNA under physiological conditions and at very low concentrations. As suggested by, from potentiometric and kinetic data, Lanznaster and co-workers the $[(L\text{-}H)(HO)Fe^{III}(\mu\text{-}OH)Cu^{II}(H_2O)]^+$ species is a potential catalyst for the hydrolysis of BDNPP and the cleavage of DNA [79]. Notably, $[(L\text{-}H)Fe^{III}(\mu\text{-}OAc)_2(\mu\text{-}PhO)Cu^{II}]^+$ is significantly more reactive than its isostructural homologues with different metal compositions ($Fe^{III}M^{II}$, where M^{II} is Zn^{II}, Mn^{II}, Ni^{II} and Fe^{II}). Of particular interest was the observation that cleavage of double-stranded plasmid DNA occurred even at very low concentrations of $[(L\text{-}H)Fe^{III}(\mu\text{-}OAc)_2(\mu\text{-}PhO)Cu^{II}]^+$ (2.5 μM), under physiological conditions (optimum pH 7.0), with a rate enhancement of 2.7·10^7 over the uncatalyzed reaction.[79]

2.5 [(L-H)FeIII(□-OAc)2(μ-PhO)FeII]ClO4 Complex

The complex $[(L\text{-}H)Fe^{III}(\mu\text{-}OAc)_2(\mu\text{-}PhO)Fe^{II}]^+$ was synthesized as a model for the active site of uteroferrin. It has been demonstrated that this complex is able to cleave DNA but also shows significant catalytic activity in the oxidation of the model substrate 3,5-di-*tert*-butylcatechol [82].

The ability of complex $[(L\text{-}H)Fe^{III}(\mu\text{-}OAc)_2(\mu\text{-}PhO)Fe^{II}]^+$ to cleave DNA phosphodiester bonds was tested in the absence and presence of oxygen, since it was previously shown that $[(L\text{-}H)Fe^{III}(\mu\text{-}OAc)_2(\mu\text{-}PhO)Fe^{II}]^+$ is capable of oxidizing catechols [82]. Data from the experiments under aerobic and anaerobic conditions suggest that a hydrolytic mechanism is responsible for the DNA cleavage since the reaction is not affected by the absence of oxygen.

Pseudo-Michaelis–Menten conditions were used in the kinetics studies of the DNA cleavage by $[(L\text{-}H)Fe^{III}(\mu\text{-}OAc)_2(\mu\text{-}PhO)Fe^{II}]^+$, and it was observed that the decrease in supercoiled DNA fitted a single exponential decay curve, and the following kinetics parameters were deduced: k_{cat} 1.91 x 10^{-3} s^{-1}; K_M 7.96 x 10^{-4} M; k_{cat}/K_M 2.40 M^{-1} s^{-1}, providing a rate enhancement of 1.91 x 10^8 over the uncatalyzed double-stranded DNA cleavage.

Thus, complex $[(L\text{-}H)Fe^{III}(\mu\text{-}OAc)_2(\mu\text{-}PhO)Fe^{II}]^+$ was found to be an important example of a synthesized promiscuous catalyst, with activity in both the oxidation of catechols and the hydrolysis of phosphodiester bonds.

2.6 Catalytic Promiscuity Promoted by Synthetic Models

In the last decade, the idea of one enzyme having one function has changed dramatically. There is a growing number of reports describing new functions for enzymes known as "moonlighting proteins", proteins that have additional functions that are generally not enzymatic, but rather structural or regulatory. On the other hand, the enzymes capable of catalyzing secondary reactions at the same active site that is specialized to catalyze a primary reaction are said to be "promiscuous" [112],[113].

The term catalytic promiscuity has been used to describe the property of biological enzymes being able to catalyze more than one chemical transformation at the active site and this promiscuity plays a natural and important role in evolution and the characterization of this activity has been reported in many examples in the literature [114],[115],[116]. A review by Karl Hult and Per Berglund, concerned with exploiting enzyme promiscuity and its impact in evolution, biological function, enzyme engineering and biocatalysis, has recently been published [117].

As an example of a promiscuous enzyme, Olguin and co-workers described a second catalytic activity of *Pseudomonas aeruginosa* arylsulfatase. Besides hydrolyzing sulfate monoesters, this enzyme catalyzes the hydrolysis of phosphate monoesters with multiple turnovers [118]. Another interesting example of catalytic promiscuity is the enzyme aminopeptidase P which performs the hydrolysis of amide bonds and also promotes phosphoester cleavage [115],[119]

A dinuclear aminopeptidase from *Streptomyces griseus* (SgAP) and its metal derivatives catalyze the hydrolysis of phosphoester bis(p-nitrophenyl)phosphate (BNPP) and a phosphonate ester (p-nitrophenyl phenylphosphonate) with extraordinary rate enhancements at pH 7.0 and 25°C, reaching 6.7 billion-fold in terms of the first-order rate constant of the di-Co(II) derivative with respect to the autohydrolytic rates [120]. Since phosphoesters are transition state-like inhibitors in peptide hydrolysis, their intrinsic hydrolysis by SgAP is rather novel. Recently, Ergan and co-workers reported the investigation of this proficient alternative catalysis of SgAP and the role of each metal ion in the dinuclear site in peptide and BNPP hydrolysis. In this study, it was shown that a divalent metal Mn(II) selectively binds to one of the dinuclear metal sites (M1), affording MnE-SgAP with an empty (E) second site for the binding of another metal (M2), which could be Mn^{II}, Co^{II}, Ni^{II}, Zn^{II}, and Cd^{II}. They demonstrated that peptide hydrolysis is controlled by M2, where the k_{cat} values for the derivatives MnM2-SgAP are different yet similar to MnCo and CoCo-SgAP and pairs of other metal derivatives. On the other hand, BNPP hydrolysis is affected by metals in both sites. Thus, the two hydrolytic catalyses must follow different mechanisms. Based on the crystal structures, docking, and biochemical data, the M1 site is close to the hydrophobic specific site and the M2 site is next to Tyr246, which is H-bonded to a coordinated nucleophilic water molecule in peptide hydrolysis; whereas a coordinated water molecule bound to M1 becomes available as the nucleophile in phosphodiester hydrolysis. Mechanistic role of each metal ion in Streptomyces dinuclear aminopeptidase: Peptide hydrolysis and $7x10^{10}$-fold rate enhancement of phosphodiester hydrolysis [121].

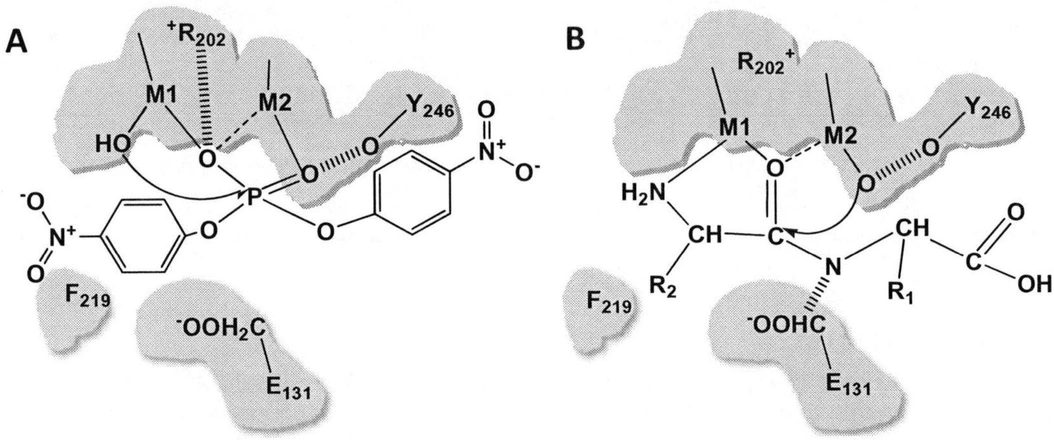

Figure 19. Putative binding modes for BNPP (A) and peptide (B) in the active site of *Sacharomyces griceus* Aminopeptise, adapted from Ercan *et al.*, 2010 [121]. The active site of the enzyme is shadowed in grey.

Inspired by Nature, scientists have reevaluated the role of promiscuity to develop promiscuous drugs and molecules, especially metal coordinated complexes that for years have been used as models of many natural enzymes. Model complexes designed to perform unique hydrolytic activities have been previously described, and although the fact that catalytic promiscuity among biological enzymes is well known, very few examples of catalytic promiscuity in model systems that mediate important cleavage reaction are known to date. The new paradigm of drug design based on artificial metalloproteases and metallonucleases should be of interest to researchers in the areas of biomimetic chemistry, as well as medicinal chemistry.

In this section we will focus on the catalytic promiscuity of metal complexes that, besides their nuclease activity, also present protease activity. Proteases (peptidases or proteolytic enzymes) are enzymes that break the peptide bonds between amino acids of proteins. Peptide bonds are stable under natural conditions, this being reflected in the half-life for spontaneous hydrolysis, which is approximately 500 years at room temperature and pH 7.0. Metal compounds that cleave peptides are among the substances that mimic enzymes which demonstrate potential use in a wide variety of different situations such as protein sequencing and proteomics. When the cleavage of proteins assisted by metals is performed under non denaturating conditions, its use can be extended to the study of protein function and structure analysis, analysis of folding structures, mapping of the active sites, binding sites for metals and ligands, conformational changes in the structure of proteins induced by ligand generation of synthetic protein and gene therapy. Lee and Suh reviewed some metal complexes that selectively catalyze hydrolytic cleavage of proteins [122],[123].

In this scenario, the development of synthetic nucleases and peptidases has been considered of great utility and importance, since it offers many possibilities for gene manipulation, the design of structural probes, and the development of novel therapeutics.

For many nucleases and for some peptidases, metal ions are essential cofactors and are directly involved in the hydrolysis of phosphodiester or amide bonds, respectively [124]. Metal compounds with the following features can be considered of interest to mimic nuclease and peptidase activities: (a) water soluble to some extent and forming cationic species in solution; (b) having an unsaturated coordination environment or labile ligands, which can be replaced by substrate molecules; (c) forming nucleophilic species near physiological pH that can promote the desired hydrolysis; and (d) possessing functional groups which can donate or remove protons as well as interact with the substrate through hydrogen bonding [124].

The first metal complex with promiscuous activity was reported by Hegg and Burstyn. In 1995 the macrocyclic copper(II) complex Cu([9]aneN$_3$)Cl$_2$ was reported to hydrolytically cleave dipeptides and proteins. Later, the same group reported that the complex has activity against RNA (phosphodiester bonds) structures [125],[126].

The interaction and recognition of dipeptides by dinuclear copper(II) OBISDIEN complexes have been determined and it has been proposed that dinuclear macrocyclic metal complexes promote hydrolysis of peptides [127].

Luiz and co-workers, 1999, using the same ligand, described the activity of a dinuclear Zn(II)-OBISDIEN complex (hexamine OBISDIEN(1,4,7,13,16,19-hexaaza-10,22-dioxacyclotetracosane)) in the hydrolysis of dipeptide glycylglycine [128].

In the same year, Meier and co-workers reported the DNA hydrolysis by Zn(II)-OBISDIEN [129].

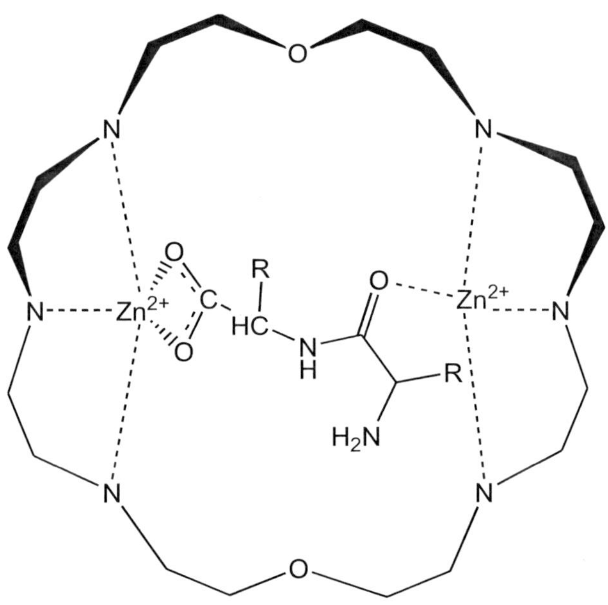

Figure 20. Schematic representation of complex glycylglycine-Zn^{II}-OBISDIEN adapted from Luiz et al., 1999 [128].

The description of crystal structures, redox and UV-vis/EPR spectroscopic properties of two new mononuclear Cu^{II} complexes, $[Cu(HL3)Cl_2]$ and [Cu(L3)Cl] (Figure 21), prepared through the reaction between copper(II) chloride and the ligand 2-[(bis(pyridylmethyl)amino)methyl]-4-methyl-6-formylphenol (HL3) under distinct base conditions, is reported along with solution studies. These Cu^{II} complexes were able to cleave peptide bonds from bovine serum albumin (BSA) and the thermostable enzyme Taq DNA polymerase at micromolar concentration, under mild pH and temperature conditions. They are active at moderate temperature (37-70°C), and the cleavage reaction is very rapid, the estimated half-life for the cleavage reaction being 30 minutes at pH 7.2. The cleavage activity seems to be specific with defined proteolytic fragments appearing after protein treatment and the location of the specific cleavage sites was assigned to solvent-accessible portions of the protein. These are two of the most active Cu^{II} complexes described to date and the most important, presenting a hydrolytic mechanism mediating protein cleavage. Later, the same group demonstrated the cleavage activity of double-stranded DNA by the mononuclear $[Cu(HL3)Cl_2]$ complex, in addition to its previously described amide bond cleavage activity.

As a nuclease the complex was able to cleave DNA in a concentration-dependent manner, as seen in the agarose gel image (Figure 21). The protease activity of cleaving peptide bonds in bovine serum albumin (BSA) and the generation of fragments can be seen in the SDS-PAGE (Figure 21). The putative cleavage sites are shown in the figure of the 3D view of BSA [130],[131].

This complex is very efficient in the hydrolysis of DNA and protein under similar experimental conditions (pH, complex concentration and temperature) and, more interestingly, its ability to act as protease or nuclease can be switched by varying the ionic strength of the medium, making this complex an interesting model for the study of catalytic promiscuity [130],[131].

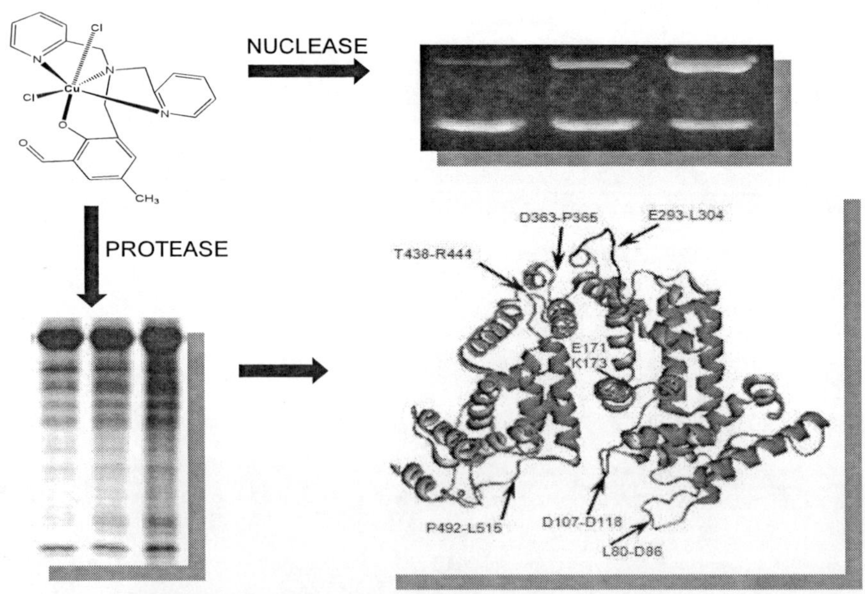

Figure 21. Protease and nuclease activity of the mononuclear $[Cu(HL3)Cl_2]$ complex. Adapted from Oliveira et al., 2005 and Oliveira et al., 2009 [130],[131].

Another good example of catalytic promiscuity is report by Rey and co-workers. Their group reported the synthesis and structural characterization of a new dicopper(II) μ-hydroxo complex containing a symmetric Schiff base derived from the facial tridentate ligand AAZ as a model for the *met* form of the active site of catechol oxidase [104]. In addition to the expected catecholase-like activity, this system was able to catalyze the hydrolysis reaction of the phosphate diester 2,4-bis(dinitrophenyl)-phosphate (BDNPP) and to cause hydrolytic damage in plasmid DNA, thus exhibiting a broad spectrum of catalytic activities. However, the catalytically active species is not the same: in the presence of catechols and a weakly basic medium, a deprotonated form of the complex acts via an oxidative mechanism. On the other hand, when the phosphate diester BDNPP is the substrate, the complex itself (which predominates in the solution from pH 6.0 to 7.5) breaks the P-O linkage following a hydrolytic pathway rather than an oxidative one [104].

Very recently, Goswami and his group reported an interesting new class of ferrocene-based amino acid reduced Schiff base ternary copper(II) complexes having planar phenanthroline bases showing efficient DNA binding propensity and significant DNA cleavage activity under red light, within the photodynamic therapy (PDT) spectral window (600-800 nm) [132]. The complexes are of copper(II) with L-methionine and phenanthroline bases, namely, [Cu(Fc-met)((phen)](NO_3), where phen is 1,10-phenanthroline and [Cu(Fc-met)((dppz)](NO_3) where dppz is dipyrido[3,2-*a*:2′,3′-*c*]phenazine. A significant role of the ferrocene moiety in enhancing the DNA photocleavage activity was observed in comparison to the mononuclear copper(II) complex. The presence of the ferrocene unit causes a change in the mechanistic pathway from the singlet oxygen to the hydroxyl radical pathway. Some features like conformational flexibility and hydrophobicity, and other physical characteristics of a molecule and its surroundings, have been implicated in mediating binding specificity or promiscuity [133]. The dppz complex shows UV-A light-induced protein cleavage activity. This opens up a hitherto unexplored area in PDT using organometallic complexes with suitable designs for cellular applications in PDT targeting both primary and secondary tumors.

It should be noted that it was only at the beginning of the new millennium that drug designers, chemists and molecular biologists begin to look beyond their target strategies. A systems biology approach offers an integrated and deeper investigation into the functioning of organelle and cellular structures in a network context and identification of much severe pathology, such as degenerative diseases, which are multifactorial, where the disease phenotype arises from the deregulation of multiple genes, pathways, and proteins [134]. Metabolic and signaling biochemical pathways have numerous steps that involve the hydrolytic cleavage of peptide or phosphate ester bonds. Likewise, the growing interest in multi-target drugs to treat complex diseases and malignancies has motivated a reassessment of the therapeutic value of catalytic promiscuity, since many successful drugs have been shown to be promiscuous [135],[136],[137],[138].

These investigations, coupled with the advancements already made regarding the molecular pathways involved, may help medicinal chemists to design new chemical entities and enable them to understand their mechanisms of action, as well as to relieve the symptoms of, or even cure, important pathologies. By combining advances in chemoinformatics, and chemogenomic and structural biology, it might be possible to rationally design the next generation of promiscuous drugs with the application of polypharmacology [139],[140],[141].

The new paradigm of drug design based on artificial metalloproteases should be of interest to researchers in the areas of biomimetic chemistry, as well as of medicinal chemistry.

REFERENCES

[1] D. Kahne, W. C. Still, Hydrolysis of a Peptide Bond in Neutral Water, *J. Am. Chem. Soc.* 110 (1988) 7529–7534.

[2] Radzicka, R. Wolfenden, Rates of uncatalyzed peptide bond hydrolysis in neutral solution and the transition state affinities of proteases, *J. Am. Chem. Soc.* 118 (1996) 6105–6109.

[3] D. W. Christianson, W.N. Lipscomb, Carboxypeptidase A, *Acc. Chem. Res.* 22 (1989) 62–69.

[4] B.W. Matthews, Structural Basis of the Action of Thermolysin and Related Zinc Peptidases, *Acc. Chem. Res.* 21 (1988) 333–340.

[5] B. K. Takasaki, J. Chin, La(III)-Hydrogen Peroxide Cooperativity in Phosphate Diester Cleavage: A Mechanistic Study, *J. Am. Chem. Soc.* 117 (1995) 8582-8585.

[6] C. A. Bunton, S. J. Farber, The Hydrolysis of Bis(2,4-dinitrophenyl) Phosphate, *J. Org. Chem.* 34 (1969) 767-772.

[7] E. L Hegg, J.N Burstyn, Toward the development of metal-based synthetic nucleases and peptidases: a rationale and progress report in applying the principles of coordination chemistry, *Coord. Chem. Rev.* 173 (1998) 133-165.

[8] D. S. Sigman, A. Mazumder, D. M. Perrin, Chemical Nucleases, *Chem. Rev.* 93 (1993) 2295-2316.

[9] F. H. Westheimer, Why Nature Chose Phosphates, *Science* 235 (1987) 1173-1178.

[10] J. Berg, J. L. Tymoczko, L. Stryer, Biochemistry, 5th edition, W. H. Freeman and Company, New York and Basingstoke (2002) p 248.

[11] N. Mitić, S. J. Smith, A. Neves, L. W. Guddat, L. R. Gahan, G. Schenk, The Catalytic Mechanisms of Binuclear Metallohydrolases, *Chem. Rev.* 106 (2006) 3338-3363.

[12] T. T. Chen, F. W. Bazer, J. J. Cetorelli, W. E. Pollard, R. M. Roberts, Purification and Properties of a Progesterone-induced Basic Glycoprotein from the Uterine Fluids of Pigs, *J. Biol. Chem.* 248 (1973) 8560-8566.

[13] H. D. Campbell, B. Zerner, A low-molecular-weight acid phosphatase which contains iron, *Biochem. Biophys. Res. Commun.* 54 (1973) 1498-1503.

[14] H. D. Campbell, D. A. Dionysius, D. T. Keough, B. E. Wilson, J. de Jersey, B. Zerner, Iron-containing acid phosphatases: Comparison of the enzymes from beef spleen and pig allantoic fluid, *Biochem. Biophys. Res. Commun.* 82 (1978) 615-620.

[15] D. B. Robinson, R. H. Glew, A Tartrate-resistant Acid Phosphatase from Gaucher Spleen, *J. Biol. Chem.* 255 (1980) 5864-5870.

[16] R. Hayman, M. J. Warburton, J. A. Pringle, B. Coles, T. J. Chambers, Purification and characterization of a tartrate-resistant acid phosphatase from human osteoclastomas, *Biochem. J.* 261 (1989) 601-609.

[17] S. H. Allen, P. R. Nuttleman, C. M. Ketcham, R. M. Roberts, Purification and characterization of human bone tartrate-resistant acid phosphatase, *J. Bone Miner. Res.* 4 (1989) 47-55.

[18] J. Janckila, T. A. Woodford, K. W. Lam, C. Y. Li, L. T. Yam, Protein-tyrosine phosphatase activity of hairy cell tartrate-resistant acid phosphatase, *Leukemia* 6 (1992) 199-203.

[19] G. Schenk, L. W. Guddat, Y. Ge, L. E. Carrington, D. A. Hume, S. Hamilton, J. de Jersey, Identification of mammalian-like purple acid phosphatases in a wide range of plants, *Gene* 250 (2000) 117-125.

[20] B. C. Antanaitis, P. Aisen, H. R. Lilienthal, Physical characterization of two-iron uteroferrin. Evidence for a spin-coupled binuclear iron cluster, *J. Biol. Chem.*258 (1983), 3166-3172.

[21] B. A. Averill, J. C. Davis, S. Burman, T. Zirino, J. Sanders-Loehr, T. M. Loehr, J. T. Sage, P. G. Debrunner, Spectroscopic and Magnetic Studies of the Purple Acid Phosphatase from Bovine Spleen, *J. Am. Chem. Soc.* 109 (1987) 3760-3767.

[22] Y.–S. Yang, J. M. McCormick, E. I. Solomon, Circular Dichroism and Magnetic Circular Dichroism Studies of the Mixed-Valence Binuclear Non-Heme Iron Active Site in Uteroferrin and Its Anion Complexes, *J. Am. Chem. Soc.* 119 (1997) 11832-11842.

[23] D. L. Wang, R. C. Holz, S. S. David, L. Que Jr., M. T. Stankovich, Electrochemical Properties of the Diiron Core of Uteroferrin and Its Anion Complexes, *Biochemistry* 30 (1991) 8187-8194.

[24] P. V. Bernhardt, G. Schenk, G. J. Wilson, Direct Electrochemistry of Porcine Purple Acid Phosphatase (Uteroferrin), *Biochemistry* 43 (2004) 10387-10392.

[25] P. R. Nuttleman, R. M. Roberts, Transfer of Iron from Uteroferrin (Purple Acid Phosphatase) to Transferrin Related to Acid Phosphatase Activity, *J. Biol. Chem.* 265 (1990) 12192-12199.

[26] W. C. Buhi, C. A. Ducsay, F. W. Bazer, R. M. Roberts, Iron Transfer between the Purple Phosphatase Uteroferrin and Transferrin and Its Possible Role in Iron Metabolism of the Fetal Pig, *J. Biol. Chem.* 257 (1982) 1712-1723.

[27] R. Hayman, S. J. Jones, A. Boyde, D. Foster, D. H. Colledge, M. B. Carlton, M .J. Evans, T. M. Cox, Mice lacking tartrate-resistant acid phosphatase (Acp 5) have disrupted endochondral ossification and mild osteopetrosis, *Development* 122 (1996) 3151-3162.

[28] N. Z. Angel, N. Walsh, M. R. Forwood, M. C. Ostrowski, A. I. Cassady, D. A. Hume, Transgenic mice overexpressing tartrate-resistant acid phosphatase exhibit an increased rate of bone turnover, *J. Bone Miner. Res.* 15 (2000) 103-110.

[29] J. C. Sibille, K. Doi, P. Aisen, Hydroxyl radical formation and iron-binding proteins. Stimulation by the purple acid phosphatases, *J. Biol. Chem.*262 (1987) 59-62.

[30] H. Kaija, S. L. Alatalo, J. M. Halleen, Y. Lindqvist, G. Schneider, H. K. Väänänen, P. Vihko, Phosphatase and oxygen radical-generating activities of mammalian purple acid phosphatase are functionally independent, *Biochem. Biophys. Res. Commun.* 292 (2002) 128-132.

[31] S.R. Räisänen, S. L. Alatalo, H. Ylipahkala, J. M. Halleen, A. I. Cassady, D. A. Hume, H. K. Väänänen, Macrophages overexpressing tartrate-resistant acid phosphatase show altered profile of free radical production and enhanced capacity of bacterial killing, *Biochem. Biophys. Res. Commun.* 331 (2005) 120-126.

[32] B. Ek-Rylander, M. Flores, M. Wendel, D. Heinegard, G. Andersson, G., Dephosphorylation of osteopontin and bone sialoprotein by osteoclastic tartrate-

resistant acid phosphatase. Modulation of osteoclast adhesion in vitro, *J. Biol. Chem.* 269 (1994) 14853-14856.

[33] G. W. Oddie, G. Schenk, N.Z. Angel, N. Walsh, L. W. Guddat, J. de Jersey, A. I. Cassady, S. E. Hamilton, D. A. Hume, Structure, function, and regulation of tartrate-resistant acid phosphatase, *Bone* 27 (2000) 575-584.

[34] M. Valizadeh, G. Schenk, K. Nash, G. W. Oddie, L. W. Guddat, D. Hume, D. A. J. de Jersey, T. R. Burke, S. Hamilton, Phosphotyrosyl peptides and analogues as substrates and inhibitors of purple acid phosphatases, *Arch. Biochem. Biophys.* 424 (2004) 154-162.

[35] P. Lång, M. Schulzberg, G. Andersson, Expression and distribution of tartrate-resistant purple acid phosphatase in the rat nervous system, *J. Histochem. Cytochem.* 49 (2001) 379-396

[36] G. Cashikar, R. Kumaresan, N. M. Rao, Biochemical characterization and subcellular localization of the red kidney bean purple acid phosphatase, *Plant Physiol.* 114 (1997) 907-915.

[37] J. Uppenberg, F. Lindqvist, C. Svensson, B. Ek-Rylander, G. Anderson, Crystal structure of a mammalian purple acid phosphatase, *J. Mol. Biol.* 290 (1999) 201-211.

[38] Y. Lindqvist, E. Johansson, H. Kaija, P. Vihko, G. Schneider, Three-dimensional structure of a mammalian purple acid phosphatase at 2.2 Å resolution with a μ-(hydr)oxo bridged di-iron center, *J. Mol. Biol.* 291 (1999) 135-147.

[39] L. W. Guddat, A. S. McAlpine, D. Hume, S. E. Hamilton, J. de Jersey, J. L. Martin, Crystal structure of mammalian purple acid phosphatase, *Structure* 7 (1999) 757-767.

[40] N. Sträter, B. Jasper, M. Scholte, B. Krebs, A. P. Duff, D. B. Langley, R. Han, B. A. Averill, H. C. Freeman, J. M. Guss, Crystal structures of recombinant human purple acid phosphatase with and without an inhibitory conformation of the repression loop, *J. Mol. Biol.* 351 (2005) 233-246.

[41] T. Klabunde, N. Sträter, R. Fröhlich, H. Witzel, B. Krebs, Mechanism of Fe(III)-Zn(II) purple acid phosphatase based on crystal structures, *J. Mol. Biol.* 259 (1996) 737-748.

[42] G. Schenk, L. R. Gahan, L. E. Carrington, N. Mitić, M. Valizadeh, S. E. Hamilton, J. de Jersey, L. W. Guddat, Phosphate forms an unusual tripodal complex with the Fe-Mn center of sweet potato purple acid phosphatase, *Proc. Natl. Acad. Sci. U.S.A.* 102 (2005) 273-278.

[43] S. K. Smoukov, L. Quaroni, X. Wang, P. E. Doan, B. M. Hoffman, L. Que Jr., Electron-nuclear double resonance spectroscopic evidence for a hydroxo-bridge nucleophile involved in catalysis by a dinuclear hydrolase, *J. Am. Chem. Soc.*, 124 (2002) 2595-2603.

[44] J. L. Beck, D. T. Keough, J. de Jersey, B. Zerner, Enzymatically active zinc, copper and mercury derivatives of the one-iron form of pig allantoic fluid acid phosphatase, *Biochim. Biophys. Acta* 791 (1984) 357-363.

[45] E. G. Funhoff, J. Ljusberg, Y. Wang, G. Andersson, B. A. Averill, Mutational Analysis of the Interaction between Active Site Residues and the Loop Region in Mammalian Purple Acid Phosphatase, *Biochemistry* 40 (2001) 11614-11622.

[46] E. G. Funhoff, C. H. W. Klaassen, B. Samyn, J. Van Beeumen, B. A. Averill, The Highly Exposed Loop Region in Mammalian Purple Acid Phosphatase Controls the Catalytic Activity, *ChemBioChem* 2 (2001) 355-363.

[47] M. Merkx, B. A. Averill, The activity of oxidized bovine spleen purple acid phosphatase is due to an Fe(III)Zn(II) 'impurity', *Biochemistry* 37 (1998) 11223-11231.

[48] M. Merkx, M. W. H. Pinkse, B. A. Averill, B. A. Evidence for nonbridged coordination of p-nitrophenyl phosphate to the dinuclear Fe(III)-M(II) center in bovine spleen purple acid phosphatase during enzymatic turnover, *Biochemistry* 38 (1999) 9914-9925.

[49] M. W. H. Pinkse, M. Merkx, B. A. Averill, Fluoride inhibition of bovine spleen purple acid phosphatase: Characterization of a ternary enzyme-phosphate-fluoride complex as a model for the active enzyme-substrate-hydroxide complex, *Biochemistry* 38 (1999) 9926-9936.

[50] E. G. Funhoff, M. Bollen, B. A. Averill, The Fe(III)Zn(II) form of recombinant human purple acid phosphatase is not activated by proteolysis, *J. Inorg. Biochem.* 99 (2005) 521-529.

[51] J. L. Beck, M. J. McArthur, J. de Jersey, B. Zerner, B. Derivatives of the purple phosphatase from red kidney bean: Replacement of zinc with other divalent metal íons, *Inorg. Chim. Acta* 153 (1988) 39-44.

[52] J. L. Beck, J. de Jersey, B. Zerner, Properties of the Fe(II)-Fe(III) derivative of red kidney bean purple phosphatase. Evidence for a binuclear Zn-Fe center in the native enzyme, *J. Am. Chem. Soc.* 110 (1988) 3317-3318.

[53] S. Gehring, P. Fleischhauer, M. Behlendorf, M. Huber, J. Lorösch, W. Haase, M. Dietrich, H. Witzel, R. Locke, B. Krebs, B. Magnetic susceptibility studies on the diiron forms of the metalloprotein purple acid phosphatase from bovine spleen and kidney bean, *Inorg. Chim. Acta*, 252 (1996) 13-17.

[54] G. Battistuzzi, M. Dietrich, R. Locke, H. Witzel, Evidence for a conserved binding motif of the dinuclear metal site in mammalian and plant purple acid phosphatases: ^{1}H NMR studies of the di-iron derivative of the Fe(III)Zn(II) enzyme from kidney bean, *Biochem. J.* 323 (1997) 593-596.

[55] G. Schenk, C. L. Boutchard, L. E. Carrington, C. J. Noble, B. Moubaraki, K. S. Murray, J. de Jersey, G. R. Hanson, S. J. Hamilton, A Purple Acid Phosphatase from Sweet Potato Contains an Antiferromagnetically Coupled Binuclear Fe-Mn Center, *J. Biol. Chem.* 276 (2001) 19084-19088.

[56] M. Jarenmark, H. Carlsson, E. Nordlander, Asymmetric dinuclear metal complexes as models for active sites in hydrolases and redox enzymes, *C. R. Chimie* 10 (2007) 433-462.

[57] P. Karsten, M. Lanznaster, A. Neves, A. J. Bortoluzzi, B. Szpoganicz, E. Schwingel, Synthesis, structure, properties, and phosphatase-like activity of the first heterodinuclear $Fe^{III}Mn^{II}$ complex with the unsymmetric ligand $H_2BPBPMP$ as a model for the PAP in sweet potato, *Inorg. Chem.* 41 (2002) 4624-4626.

[58] M. A. S. Aquino, J. S. Lim, A. G. Sykes, Mechanism of the reaction of different phosphates with the iron(II)iron(III) form of purple acid phosphatase from porcine uteri (uteroferrin), *J. Chem. Soc., Dalton Trans.* 4 (1994) 429-436.

[59] X. Wang, L. Que Jr., Extended X-ray absorption fine structure studies of the anion complexes of FeZn uteroferrin, *Biochemistry* 37 (1998) 7813-7821.

[60] X. Wang, C. R. Randall, A. E. True, L. Que Jr., X-ray absorption spectroscopic studies of the FeZn derivative of uteroferrin, *Biochemistry* 35 (1996) 13946-13954.

[61] E. True, R. C. Scarrow, C. R. Randall, R. C. Holz, L. Que Jr., EXAFS studies of uteroferrin and its anion complexes, *J. Am. Chem. Soc.* 115 (1993) 4246-4255.

[62] X. Wang, R. Y. N. Ho, A. K. Whiting, L. Que Jr., Spectroscopic characterization of a ternary phosphatase-substrate- fluoride complex. Mechanistic implications for dinuclear hydrolases, *J. Am. Chem. Soc.* 121 (1999) 9235-9236.

[63] L. Gavrilova, B. Bosnich, Principles of mononucleating and binucleating ligand design, *Chem. Rev.* 104 (2004) 349-383.

[64] R. Than, A. A. Feldmann, B. Krebs, Structural and functional studies on model compounds of purple acid phosphatases and catechol oxidases, *Coord. Chem. Rev.* 182 (1999) 211-241.

[65] M. Suzuki, H. Furutachi, H. Ōkawa, Bimetallic dioxygen complexes derived from 'end-off' compartmental ligands, *Coord.Chem. Rev.* 200 (2000) 105-129.

[66] S. R. Collinson, D. E. Fenton, Metal complexes of bibracchial Schiff base macrocycles, *Coord. Chem. Rev.* 148 (1996) 19-40.

[67] P. Guerriero, S. Tamburini, P. A. Vigato, From mononuclear to Polynuclear Macrocyclic or Macroacyclic Complexes, *Coord. Chem. Rev.* 139 (1995) 17-243.

[68] D. E. Fenton, Metallobiosites and their synthetic analogues - a belief in synergism - 1997-1998 Tilden Lecture, *Chem. Soc. Rev.* 28 (1999) 159-168.

[69] C. Belle, J. L. Pierre, Asymmetry in bridged binuclear metalloenzymes: Lessons for the chemist, *Eur. J. Inorg. Chem.* 23 (2003) 4137-4146 and references therein.

[70] L. R. Gahan, S. J. Smith, A. Neves, G. Schenk, Phosphate ester hydrolysis: metal complexes as purple acid phosphatase and phosphotriesterase analogues, *Eur. J. Inorg. Chem.* (2009) 2745-2758.

[71] P. Hendry, A. M. Sargeson, Metal-ion Promoted Phosphate Eser Hydrolysis – Intramolecular Attack of Coordinated Hydroxide Ion, *J. Am. Chem. Soc.* 111 (1989) 2521-2527.

[72] P. Hendry, A. M. Sargeson, Metal-Ion Promoted Reactions of Phosphate Derivatives, *Prog. Inorg. Chem.* 38 (1990) 201-258.

[73] Neves, M. A. de Brito, V. Drago, K. Griesar, W. Haase, A New Asymmetric N_5O_2-Donor Binucleating Ligand and its First Fe(II)Fe(III) Complex as a Model for the Redox Properties of Uteroferrin, *Inorg.Chim. Acta* 237 (1995) 131–135.

[74] E. Bernard, W. Moneta, J. Laugier, S. Chardon-Noblat, A. Deronzier, J.-P. Tuchagues, J.-M. Latour, A Mixed-valent, Unsymmetrical Fe^{III}-Fe^{II} Complex with a Terminal Phenolato Ligand as a Model for the Active-site of Purple Acido-Phosphatases, *Angew. Chem. Int. Ed.* 33 (1994) 887–889.

[75] E. Lambert, B. Chabut, S. Chardon-Noblat, A. Deronzier, G. Chottard, A. Bousseksou, J.-P. Tuchagues, J. Laugier, M. Bardet, J.-M. Latour, Synthesis, structural, magnetic, and redox properties of asymmetric diiron complexes with a single terminally bound phenolate ligand. Relevance to the purple acid phosphatase enzymes, *J. Am. Chem. Soc.* 119 (1997) 9424–9437.

[76] M. Lanznaster, A. Neves, A. J. Bortoluzzi, B. Szpoganicz, E. Schwingel, New $Fe^{III}Zn^{II}$ complex containing a single terminal Fe-O-phenolate bond as a structural and functional model for the active site of red kidney bean purple acid phosphatase *Inorg. Chem.* 41 (2002) 5641–5643.

[77] P. Karsten, A. Neves, A. J. Bortoluzzi, J. Strähle, C. Maichle-Mössmer, Synthesis, structure and catalase-like activity of a new dinuclear mixed valence $Mn^{III}Mn^{II}$ complex containing an unsymmetric N_5O_2 donor ligand, *Inorg. Chem. Commun.*5 (2002) 434-438.

[78] S. C. Batista, A. Neves, A. J. Bortoluzzi, I. Vencato, R. A. Peralta, B. Szpoganicz, V. V. E. Aires, H. Terenzi, P. C. Severino, Highly efficient phosphate diester hydrolysis and DNA interaction by a new unsymmetrical $Fe^{III}Ni^{II}$ model complex, *Inorg. Chem. Commun.* 6 (2003) 1161-1165.

[79] M. Lanznaster, A. Neves, A. J. Bortoluzzi, V. V. E. Aires, B. Szpoganicz, H. Terenzi, P. C. Severino, J. M. Fuller, L. R. Gahan, G. R. Hanson, M. J. Riley, G. Schenk, A new heterobinuclear $Fe^{III}Cu^{II}$ complex with a single terminal Fe^{III}-$O_{(phenolate)}$ bond. Relevance to purple acid phosphatases and nucleases, *J. Biol. Inorg. Chem.* 10 (2005) 319-332.

[80] F. R. Xavier, A. Neves, A. Casellato, R. A. Peralta, A. J. Bortoluzzi, B, Szpoganicz, P. C. Severino, H. Terenzi, Z. Tomkowicz, S. Ostrovsky, W. Haase, A. Ozarowski, J. Krzystek, J. Telser, G. Schenk, L. R. Gahan, Unsymmetrical $Fe^{III}Co^{II}$ and $Ga^{III}Co^{II}$ Complexes as Chemical Hydrolases: Biomimetic Models for Purple Acid Phosphatases (PAPs), *Inorg. Chem.* 2009, 48, 7905–7921.

[81] G. Schenk, R. A. Peralta, S. C. Batista, A. J. Bortoluzzi, B. Szpoganicz, A. K. Dick, P. Herrald, G. R. Hanson, R. K. Szilagyi, M. J. Riley, L. R. Gahan, A. Neves, Probing the role of the divalent metal ion in uteroferrin using metal ion replacement and a comparison to isostructural biomimetics, *J. Biol. Inorg. Chem.* 13 (2008) 139-155.

[82] Neves, A. J. Bortoluzzi, R. Jovito, R. A. Peralta, B. de Souza, B. Szpoganicz, A. C. Joussef, H. Terenzi, P. C. Severino, F. L. Fischer, G. Schenk, M. J. Riley, S. J. Smith, L. R. Gahan, Catalytic Promiscuity: Catecholase-like Activity and Hydrolytic DNA Cleavage Promoted by a Mixed-Valence $Fe^{III}Fe^{II}$ complex, *J. Braz. Chem. Soc.* 2010, in press.

[83] E. P. Day, S. S. David, J. Peterson, W. R. Dunham, J. J. Bonvoisin. R. H. Sands, L. Que, Magnetization and electron paramagnetic resonance studies of reduced uteroferrin and its 'EPR-silent' phosphate complex, *J. Biol. Chem.* 263 (1988) 15561-15567.

[84] N. S. Goncalves, A. Horn Jr, M. Lanznaster, L. K. Noda, A. Neves, Resonance Raman spectroscopy of $Fe^{II}Fe^{III}$ and $Fe^{III}Fe^{III}$ model complexes containing an unsymmetrical dinucleating ligand: A biomimetic redox pair for Uteroferrin, *J. Braz. Chem. Soc.* 17 (2006) 1658-1663.

[85] R. W. Z. de Oliveira, A. Neves, I. C. Vieira, Development of a new biomimetic sensor based on an $Fe^{III}Fe^{II}$ complex for the determination of phenolic compounds, *Sensors Actuators B, Chemical,* 129 (2008) 424–430.

[86] C. Belle, I. Gautier-Luneau, L. Karmazin, J.-L. Pierre, S. Albedyhl, B. Krebs, M. Bonin, Regio-directed synthesis of a $Zn^{II}Fe^{III}$ complex from an unsymmetrical ligand and its relevance to purple acid phosphatases, *Eur. J. Inorg. Chem.*12 (2002) 3087–3090.

[87] C. Belle, I. Gautier-Luneau, G. Gellon, J.-L. Pierre, I. Morgenstern-Badarau, E. Saint-Aman, Directed syntheses and structural studies of mononuclear copper(II) and heterodinuclear copper(II)-iron(III) complexes from the same unsymmetrical dinucleating ligand, *J. Chem. Soc., Dalton Trans.* 19 (1997) 3543–3546.

[88] M. B. Robin, P. Day, Mixed valence chemistry - A survey and classification, *Adv. Inorg. Chem. Radiochem.* 10 (1967) 247-422.

[89] L. W. Paes, R. B. Faria, J. O. Machuca-Herrera, A. Neves, S. P. Machado, Molecular modeling utilizing purple acid phosphatase biomimetic models *Can. J. Chem.* 82 (2004) 1619-1624.

[90] M. A. de Brito, Análogos sintéticos para a estrutura do sítio ativo das Fosfatases Ácidas Púrpuras *PhD Thesis* (1994) 74p.

[91] P. G. DeBrunner, M. P. Hendrich, J. de Jersey, D. T. Keough, J. T. Sage, B. Zerner, Mössbauer and electron-paramagnetic-res study of the binuclear iron center in Purple Acid Phosphatasc, *Biochim. Biophys. Acta* 745 (1983) 103-106.

[92] Neves, M. Lanznaster, A. J. Bortoluzzi, R. A. Peralta, A. Casellato, E. E. Castellano, P. Herrald, M. J. Riley, G. Schenk, An unprecedented $Fe^{III}(\mu\text{-OH})Zn^{II}$ complex that mimics the structural and functional properties of purple acid phosphatases, *J. Am. Chem. Soc.* 129 (2007) 7486-7487.

[93] M. Scarpellini, A. Neves, R. Horner, A. J. Bortoluzzi, B. Szpoganicz, C. Zucco, R. A. N. Silva, V. Drago, A. S. Mangrich, W. A. Ortiz, W. A. C. Passos, M. C. B. de Oliveira, H. Terenzi, Phosphate diester hydrolysis and DNA damage promoted by new cys aqua/hydroxy copper(II) complexes containing tridentate imidazole-rich ligands *Inorg. Chem.* 42 (2003) 8353-8365.

[94] S. J. Smith, A. Casellato, K. S. Hadler, N. Mitic, M. J. Riley, A. J. Bortoluzzi, B. Szpoganicz, G. Schenk, A. Neves, L. R. Gahan, The reaction mechanism of the Ga(III)Zn(II) derivative of uteroferrin and corresponding biomimetics, *J Biol Inorg Chem* 12 (2007) 1207–1220.

[95] S. J. Smith, R. Jovito, A. Horn Jr, C. J. Noble, G. R. Hanson, R. Stranger, V. Jayaratne, G. Cavigliasso, L. R. Gahan, G. Schenk, A. Neves, M. J. Riley. Manuscript in preparation.

[96] M. Merkx, B. A. Averill, Ga^{3+} as a functional substitute for Fe^{3+}: Preparation and characterization of the $Ga^{3+}Fe^{2+}$ and $Ga^{3+}Zn^{2+}$ forms of bovine spleen purple acid phosphatase *Biochem.* 37 (1998) 8490-8497.

[97] R. A. Peralta, Novos complexos binucleares de cobre(II) e de ferro(III) zinco(II): biomiméticos sintéticos para catecol oxidase e para fosfatases ácidas púrpuras, *PhD Thesis*, (2005).

[98] J. Kirby, M. Younas, Reactivity of phosphate esters – Reactions of diesters with nucleophiles *J. Chem. Soc., Section B* (1970) 1165-1172.

[99] S. A. Khan, A. J. Kirby, Reactivity of phosphate esters - Multiple structure reactivity correlations for reactions of triesters with nucleophiles, *J. Chem. Soc., Section B* (1970) 1172-1182.

[100] S. J. Franklyn, Lanthanide mediate DNA hydrolysis, *Curr. Opin. Chem. Biol.* 5 (2001) 201-208.

[101] H. Chao, W-J. Mei, Q.-W. Huang, L.-N. J. Ji, DNA binding studies of ruthenium(II) complexes containing asymmetric tridentate ligands, *J. Inorg. Biochem.* 92 (2002) 165-170.

[102] L. Otero, P. Smircich, M. Vieites, M. Ciganda, P. C. Severino, H. Terenzi, H. Cerecetto, D. Gambino, B. Garat, DNA conformational changes and cleavage by ruthenium(II) nitrofurylsemicarbazone complexes, *J. Inorg. Biochem.* 101 (2007) 74-79.

[103] R. Tamilarasan, D. R. McMillin, Photophysical studies of copper phenanthrolines bound to DNA, *Inorg. Chem.* 29 (1990) 2798-2802.

[104] N. A. Rey, A. Neves, A. J. Bortoluzzi, C. T. Pich, H. Terenzi, Catalytic Promiscuity in Biomimetic Systems: Catecholase-like Activity,Phosphatase-like Activity, and

Hydrolytic DNA Cleavage Promoted by a New Dicopper(II) Hydroxo-Bridged Complex, *Inorg. Chem.* 46 (2007) 348-350.
[105] H.-L. Chan, H.-Q. Liu, B.-C. Tzeng, Y.-S. You, S.-M. Peng, M. Yang, C.-M. Che, Syntheses of ruthenium(II) quinonediimine complexes of cyclam and characterization of their DNA-binding activities and cytotoxicity, *Inorg. Chem.* 41 (2002) 3161-3171.
[106] S.-S. Zhang, S.-Y. Niu, B. Qu, G.-F. Jie, H. Xu, C.-F. J. Ding, Studies on the interaction mechanism between hexakis(imidazole) manganese(II) terephthalate and DNA and preparation of DNA electrochemical sensor, *Inorg. Biochem.* 99 (2005) 2340-2347.
[107] S. Mahadevan, M. Palaniandavar, Spectral and Electrochemical Behavior of Copper(II)−Phenanthrolines Bound to Calf Thymus DNA. [(5,6-dimethyl-OP)$_2$Cu]$^{2+}$(5,6-dimethyl-OP=5,6-Dimethyl-1,10-phenanthroline) Induces a Conformational Transition from B to Z DNA, *Inorg. Chem.* 37 (1998) 3927-3934.
[108] P. R. Reddy, K. S. Rao, B. Satyanarayana, Synthesis and DNA cleavage properties of ternary Cu(II) complexes containing histamine and amino acids, *Tetrahedron Lett.* 47 (2006) 7311-7315.
[109] Sreedhara, J. D. Freed, J. A. Cowan, Efficient Inorganic Deoxyribonucleases. Greater than 50-Million-Fold Rate Enhancement in Enzyme-Like DNA Cleavage, *J. Am. Chem. Soc.* 122 (2000) 8814–8824.
[110] Sreedhara, J. A. Cowan, Targeted site-specific cleavage of HIV-1 viral Rev responsive element by copper aminoglycosides, *J. Biol. Inorg. Chem.* 6 (2001) 337-347.
[111] J. A. Cowan, Chemical Nucleases, *Curr. Op. Chem. Biol.* 5 (2001) 634–642.
[112] S. D. Copley, Enzymes with extra talents: moonlighting functions and catalytic promiscuity, *Curr. Op. Chem. Biol.* 7 (2003) 265–272.
[113] O. Khersonky, C. Roodveldt, D. S. Tawfik, Enzyme promiscuity: evolutionary and mechanistic aspects, *Curr. Op. Chem. Biol.* 10 (2006) 498-508.
[114] P. J. O'Brien, D. Herschlag, Catalytic promiscuity and the evolution of new enzymatic activities, *Chemistry & Biology* 6 (1999) 91-105.
[115] U. T. Bornscheuer, R. J. Kazlauskas, Catalytic promiscuity in biocatalysis: using old enzymes to form new bonds and follow new pathways, *Angew. Chem. Int. Ed.* 43 (2004) 6032-6040.
[116] G. J. Poelarends, V. Puthan Veetil, C. P. Whitman, The chemical versatility of the β–α–β fold: Catalytic promiscuity and divergent evolution in the tautomerase superfamily, *Cel. Mol. Lif. Sci.* 65 (2008) 3606-3618.
[117] K. Hult, P. Berglund, Enzyme promiscuity: mechanism and applications, *Trends Biotech.* 25 (2007) 231-238.
[118] L. F. Olguin, S. E. Askew, A. C. O'Donoghue, F. Hollfelder, Efficient Catalytic Promiscuity in an Enzyme Superfamily: An Arylsulfatase Shows a Rate Acceleration of 10^{13} for Phosphate Monoester Hydrolysis, *J. Am. Chem. Soc.* 130 (2008) 16547–16555.
[119] R. J. Kazlauskas, Enhancing catalytic promiscuity for biocatalysis *Curr. Opin. Chem. Biol.* 9 (2005) 195-201.
[120] Ercan, H. I. Park, L.-J. Ming, A "Moonlighting" Dizinc Aminopeptidase from *Streptomyces griseus*: Mechanisms for Peptide Hydrolysis and the 4 x 10^{10}-Fold Acceleration of the Alternative Phosphodiester Hydrolysis, *Biochemistry* 45 (2006) 13779–13793.

[121] Ercan, W. M. Tay, S. H. Grossman, L.-J. Ming, Mechanistic role of each ion in Streptomyces dinuclear aminopeptidase: Peptide hydrolysis and 7x10(10)-fold rate enhancement of phosphodiester hydrolysis, *J. Inorg. Biochem.* 104 (2010) 19–29.

[122] J. Suh, W. S. Chei, Metal complexes as artificial proteases: toward catalytic drugs, *Curr. Op. Chem. Biol.* 12 (2008) 207–213.

[123] T. Y. Lee, J. Suh, Target-selective peptide-cleaving catalysts as a new paradigm in drug design, *Chem. Soc. Rev.* 38 (2009) 1949–1957.

[124] R. Krämer, Bioinorganic models for the catalytic cooperation of metal ions and functional groups in nuclease and peptidase enzymes, *Coord. Chem. Rev.* 182 (1999) 243–261.

[125] E. L. Hegg, J. N. Burstyn, Hydrolysis of Unactivated Peptide Bonds by a Macrocyclic Copper(II) Complex: $Cu([9]aneN_3)Cl_2$ Hydrolyzes both Dipeptides and Proteins. *J. Am. Chem. Soc.* 117 (1995) 7015-7016.

[126] E. L. Hegg, K. A. Deal, L. L. Kiessling, J. N. Burstyn, Hydrolysis of Double-Stranded and Single-Stranded RNA in Hairpin Structures by the Copper(II) Macrocycle $Cu([9]aneN_3)Cl_2$, *Inorg. Chem.* 36 (1997) 1715-1718.

[127] M. T. B. Luiz, B. Szpoganicz, M. Rizzoto, A. E. Martell, M. G. Basallote, Equilibrium studies and molecular recognition in the glycylglycine, dl-alanyl-dl-alafflne and glycyl-/-leucine Cu(II)-1,4,7,13,16,19-hexaaza-10,22-dioxacyclotetracosane complexes, *Inorg. Chim. Acta* 254 (1997) 345-351.

[128] M. T. B. Luiz , B. Szpoganicz , M. Rizzoto , M. G. Basallote , A. E. Martell. Molecular recognition of dipeptides. Catalysis of deuteration and hydrolysis of glycylglycine by dinuclear OBISDIEN Zn(II) complexes, *Inorg. Chim. Acta* 287 (1999) 134–141.

[129] M. M. Meier, P. Karloh, H. Terenzi, B. Szpoganicz, Equilibrium and hydrolysis studies of phosphate esters model molecules and DNA catalyzed by OBISDIEN-Zn(II) complexes. *South. Brazil. J. Chem.* 7 (1999) 11-23.

[130] M. C. B. Oliveira, M. Scarpellini, A. Neves,H Terenzi, A. J. Bortoluzzi, B. Szpoganics, A. Greatti, A. S. Mangrich, E. M. Souza, P. M. Fernandez, M. R. Soares, Hydrolytic Protein Cleavage Mediated by Unusual Mononuclear Copper(II) Complexes: X-ray Structures and Solution Studies, *Inorg. Chem.* 44 (2005) 921-929.

[131] M. C. B. Oliveira, D. Mazera, M. Scarpellini, P. C. Severino, A. Neves, H. Terenzi, Mononuclear Cu^{II}-phenolate Bioinspired Complex is Catalytically Promiscous: Phosphodiester and Peptide Amide Bond Cleavage, *Inorg. Chem.* 48 (2009) 2711-2713.

[132] T. K. Goswami, M. Roy, M. Nethaji, A. R. Chakravarty, Photoinduced DNA and Protein Cleavage Activity of Ferrocene-Appended L-Methionine Reduced Schiff Base Copper(II) Complexes of Phenanthroline Bases, *Organometallics* 28 (2009) 1992–1994.

[133] M. L. Radhakrishnan, B. Tidor, Specificity in Molecular Design: A Physical Framework for Probing the Determinants of Binding Specificity and Promiscuity in a Biological Environment, *J. Phys. Chem. B* 111, (2007) 13419-13435.

[134] M. L Bolognesi, R. Matera, A. Minarini, M. Rosini, C. Melchiorre. Alzheimer's disease: new approaches to drug discovery, *Curr. Op. Chem. Biol. 13* (2009) 303–308.

[135] D. E. Wilcox, Binuclear Metallohydrolases, *Chem. Rev.* 96 (1996) 2435-2458.

[136] X. Zhang, A. Crespo, A. Fernández, Turning promiscuous kinase inhibitors into safer drugs, *Trends Biotech.* 26 (2008) 295-301.

[137] S. K. Mencher, L. G. Wang, Promiscuous drugs compared to selective drugs (promiscuity can be a virtue), *BMC Clin. Pharm.* 5 (2005) 3-7.
[138] T. Hampton, Promiscuous anticancer drugs that hit multiple targets may thwart resistance, *JAMA* 292 (2004) 419-422.
[139] F. Fliri, W. T. Loging, P. F. Thadeio, R. A. Volkmann, Biological spectra analysis: Linking biological activity profiles to molecular structure *Proc. Nat. Acad. Sci.* 102 (2005) 261-266.
[140] L Hopkins, J. S Mason, J. P. Overington, Can we rationally design promiscuous drugs? *Curr. Op. Struct. Biol.* 16 (2006) 127-136.
[141] D. Rognan, Chemogenomic approaches to rational drug design. *Brit. J. Pharm.* 152 (2007) 38-52.

In: Molecular and Supramolecular Bioinorganic Chemistry ISBN: 978-1-61728-785-5
Ed: A. L. R. Mercê, J. Felcman and M. A. L. Recio

Chapter 3

PHOSPHATES IN BIOLOGY: SOME INTERACTIONS WITH METAL IONS AND OTHER BIOLOGICAL COMPOUNDS

***Judith Felcman**[*1] **and Ana Lucia Ramalho Mercê**[≠2]*
[1]Pontifícia Universidade Católica do Rio de Janeiro, Brazil
[2] Universidade Federal do Paraná, Curitiba – Paraná – Brazil

1. INTRODUCTION

Phosphorus is an essential element that plays a crucial role in cell structure and metabolism. The importance of phosphorus in the metabolism of human osseous and non-osseous tissues has been well established long ago.

In living organisms, phosphorus exists surrounded by four oxygen atoms constituting the phosphate ion (PO_4^{3-}). Inside cells, PO_4^{3-} regulates enzymatic activity, and is an essential component of nucleic acids, adenosine triphosphate, and phospholipid membranes. Outside cells, PO_4^{3-} resides primarily in bone and teeth as hydroxyapatite.

A nucleic acid is a macromolecule composed of nucleotide chains. In biochemistry these molecules carry genetic information or form structures within cells. The most common nucleic acids are deoxyribonucleic acid (DNA) and ribonucleic acid (RNA). Nucleic acids are universal in living beings as they are found in all cells, even in viruses.

Adenosine 5' triphosphate (ATP) is a nucleotide that performs many essential roles in the cell. It is the major energy currency of the cell, providing the energy for most of its energy-consuming activities. ATP is an organic compound composed of adenine, the sugar ribose, and three phosphate groups. It is one of the monomers used in the synthesis of RNA and, after conversion to deoxy-ATP (dATP), of DNA. It regulates many biochemical pathways. ATP serves as the main energy source within the cell to drive a number of biological processes

* Department of Chemistry, Pontifícia Universidade Católica do Rio de Janeiro – PUC – Rio – RJ – Brazil. Postal Adress: R. Marquês de S. Vicente, 225 – Gávea 22453-900 – Rio de Janeiro – RJ – Brazil, felcman@puc-rio.br

≠ Universidade Federal do Paraná – UFPR CP 19081 – LEQ – DQ – , Centro Politécnico, Jardim das Américas - 81531-980 – Curitiba – Paraná – Brazil, anamerce@ufpr.br

such as photosynthesis, muscle contraction, and protein synthesis. It is cleaved by hydrolysis to yield adenosine diphosphate (ADP), inorganic phosphorus, and energy. ADP can be further cleaved to yield adenosine monophosphate (AMP), additional phosphorus, and more energy. When the phosphorus and energy are immediately used to drive other reactions, such as the synthesis of uridine diphosphate (UDP), an RNA precursor, from uridine monophosphate (UMP), the pair of reactions are said to be coupled. New ATP is produced from AMP using the energy released from the breakdown of fuel molecules, such as fats and sugars. Extracellularly, ATP has been found to act as a neurotransmitter.

ATP receptors are widespread through the body. On its own, ATP is known to have effects on arteries, intestines, lungs, and bladder. It is also often released in tandem with other neurotransmitters, perhaps to add chemical stability.

Phospholipids are fat derivatives in which one fatty acid has been replaced with a phosphate group and one of several nitrogen-containing molecules. A phospholipid molecule consists of a hydrophilic polar head group and a hydrophobic tail. The polar head group contains one or more phosphate groups. The hydrophobic tail is made up of two fatty acid chains. When many phospholipid molecules are placed in water their hydrophilic heads tend to face the water and the hydrophobic tails are forced to stick together, forming a bilayer. The cell membrane or "phospholipid bilayer" is a semipermeable lipid bilayer of proteins and lipids which are involved in a broad range of cellular processes, and also serves as the attachment point for both the intracellular cytoskeleton and, if present, the cell wall.

Phosphorylated compounds perform many biological functions such as participating in hydrolytic enzymatic biochemical processes. The cleavage that occurs in natural compounds that release phosphates, provides the energy required by living organisms for muscular movement, synthesis of complex biological molecules and transport of solutes against concentration gradients [1].

ATPases are a class of enzymes that catalyze the decomposition of ATP into ADP and a free phosphate ion. This dephosphorylation reaction releases energy, which the enzyme - in most cases - harnesses to drive other chemical reactions that would not occur otherwise. This process is very common in all known forms of life. Some of such enzymes, called "transmembrane ATPases" are integral membrane proteins (anchored within biological membranes), which move solutes across the membrane.

Transmembrane ATPases import many of the metabolites necessary for cell metabolism and export toxins, waste, and solutes that can hinder cellular processes. An important example is the sodium-potassium exchanger (or Na^+/K^+ATPase), which establishes the ionic concentration balance that maintains the cell potential. Another example is the hydrogen potassium ATPase or H^+/K^+ATPase or gastric proton pump which acidifies the contents of the stomach.

Besides exchangers, other categories of transmembrane ATPase include co-transporters and pumps (although some exchangers are also pumps). Some of these, like the Na^+/K^+ATPase, cause a net flow of charge, but others do not. These are called "electrogenic" and "nonelectrogenic" transporters, respectively.

Phosphotransferases are a large group of enzymes comprising those that transfer not only phosphate but also diphosphate, nucleotidyl residues, and others. They are subdivided according to the acceptor group.

Phospholipids and sphingolipids play critical roles in signal transduction, intracellular membrane trafficking, and control of cell growth and survival. LPTs (lipid phosphatases/

phosphotransferases) are a family of integral membrane proteins with central roles in bioactive lipid metabolism and signaling. They are characterized by a core domain containing six transmembrane spanning α-helices connected by extramembrane loops, two of which interact to form the catalytic site. LPT family members are localized to all major membrane compartments of the cell. The transmembrane topology of these proteins places their active site facing the lumen of endomembrane compartments or the extracellular face of plasma membranes. Sequence conservation between the active site of LPPs (lipid phosphate phosphatases), SPPs (sphingosine phosphate phosphatases) and the recently identified SMSs (sphingomyelin synthases) with vanadium-dependent fungal oxidases providing a framework for understanding their common catalytic mechanism. LPPs hydrolyses LPA (lysophosphatidic acid), S1P (sphingosine 1-phosphate) and structurally related substrates. Although LPPs can dephosphorylate intracellularly generated substrates to control intracellular lipid metabolism and signaling, being their best understood function to regulate cell surface receptor-mediated signaling by LPA and S1P by inactivating these lipids at the plasma membrane or in the extracellular space.

It is also known that the human genome remarkably has 518 protein kinases [2]. The kinases phosphorylate about 30% of mammalian cellular proteins on the amino acids serine, threonine and tyrosine with an approximate ratio of 1800: 200: 1 respectively [3,4]. Phosphorylated proteins mediate cell division, cellular differentiation, hormonal signal transduction and much more. While tyrosine phosphorylation often initiates major cellular events, the phosphorylation of residues of serine and threonine on proteins is key to the cascade reactions that follow, which are of great interest.

Phosphocreatine, also known as creatine phosphate or PCr, is a phosphorylated creatine molecule that is an important energy store in skeletal muscle. It is used to anaerobically generate ATP from ADP, forming creatine for the 2 to 7 seconds following an intense effort. This occur when PCr donates a phosphate group through a reaction catalyzed by creatine kinase (the presence of creatine kinase in plasma is indicative of tissue damage and is used in the diagnosis of myocardial infarction) [5]. This reaction is reversible, and therefore acts as a spatial and temporal buffer of ATP concentration. Phosphocreatine is part of a coupled reaction; the energy given off from one reaction is used to regenerate the other compound - be it ATP or PC. Phosphocreatine plays a particularly important role in tissues that have high, fluctuating energy demands such as muscle and brain. It is synthesized in the liver and transported to the muscle cells, for storage, via the bloodstream.

Phosphoserine is a component of many proteins. The phosphorylation of the alcohol functional group in serine to produce phosphoserine is catalyzed by various types of kinases. The serine/threonine kinase PKB/Akt plays an essential role in various cellular processes including cell growth and proliferation, metabolism and cell survival. The inherited deficiency of the enzyme phosphoserine aminotransferase manifests as uncontrollable seizures, psychomotor retardation, reduced head circumference and increased muscle tone.

Phosphorylation by protein kinases is a central theme in biological systems and aberrant protein kinase activity has been implicated in a variety of human diseases. For this reason, Eldar-Finkelman et al., [6] considering that modulation of kinase activity represents an attractive therapeutic approach for the treatment of human illnesses reviewed the strategies to design such peptide inhibitors, focusing mainly on the serine/threonine protein kinase family.

Acetylcholinesterase (AchE) is found in the synapse between nerve cells and muscle cells. It is an enzyme that catalyzes the hydrolysis of the neurotransmitter acetylcholine into choline and acetic acid, a necessary reaction to allow a cholinergic neuron to return to its resting state after activation [7]. This reaction effectively stops the signal, allowing the pieces to be recycled and rebuilt into new neurotransmitters for the next message. AchE has one of the fastest reaction rates among our enzymes, breaking up each molecule in about 80 microseconds. To perform this reaction, one serine of this enzyme interacts with a phosphate group. This is the reason for the symptoms due to the pesticides containing phosphates.

Alzheimer's disease (AD) is a neurodegenerative disease characterized by cognitive impairment and personality changes. The development of drugs to treat the cognitive deficits of AD has focused on agents which counteract loss in cholinergic activities. These symptoms of AD have been successfully treated with AchE inhibitors.

Gliotransmission is a process in which astrocytes are dynamic elements that influence synaptic transmission and synaptogenesis. The best-known gliotransmitters are glutamate and ATP. However, in the past decade, it has been demonstrated that D-serine, a D-amino acid, acts as a gliotransmitter in glutamatergic synapses. The physiological relevance of D-serine is sustained by the way in which it modulates the action of glutamatergic neurotransmission, neuronal migration and long-term potentiation (LTP). In addition, the synthesis and degradation mechanisms of D-serine have been proposed as potential therapeutic targets for the treatment of Alzheimer's disease, schizophrenia and related disorders [8].

These examples are merely a small sample of important biological molecules and mechanisms involving phosphates.

Since it is vast to unravel all possible interactions in the living systems due to the great number of existing potential ligands and metal ions involved in biological processes, this work intends to describe some of the interactions of phosphorylated ligands with other biological compounds and with metal ions that are present in living organisms.

2. PHOSPHATES AND COPPER

2.1. Experimental Evidences

Several metabolic impairments and diseases are caused by the involvement of copper, adenosine 5'triphosphate (ATP) or phosphorylated amino acids such as phosphocreatine (PCr), as well as other ligands of biological importance.

Van Hoof et al. [9] studied copper uptake in inside-out plasma membrane vesicles derived from roots of copper-sensitive, moderately copper-tolerant and highly copper-tolerant populations of *Silene vulgaris* (Amsterdam, Marsberg and Imsbach, respectively). They verified that vesicles derived from Imsbach plants accumulated two to three times more copper than those derived from Marsberg and Amsterdam plants, respectively. This accumulation was found to be ATP-dependent. Also, 9-amino-6-chloro-2-methoxyacridine fluorescence quenching rates upon copper addition decreased in the order Imsbach>Marsberg>Amsterdam. These results support the hypothesis that the efflux of copper across the root´s plasma membrane plays a role in the copper tolerance mechanism in *S. vulgaris*.

In human cells copper modulates the phosphorylation of a key copper transporter, the Wilson's disease protein (WNDP). The basal phosphorylation site is located in the C-terminal

Ser(796)-Tyr(1384) region of WNDP and the copper-induced phosphorylation appears to require the presence of the functional N-terminal domain of this protein. The role of copper as a modulator of protein phosphorylation can help to understand how copper transport is regulated in mammalian cells [10].

There are several evidences that Cu^{2+} plays a physiological role in regulating nervous system functions. Li et al. [11] verified that Cu^{2+} greatly enhanced current activated by 10 μM ATP in the majority of rat nodose ganglion neurons. However, Cu^{2+} did not enhance ATP-activated current in the presence of a maximally-effective concentration of Zn^{2+}. The observations suggest that Cu^{2+} increases the affinity of the receptor for ATP by acting at the Zn^{2+} modulatory site. In addition, a subset of neurons in the nodose ganglion expresses that ATP-gated receptor-channels are insensitive to modulation by physiological concentrations of Cu^{2+}, Zn^{2+} and protons.

Isolated rat hearts were perfused with glucose, octanoate or glucose + octanoate and different concentrations of the copper-based antineoplastic drugs casiopeina II-gly (CSII) or casiopeina III-i-a (CSIII). Both casiopeinas induced diminution in cardiac work and O_2 consumption. There was a strong inhibition of the pyruvate and 2-oxoglutarate dehydrogenases as well as the total creatine kinase. These observations suggest that these drugs target ATP generation by oxidative phosphorylation and its transfer towards myofibrils. As a consequence, the cellular contents of ATP and phosphocreatine were also lowered by the casiopeinas [12].

Male weanling rats were made copper deficient with a purified diet containing all known essential dietary nutrients except copper. Copper deficiency was verified by indirect (anemia, growth retardation, hypercholesterolemia, gross pathology, and abnormal electrocardiograms) and direct (tissue copper analysis) criteria. Phosphorus-31 nuclear magnetic resonance spectroscopic analysis of perchloric acid extracts of cardiac, renal, and hepatic tissue revealed significant metabolic changes associated with the dietary copper deficiency, including a generalized marked decrease in ATP and phosphocreatine levels and a corresponding increase in inorganic orthophosphate and ADP levels in these tissues. Tissue-specific changes consisting of elevated levels of ribose 5-phosphate (heart), phosphocholine (heart), and inosine monophosphate (kidney) and decreased glycerol 3-phosphorylethanolamine (liver) and glycerol 3-phosphorylcholine (liver) were detected in the membranes of copper-deficient rats, which resulted in pronounced vacuolization throughout the tissues. Although the physiological and metabolic disturbances manifested in the hearts of copper-deficient animals generally mimic myocardial responses to chronic ischemia, the changes observed were interpreted in a broader context to represent the appearance of a copper-dependent cardiomyopathy [13].

In another study, rats were randomly assigned either copper-adequate (6.0 mg Cu/kg feed) or copper-deficient (< 1.0 mg Cu/kg feed) diets for a 5 week period. Their cardiac ATP levels and energy charges did not differ between the two groups. Levels of nicotinamide adenine dinucleotide (NAD) and phosphocreatine as well as the sum of pyridine nucleotides were greater in copper-deficient rats compared with copper-adequate rats using HPLC analysis. Mitochondrial respiratory control ratios and oxygen consumption rates in copper-deficient rat hearts were depressed, although their ADP:O ratios were similar to copper-adequate rat hearts [14].

2.2. Binary Complex of Cu(II) and ATP

There are many suggestions about the bonding sites of ATP in Cu(II) complexes [15]. They can be the oxygen atoms of the phosphates, the nitrogen atoms (N7 and NH_2) of the adenosine or one oxygen atom of the//phosphate and the N7 atom of the adenosine (Ado) molecule.

The value of log K/6.27 for the CuATP//complex, indicates that ATP behaves as a bidentate ligand. In comparison, the values of log K for the CuAMP complex, where AMP acts as a monodentate ligand with one oxygen atom of the phosphate as donor atom, and for CuAdo where the donor atom is one nitrogen atom, are 3.02 and 2.88, respectively [16]. From these values it can be suggested that the donor atoms in the CuATP complex are probably two oxygen atoms of phosphates.

In the case of the Cu:ATP species the maximum absorption band in the electronic spectrum occurs at 800 - 748 nm. These high values indicate that the electron donors in this case are not the nitrogen atoms and therefore, must be two oxygen atoms of the phosphates. This result reinforces the suggestion in the late paragraph above [17].

2.3. Binary Complexes of Cu(II) and PCr

The phosphocreatine, (PCr) molecule has three possible donor atoms: one oxygen of the phosphate group, the nitrogen atom of the guanidine group and the oxygen of the carboxylate group.

Although there is a possibility that PCr may act as a tridentate ligand, this hypothesis is not probable in solution because the value of log K_{CuPCr} (7.89) [18] is much lower than the value for the corresponding complex with phosphoserine (9.38), where this last ligand acts as tridentate [19]. By comparing the corresponding values of log K_{CuL} where L = diisopropylphosphorylguanidine (DPG), which does not have the carboxylate group (log K_{CuDPG} = 8.63 [20]), or creatine (Cr), which does not have the phosphate group (log K_{CuCr} = 7.20 [21]), it can be concluded that the value of log K_{CuPCr} can be either by coordination through the oxygen atom of the phosphate group and the nitrogen atom of the guanidine group, or through the oxygen atom of the carboxylate group and the nitrogen atom of the guanidine group. Since NMR analysis is not very useful for Cu(II) ion and far IR spectra cannot differentiate a M-O band corresponding either to an oxygen of phosphate groups or to an oxygen of carboxylate groups, these complexes were synthesized in the solid state in order to try to elucidate the coordination sites. One of these obtained complexes with Cu(II), PCr and Tyr, after its characterization, it was concluded that it was a Cu:Tyr:$PO_3H_2.H_2O$ complex [22] where PCr lost the creatine molecule. Therefore, it is more probable that the oxygen atom bound to Cu(II) in the Cu-PCr complex is from the phosphate group.

As mentioned before, in solution, it was possible to see that PCr acts as a bidentate ligand, the nitrogen atom of the guanidine group being one of the donor atoms. However, it was not possible to conclude whether the second donor atom was the oxygen atom of the phosphate or of the carboxylate group.

In order to try to solve this problem a Cu(II)-PCr complex from $CuSO_4.5H_2O$ and PCr was synthesized in aqueous media in a 1:1 ratio. A complex of Cu(II) and PCr of stoichiometry [Cu(PCr)(H_2O)] was prepared from aqueous solution at pH 6. The synthesis and characterization employing thermogravimetric, vibrational spectroscopy and electron paramagnetic resonance analyses were described [18] and the results suggested that PCr in solid state, acts as a tridentate ligand (the nitrogen atom of the guanidine group and the oxygen atoms of the phosphate and of the carboxylate groups being the donor atoms). The fourth position is occupied by a water molecule. These results were confirmed through computational calculations (DFT/B3LYP:311G theoretical procedure). The tetra coordination of Cu(II) arranged in a quadratic planar geometry was found to be more stable in the DFT calculations. Based on these calculations the wave numbers were attributed. Band deconvolution analysis (Peak Solve) was performed in order to verify the normal stretching modes of the -NH, -OH and -CH bonds [23].

The calculated vibrational spectrum is in agreement with the FT-IR experimental obtained spectrum. From these results it can be concluded that there is no similarity in the coordination mode of PCr with Cu(II) in solid state and in solution studies [23].

Recently, by using Raman spectroscopy in an aqueous solution of Cu(II)-putrescine (Put) it could be shown that the PO and NH bands are displaced when compared to the bands of Put alone. This suggests that Cu(II) binds to Put through the oxygen of the phosphate [24].

2.4. Ternary Complexes

2.4.1. Polyamines

Natural polyamines (PA) such as putrescine (Put, $NH_2(CH_2)_4NH_2$), spermidine (Spd, $NH_2(CH_2)_3NH(CH_2)_4NH_2$) and spermine (Spm, $NH_2(CH_2)_3NH(CH_2)_4NH(CH_2)_3NH_2$), are found in most living organisms and are involved in many biological reactions, particularly in metalation reactions, when they act as substrates for hormone synthesis [25-27] and react with nucleic acid, leading to structural changes in the latter. These reactions determine the role of PA in transfer processes of genetic information [26 - 28].

It has been observed that natural polyamines occur in relatively high concentration in young cells, including cancer cells [29]. It was also found out that polyamines regulate the stability of activating transcription factor-2-mRNA through RNA-binding proteins in intestinal epithelial cells and that polyamine-modulated ATF-2 expression plays a critical role in regulating epithelial apoptosis [30]. Moreover, it was reported that a novel class of quinine-bearing polyamines was recently proposed as multi-target-directed ligands to combat Alzheimer′s disease, with various factors in its etiopathology [31].

Copper(II) ion is involved in numerous enzymatic processes, including Cu-dependent amine oxidases (CAO). It has been reported that the DNA-bound to CAO is enzymatically active and that the oxidation of DNA-bound polyamines leads to the degradation of DNA [32]. In addition, complex formation of the naturally occurring polyamines with copper(II) inhibit the growth processes of these polyamines in cancer cells [33].

In physiological pH (≈ 7), Put, Spd and Spm are found in the protonated cation form. Due to their protonated – NH_3^+ groups, and mainly in this form, they interact with negatively charged fragments of other bioligands such as amino acids, proteins, fatty acids, phospholipids, nucleotides and nucleic acids [25]. The polyamine – bioligand interactions are observed in the range of pH in which the polyamine is protonated while the bioligand is deprotonated. The tendency towards the formation of molecular complexes depends not only on the number of protonated amine groups but also on the length of the polyamine methylene chain [34]. The potential sites of metal-ligand bonds are also the active centers of polyamine interactions with bioligands [17,34].

Bunce and Kong [35] proposed that Spd interacts with adenosine 5' monophosphate through two protonated amine groups (those separated by the three carbon chain) with the phosphate oxygen negative centers of the nucleotide.

Metal ion complexes of nucleosides or nucleotides interact with polyamines that are present in living cells. The determination of the structure of these complexes in solution is crucial for the understanding of some biological processes including the participation of metals in the processes of replication and transcription of nucleic acid [36,37]. The coordination properties of polyamines present in biological systems are important for the role they play in various cell processes [38]. Antitumor activity in the presence of plasma amineoxidase requires a terminal NH_2-$(CH_2)_3$-NH group and at least three carbon atoms between the primary and secondary amino groups [39], while antimicrobial activity is based on the ability to serve as an effective chelating agent [40].

2.4.2. Cu (II) Ternary Complexes with ATP and Polyamines

In order to a better understanding of the interactions of phosphates and PAs and the role of Cu(II), solution equilibria and spectral data of the binary and ternary complexes of Cu(II) ions with ATP and the PAs - ethylenediamine (en), 1,3-diaminopropane (tn), spermidine (Spd), and spermine (Spm) - were studied [16]. The system Cu(II):ATP:bis-[(2S)-2-pyrrolidinylmethyl]ethylenediamine (tetraamine) was also studied [16].

The decrescent order of the stability constant values of the ternary Cu(II) complexes with the polyamines and ATP was CutetraamineATP > CuSpdATP > CuenATP CutnATP. It is the same order for the corresponding binary Cu(II) complexes with the polyamines. This behaviour was confirmed by an increase in the maximum wavelength of the absorption spectra and the g_0 parameter and the decrease in the A_0 parameter of the EPR analyses as the stability constants decrease. This order of stability corresponds to the number of nitrogen donor atoms of the PA (tetraamine = 4, Spd = 3 and en = tn = 2). However, when the PA was Spd or tetraamine the protonated ligands favoured more stable ternary complexes. In fact, CuHSpdATP is the only protonated species that predominates in these ternary systems studied and this occurs at physiological pH ($\approx$ 7). This is a very important fact. Only when the polyamine is protonated it can interact with the deprotonated ATP. Since Spd is a naturally occurring polyamine, it can be concluded that its coordination to Cu(II) does not hinder its interaction with the phosphate groups of the nucleotides [16].

Studies in the solid state of the same complexes led to the proposition that the polyamines coordinate to the metal ion using two amine groups from each diamine, and the nucleotide interacts with the Cu(II) ion through the phosphate chain. In Spd one amine group and in Spm two amine groups remain free. The two remaining positions are occupied by two oxygen atoms of the phosphates [41].

The destabilizing effect of copper ions on DNA is generally explained by the disruption of the complementary hydrogen bonding scheme between the 6-amino group and the phosphate moiety [42]. Solution studies involving ATP, polyamines and copper ions, report that the main centers of metalation are oxygen atoms of the phosphate group of the nucleotide and deprotonated nitrogen atoms of the polyamines [16-17,34,43-45].

From those observations it can be suggested that the second ligand plays an important role, affecting the stability of ATP regarding dephosphorylation and thus facilitating the hydrogen bonds, even far from the site of metal attachment. The mechanisms behind many processes involving nucleotides, polyamines and metal ions have not been explained yet. Thus the knowledge on the coordination modes of ATP and polyamines can be useful to a better understanding of their behavior *in vivo* [41].

2.4.3. Cu (II) Ternary Complexes with ATP or Phosphocreatine and Amino Acids Containing a Hydroxyl Group

Several metabolic impairments and diseases are caused by the involvement of copper, ATP or phosphorylated amino acids containing a hydroxyl group.

Patients with gout present iron accumulation and overproduction of uric acid. These symptoms result from ATP degradation, decreased uric acid elimination caused by molybdenum and sulfur binding to copper in the kidneys, precipitation of sodium urate-iron crystals in the joints due to high ferritin, saturated transferrin and low CuZn-SOD as well as Cu-thionein in the joints and also development of inflammation triggered by tyrosine bonding to the sodium urate-iron crystals and being transformed by tyrosine kinase [46].

An increase in myocellular creatine was also observed after starvation associated with reduced serine phosphorylation [47]. As was already described, copper regulates the intracellular phosphorylation of the Wilson's disease protein, a human copper-transporting ATPase [10].

The autoregulatory process of glucose transport occurs independently of insulin. The activation of the energy-sensing enzyme AMP-activated protein kinase (AMPK) has been shown to increase insulin-independent glucose transport into skeletal muscle in response to exercise and hypoxia. Studies with incubated rats suggest that glucose autoregulates its own transport into skeletal muscle by a mechanism involving AMPK. The concentration of malonyl CoA increased and acetyl CoA carboxylase phosphorylation at serine 79 decreased when the concentration of glucose in the medium was raised [48].

Studies [17] were performed to assess the interactions of a phosphorylated ligand, an amino acid containing a hydroxyl group and the metal ion Cu(II). The authors report the results of aqueous solution equilibria and spectral studies of ternary copper(II) ion complexes with ATP or PCr and with the amino acids glycine, serine, tyrosine and threonine. Glycine is the only amino acid that does not have a hydroxyl group, and was used for comparison with the others in order to establish possible hydroxyl-phosphate interactions. These amino acids were chosen because glycine is the amino acid synthesized from serine [49]. Serine is involved in the synthesis of sphingol and therefore in the formation of sphingomyelins (phospholipids) in the brain [50]. In phosphoproteins, serine is present as phosphoserine [49] and it is part of the active centre in the enzyme acetylcholinesterase. The synthesis of acetylcholine is made through the choline-acetylase enzyme in the presence of ATP. PCr can replace ATP in this reaction [51]. The behavior of threonine and its importance is similar to that of serine. Finally, tyrosine is the precursor of the thyroid hormones and of melanin [15].

There are many possibilities regarding the bonding sites of ATP in Cu(II) complexes, but based on the values here obtained it can be suggested that the donor atoms in the Cu:ATP complex are probably two oxygen atoms of the phosphates [17].

The values of the stability constants of the ternary Cu(II) complexes with the amino acids and ATP or PCr indicate that these are stable complexes [18]. ΔlogK ($\log\beta_{CuaaL} - (\log\beta_{Cuaa} + \log\beta_{CuLATP})$) values with aa = Gly, Ser, Tyr and Thre, and L = ATP or PCr, showed that in all cases but specially when aa = Ser or Tyr and L=PCr, the ternary complexes are less stable than the binary ones, suggesting that no interaction occurs between the ligands in these ternary complexes. Since PCr acts as a bidentate ligand, there is one non-coordinated position and both Ser and Tyr have an OH group free. Since the possibility of interaction does not occur when the amino acid is Gly, which does not have an OH group, it seems that the interaction in the ternary complexes of PCr with Ser and Tyr is between a phosphate or carboxylate (probably the latter) of PCr and the OH group of the amino acids. These results can help understanding biological interactions among a phosphorylated ligand, an amino acid containing an hydroxyl group and the metal ion Cu(II) [18].

3. Phosphates and Cobalt

Cobalt is an essential trace element well known as a constituent of vitamin B_{12}, but it is also described as toxic for living organisms or for the environment. Cobalt causes DNA strand to break due to its capacity to produce active oxygen species (ROS) known to be highly reactive against DNA and other biomolecules. The use of the long-acting cobalt-ATP complex (Co-ATP) allows the observation of the effects of ATP without the influence of its metabolites.

Hegedüs et al. [53] compared the effects of intravenous Co-ATP on the cerebral blood flow (CBF), polarographically detected oxygen currents (O_2a), mean arterial blood pressure (MABP), heart rate, respiration rate, cerebral electrical activity, arterial blood gases, pH, and glucose in 13 normotensive (NT) rabbits to those in 14 stroke-prone spontaneously hypertensive (HT) animals. CBF was measured by the hydrogen and heat clearance methods. In response to Co-ATP, MABP decreased and CBF increased significantly in both groups. The decrease in MABP was greater in HT rabbits, while CBF response was 25% smaller than in NT animals. The ratio of O_2a to CBF diminished moderately and simultaneously with the CBF increase in NT rabbits. In HT rabbits, the decrease in O_2a/CBF was larger and began when CBF response reached its maximum. These results suggested that despite the restricted CBF response, long-acting ATP should still be taken into consideration as a supplementary treatment of hypertensive encephalopathy because of its beneficial effects on cerebral metabolism and hypertension [53].

The mechanism of long-term anoxic damage in brain tissue was investigated by Kass et al. [54] using rat hippocampus slices as a model system. The effects of short-duration of anoxia on subsequent transmission through two neural pathways were studied. 10 min of anoxia irreversibly abolishes the transmission between the perforant path and the dentate granule cells while only 7 min of anoxia irreversibly abolishes transmission between the Schaeffer collaterals and the CA1 pyramidal cells. The inclusion of 2 mM-cobalt in the buffer reduces the normoxic uptake of $^{45}Ca^{2+}$ so that the uptake during anoxia is no greater than the normoxic uptake in the absence of cobalt. This is associated with total protection against long-term transmission damage following 10 min of anoxia in the dentate gyrus [54].

Saccharomyces cerevisiae Cet1p catalyzes the first step of mRNA capping, the hydrolysis of the phosphate of triphosphate-terminated RNA to form a 5' diphosphate end. The RNA triphosphatase activity of Cet1p is magnesium-dependent and has a turnover number of 1 s^{-1}. Kiong Ho et al. [55] showed that purified recombinant Cet1p has a robust ATPase activity (K_m = 2.8 μM; V_{max} = 25 s^{-1}) in the presence of manganese. Cobalt is also an effective cofactor, while magnesium, calcium, copper, and zinc are not.

Wells et al. [56] studied the mechanism of inhibition of the myosin subfragment 1 (SF,) ATPase activity by Co(III)-phenATP.

Pecoraro et al. [57] investigated the substrate specificity of creatine kinase using chromium(III) and cobalt(III) complexes of adenosine 5'-diphosphate. They concluded that in general, chromium nucleotides are bound more strongly than cobalt complexes, and binding affinity increases as water replaces ammonia in the first coordination sphere of the metal. Both trends are a result of stronger hydrogen-bond interactions between the metal complex and protein.

Cobalt was used to study the YrbI protein of *Haemophilus influenzae* (HI1679). The crystal structure was determined by Parsons et al [58]. They reported that the protein exhibits the alpha/beta-hydrolase fold and resembles most closely the fold of the L-2-haloacid dehalogenase (HAD) superfamily. A detailed sequence analysis revealed remote homology to two members of the HAD superfamily, the P-domain of Ca^{2+} ATPase and phosphoserine phosphatase. The 19-kDa chains of HI1679 form a tetramer both in solution and in the crystalline form. The four monomers are arranged in a ring such that four beta-hairpin loops, each inserted after the first beta-strand of the core alpha/beta-fold, form an eight-stranded barrel at the center of the assembly. Four active sites are located at the subunit interfaces. Each active site is occupied by a cobalt ion. The cobalt is octahedrally coordinated to two aspartate side-chains, one backbone oxygen, and three solvent molecules, indicating that the physiological metal may be magnesium. HI1679 hydrolyzes a number of phosphates, including 6-phosphogluconate and phosphotyrosine, suggesting that it functions as a phosphatase *in vivo*. The physiological substrate is yet to be identified, although the location of the gene on the yrb operon suggests involvement in sugar metabolism [58].

The endogenous of the adenovirus protein kinase specifically phosphorylates the capsid protein IIIa in the presence of Mg^{2+}, utilizing either ATP or GTP as a phosphate donor. When Mn^{2+} is substituted for Mg^{2+} or Co^{2+} in the presence of ATP, phosphorylation of IIIa is enhanced by 2-3 fold. However, in addition to IIIa phosphorylation, the core proteins V and VII are now phosphorylated. Further, when these two ions are substituted for Mg^{2+}, the profile of the phosphoamino acid residue changes: instead of only phosphoserine being labeled, phosphothreonine is now extensively labeled. These results are specific for the above endogenous viral proteins, since when the viral protein kinase is used to phosphorylate added exogenous substrates such as casein, neither enhancement of phosphorylation nor any change in phosphoamino acid profile are achieved by substituting Mn^{2+} or Co^{2+} for Mg^{2+}. It is thus likely that MnATP or CoATP somehow interacts differently with the structural proteins of adenovirus than does MgATP and this facilitates their accessibility to the enzyme [59].

Using Thr(P)-inhibitor-1 and Ser(P)-casein as substrates, studies on the activation of calcineurin purified from bovine brain demonstrated that phosphatase requires the synergistic action of Ca^{2+}, calmodulin and another divalent cation (Mg^{2+}, Mn^{2+}, Co^{2+} or Ni^{2+}, but not Zn^{2+}) for the full expression of its activity [60].

Adenylate kinase (AK) is involved in the reversible transfer of the terminal phosphate group from ATP to AMP and contributes to the maintenance of a constant level of cellular adenine nucleotides. Cobalt, zinc and iron (II) have been reported to be present in the AKs of some Gram-negative bacteria. The AK of two sulfate-reducing bacteria, *Desulfovibrio gigas* and *Desulfovibrio desulfuricans* was purified from crude extracts and was biochemically and spectroscopically characterized in the native and full cobalt and zinc-substituted forms. The involvement of three cysteine residues and one histidine were confirmed [61]. The purified native zinc-containing AK was crystallized and the cobalt- and iron-containing forms of recombinant AK were also obtained. A monomer was present in the asymmetric unit for both Zn^{2+}-AK and Fe^{2+}-AK forms, and a dimer was present in the Co^{2+}-AK form [62].

Studies in solution of the interactions between ATP and ADP with Co (II) and with spermine (Spm) and 1,11-diamine-4,8-diazaundecane (3,3,3-tet) showed that in the systems: Co(ADP)H_4(Spm), Co(ATP)H_4(Spm), Co(ADP)H_4(3,3,3-tet) and Co(ATP)H_4(3,3,3-tet), there was the molecular complexes formation of the type ML…L', in which the N(7) atom and oxygen atoms of the phosphate group are involved in coordination and the fully protonated polyamine is engaged in noncovalent interactions with the nucleotide N(1) [63].

4. Phosphates and Aluminum

Until recently, aluminum was considered to be harmless to humans as it would reportedly be kept out of the body by natural barriers [64]. Nowadays elevated levels of aluminum are associated with neurological disorders. In the last years aluminum was detected in the brain of patients with Alzheimer's [65] and Dialysis Encephalopathy [66] and was also linked to microcytic anemia and osteomalacia [67,68]. Patients submitted to hemodialysis, in which antacids containing aluminum were used to diminish the absorption of phosphates, present symptoms of neurological dysfunctions such as impairment of speech and movements, personality alterations, altered encephalograms, dementia and psychosis which can lead to death after some months. These patients showed high aluminum concentrations in the blood, bones and brain (mainly in gray matter) [69]. Studies show that after a 90-day oral exposure of rats to Al-lactate there was a decrease in adenosinetriphosphatase (ATPase) and that aluminum contribute to oxidative stress and this constitutes a major risk related to Alzheimer's disease and other degenerative disorders. Chronic exposure to aluminum accelerates and enhances the production of the aberrant splicing isoform of preselin-2, and promotes the development of Alzheimer's disease [71]. It seems that there is a tendency to accumulate aluminum with aging due to a diminished homeostatic control [72]. Aluminum binds to biological phosphates including nucleotides, phosphorylated proteins, and phospholipids. In all cases the phosphate group is probably the target of the biochemical action of Al(III), due to its high affinity to both inorganic and organic phosphates [73]. The study of these interactions is very important to the understanding of the role of aluminum in living organisms.

It must be taken into account that the cognitive impairments that occur in Alzheimer's disease and other neurological dysfunctions seem to involve protein phosphatases [74, 75]. It is not entirely established if aluminum plays a role in these processes directly or indirectly due to its competition with calcium [76] and other metal ions but it is very probable that its affinity to phosphates may affect neuronal communication. Aluminum (III) may also interfere in bone formation and other biological processes. For instance, protein phosphorylation, which is very important to many regulatory processes, can be affected by the presence of this element. In particular, the complexation of aluminum with phosphates is very harmful because it causes serious damages to the processes where phosphates are present.

To solve this problem and to provide new treatments it is necessary to understand what happens at the molecular level. The stability of Al(III) complexes with mono, di and triphosphates [73-77,78] as well as NMR spectroscopic studies of nucleotides [79] have been reported.

However, as aluminum is easily hydrolyzed [80], a great part of the ingested aluminum is eliminated as insoluble phosphate, $AlPO_4$, or hydroxy-phosphates [73]. The absorbed Al(III) must be protected from hydrolysis in order not to be able to circulate in the organisms and reach the brain. That means that it must have its charge neutralized by complexation with negatively charged compounds with low molecular weight [69].

Aluminum (III) binds very easily to O-donor groups [81] and phosphates are among the preferred ones [82]. Aluminum reacts with membrane phospholipids [83] and biophysical studies suggest the direct interaction of this metal ion with the phosphate binding site [84]. NMR studies of phosphate-Al complexes have shown that phosphate takes part in the coordination site [85]. This is one of the reasons why it is necessary to understand the way this ion interacts with ligands containing phosphates.

4.1. Binary Complexes of Al(III) and Phosphoserine.

Kiss et al. [86], studying the system Al(III):phosphoserine (Pser) suggest that the formation of the complex begins with the binding to the phosphate group resulting in the species AlPserH and with the amine group still protonated and ends with the ligand acting as tridentate (OPO_3^{2-}, NH_2, CO_2^-). Although the interaction Al(III) – NH_2 is not favorable, the amino group of the phosphoserine is beside two binding sites where the oxygen is the donor atom, thus favoring a tridentate coordination.

4.2. Binary Complexes of Al(III) and PCr

The complexes of Al(III) and PCr were studied by Silva [87]. Pser and PCr have the same donor atoms. The value of the formation constant of AlPCr was found to be 11.37 [87] and it is very similar to the value for the formation constant for AlPser – 11.48 [86]. For this reason it was suggested that in the complex Al(III) PCr, the ligand acts as tridentate, and that the donor atoms in the AlPCr complex are the oxygen atom of the phosphate, the nitrogen atom of the guanidine group and the oxygen atom of the carboxylate.

Comparing the ^{31}P NMR spectrum of PCr with those of the AlPCr complex, confirms the complexation through the phosphate [20]. The ^{27}Al NMR spectra of the complexes indicate tetrahedral molecules [88]. The spectrum of PCr shows a peak at 48.923 ppm [20]. Peaks around 50 ppm are typical of Al-phosphates [88].

These results were confirmed through computational calculations (DFT/B3LYP:311G theoretical procedure). The tetra coordination of Al(III) arranged in a tetrahedral geometry was found to be more stable in the DFT calculations. Besides the three positions occupied by PCr, the forth position is occupied by a water molecule. Based on these calculations the wave numbers were attributed. The calculated vibrational spectrum agrees well with the RAMAN experimental spectrum [89].

4.3. Binary Complexes of Al(III) and ATP

Kiss et al. [78], Wang et al. [79] and Jackson and Vouyi [90] suggest the coordination of Al(III) with ATP through the terminal phosphates (β and γ). Mao et al. [91] suggest that the sites of coordination are the N-7 from adenine besides the phosphates.

Potentiometric studies performed with ATP and Al(III), varying the concentrations of the ligands and the ligand:metal ratio indicated the values for the log of the complex stability constants of 11.30, 7.82 and 2.46 for the species AlATPH, AlATP and $AlATPH_{-1}$, respectively [78].

Silva [87] found similar values for these species. However, when the calculated and experimental curves were compared, it could be seen that after pH 6 they did not coincide.

According to Wang et al. [79], studies of ^{1}H NMR with Al:ATP system detected the species Al_2ATP_2OH ($Al_2ATP_2H_{-1}$) in neutral pH together with the species AlATP, besides some hydrolyzed species. Studies with ^{31}P indicate that in the species $Al_2ATP_2H_{-1}$ each Al(III) ion is coordinated through the phosphates β and γ to two different ATP molecules. This dimer seems to be formed by bases attraction, so repeating the determination of the formation constants with the introduction of the species $Al_2ATP_2H_{-1}$ lead to different results which are shown in Table 1 [87].

Table 1. Stability constants (log β) of the complexes formed with aluminum (III) and ATP.[87]

Species	log β
$AlATPH_2$	15.68 ± 0.02
AlATPH	13.04 ± 0.01
AlATP	9.21 ± 0.01
$AlATPH_{-1}$	2.80 ± 0.02
$Al_2ATP_2H_{-1}$	16.16 ± 0.01

With these results, the experimental and calculated curves are the same.

Computational calculations (DFT:B3LYP/6-311G theoretical procedure) show that the structure of AlATP is a distorted octahedron and the coordination takes place through the two terminal phosphates (ß and γ). The calculated vibrational spectrum agrees well with the RAMAN experimental spectrum [89].

4.4. Al(III) Ternary Complexes with ATP or Phosphocreatine and Amino Acids Containing a Hydroxyl Group

The calculated complexation constant values of the ternary complexes with ATP do not suggest any interaction besides complexation. However, in the case of phosphocreatine, all results indicate some additional interaction besides complexation. This interaction cannot be related to the hydroxyl group since it occurs also with glycine. Perhaps the coordinated water molecule or the amino group can be the site of interaction [87]. These possibilities are still to be checked by theoretical calculations.

5. Phosphates and Vanadium

Vanadium plays an exceptional role among the bioelements because both anionic and cationic forms can participate in biological processes [92]. Dietary vanadium occurs mainly as $H_2VO_4^-$ and enters cells probably through the phosphate transport mechanism and undergoing a rapid one-electron reduction [93]. Therefore, outside the cells vanadium is present as vanadate and inside the cells as vanadyl. In its anionic forms (vanadates (V)), they resemble phosphates, but in its cationic forms - mainly as VO^{2+} - it behaves like a typical transition metal ion, competing with other metal cations in coordination with biogenic ligands or compounds. This duality, added to the facility with which it changes oxidation states and coordination environments, may be responsible for the very peculiar and somewhat unparallel behavior of this biometal [92].

Vanadate has been proposed as substitute for phosphate in a number of phosphate binding enzymes in view of the similarity of PO_4^{3-} and VO_4^{3-} [94]. Vanadium compounds have been shown to cause a variety of biological and metabolic effects, including the inhibition of certain enzymes, alteration of the contractile function, and as an insulin-like regulator of glucose metabolism. The similarity of phosphate and vanadate is an important factor to be investigated when these effects are studied. This similarity must be the reason why bone is an active accumulator of vanadium. In order to investigate this rapid exchange of vanadate with bone phosphate, calcium hydroxylapatite, $Ca_{10}(PO_4)_6(OH)_2$, was used as a model for the inorganic phase of bone. Under physiological conditions, the exchange was only observed with amorphous material. These model studies showed that the incorporation of small amounts of vanadium into the phosphate sites only produces weak distortions at macroscopic (crystallographic parameters, crystal ordering) and microscopic (local distortions, weakening of chemical bonds) levels in the apatite lattice [95].

It has been shown that superoxide and several xenobiotic free radicals can release iron from ferritin through a reductive mechanism. Vanadyl reacts with oxygen to generate superoxide and vanadate. Monteiro et al. [96] investigated the ability of vanadyl and vanadate to release iron from ferritin. Iron release was measured by monitoring the generation of the $Fe^{(2+)}$-ferrozine complex. It was found that vanadyl, but not vanadate, was able to mobilize ferritin iron in a concentration dependent manner. Initial rates, and iron release over 30 minutes, were unaffected by the addition of superoxide dismutase. Glutathione or vanadate added in relative excess to the concentration of vanadyl, inhibited iron release up to 45%. The addition of ferritin at the concentration used for measuring iron release prevented vanadyl-induced NADH oxidation. Vanadyl promoted lipid peroxidation in phospholipid liposomes. Addition of ferritin to the system stimulated lipid peroxidation up to 50% above that with vanadyl alone. Ferritin alone did not promote significant levels of lipid peroxidation. [96]

Plasma taken from 9 normal control and 8 manic-depressive subjects was analysed for vanadium by atomic absorption spectrometry. The mean plasma vanadium concentrations were 0.15 microM in normal control, 0.34 microM in manic, 0.28 microM in depressed subjects, and 0.23 microM in manic-depressive subjects after recovery. The differences between normal, manic and recovered subjects were statistically significant. Significant negative correlations were found between plasma vanadium concentration and the ratio of Na-K-Mg ATPase to Mg-ATPase in 2 manic-depressive subjects, but not in normal subjects. The results suggest that vanadium may be a cause behind the variations in Na-K-Mg ATPase and sodium pump activity which are associated with manic-depressive illness [97]. Vanadium content of hair from manic patients, recovered patients, and normal controls was measured by neutron activation analysis. The mean vanadium content of the hair of the manic group was significantly higher than that of the control group and that of the group of recovered manic patients [98]. These high levels of vanadium reported in plasma in mania and depression and in hair in mania are evidence of the involvement of vanadium in the etiology of manic depressive psychosis. Lithium has been reported to reduce the inhibition of Na-K ATPase by vanadate. Several groups of psychotropic drugs (e.g. phenothiazines, monoamine oxidase inhibitors) have been shown to catalyze the reduction of vanadate to the less active vanadyl ion. Therapies based on decreasing vanadate levels in the body (e.g. ascorbic acid, EDTA, methylene blue) have been reported to be effective in both depression and mania [99].

Evidence that elevated elemental vanadium levels may play an etiological role in at least some forms of manic-depression and the fact that lithium ion, which inhibits the hydrolytic degradation of inositol monophosphates is the most common therapeutic agent used in the control of bipolar disorder, leads to studies that demonstrated that vanadate treatment of intact cells from several different clonal lines synergistically induces a substantial augmentation in neurotransmitter receptor-mediated or growth factor receptor-triggered inositol trisphosphate accumulation *in situ*. Furthermore, studies using cellular extracts indicate that the effects of vanadate treatment *in situ* may be due to its ability to inhibit hydrolysis of inositol 1,4,5-trisphosphate inositol 1,3,4-trisphosphate, and inositol 1,3,4,5-tetrakisphosphate *in vitro*. These results suggest that vanadate treatment may facilitate the characterization of inositol phosphate metabolism and intracellular signaling [100].

Liochev et al. [101] studied extensively the oxidation of NADH by vanadium. Sugars and sugar phosphates enable vanadate to catalyze the oxidation of NADH. Superoxide dismutase inhibits this oxidation. The incubation of sugars with vanadate, prior to addition of NADH, accelerates this oxidation of subsequently added NADH and eliminates the lag phase otherwise noted. Incubation of sugars with vanadate also results in the reduction of vanadate to vanadyl, with the appearance of a blue-green color probably associated with a vanadyl-vanadate complex. It appears that sugars reduce vanadate to vanadyl which, in turn, reduces O_2 to O^{2-} and that vanadate plus O^{2-} then catalyzes the oxidation of NAD(P)H by means of a free-radical chain reaction. Such oxidation of NAD(P)H may account for several of the biological effects of vanadate [101].

Vanadyl (V(IV)) was found to induce rapidly developing lipid peroxidation in intact and sonicated mitochondria as well as in phosphatidylcholine suspension. The ability of vanadate (V(V)) to induce lipid peroxidation was much less pronounced compared to that of vanadyl.

The peroxidative action of vanadate on phosphatidylcholine increased significantly in the presence of NADH and ascorbate. Preincubation of vanadate with glucose had the same effect. Vanadyl-induced lipid peroxidation was not essentially influenced by SOD, catalase and ethanol but was completely inhibited by butylated hydroxytoluene [102]. The mechanism of vanadate (V(V))-dependent oxidation of NADH was different in phosphate buffers and in phosphate-free media. In the latter vanadyl generated by direct V(V)-dependent oxidation of NADH formed a complex with V(V). In phosphate buffers V(IV) autoxidized instead of forming a complex with V(V). The generated superoxide radical initiated a high-rate free radical chain oxidation of NADH. Phosphate did not stimulate the V(V)-dependent NADH oxidation catalyzed by O^{2-}-generating systems. Monovanadate proved to be a stronger catalyzer of NADH oxidation compared to polyvanadate [103]. Vanadate (V(V)) stimulates the oxidation of NADH by xanthine oxidase and superoxide dismutase eliminates the effect of V(V). Paraquat stimulates both the oxidation of NADH by xanthine oxidase and the V(V) enhancement of that oxidation. Xanthine, which is a better substrate for xanthine oxidase than is NADH, causes a V(V)-dependent co-oxidation of NADH which is transient and eliminated by SOD. Urate inhibits the V(V)-stimulated oxidation of NADH by xanthine oxidase or by Rose Bengal plus light. Measurement of rates of both O^{2-} production and V(V)-stimulated NADH oxidation showed that many molecules of NADH were oxidized per O^{2-} produced. These chain lengths were an inverse function of the overall reaction rate. Minimum chain lengths, calculated on the basis of 100% univalent reduction of O_2 to O^{2-}, were smaller than average chain lengths measured by a factor of five. All of these results are in accordance with the view that V(V) does not directly affect the activity of the enzyme, but rather catalyzes the free-radical chain oxidation of NADH by O^{2-}. It was further shown that phosphate was not involved and that the active form of V(V) was orthovanadate, rather than decavanadate [104]. As it can be seen, not all effects of vanadium are related to its similarity to phosphate although this is an important factor and affects several mechanism. Furthermore, all these effects of vanadyl and vanadate are important to understand the insulin-like and other biological actions of vanadium.

One of the leading causes of mortality among diabetics is myocardial disease. Although originally thought to occur as a result of atherosclerosis, various studies have shown that heart disease can occur in the absence of atherosclerosis, suggesting a diabetic cardiomyopathy. Using diabetic animals, it has been possible to characterize diabetes-induced myocardial abnormalities. Diabetic rat hearts do not respond to conditions of high stress as well as do controls. Functional depression is accompanied by altered cardiac enzyme systems. A decrease in myosin ATPase activity, which appears to be a result of diabetes-induced hypothyroidism, can be observed. Also, a depression of sarcoplasmic reticular calcium ATPase, along with a depression of calcium uptake by the SR, is seen in diabetic rat hearts. Na+, K+ ATPase activity has also been shown to be depressed, and this depression appears to correlate with depressed atrial contractility. Insulin treatment has been relatively successful at reversing or preventing myocardial changes in diabetic rats. Among a great variety of treatments, vanadate treatment has proved to be encouraging in that normalizing both function and hyperglycemia [105].

Vanadate mimics the metabolic actions of insulin. In diabetic rodents, vanadate also sensitizes peripheral tissues to insulin. Sun et al. [106] analyzed whether this latter effect is brought about by a mechanism other than the known insulinomimetic actions of vanadium *in vitro*. They reported that the levels of glucose 6-phosphate (G-6-P) in adipose tissue, liver, and muscle of streptozotocin-treated (STZ)-hyperglycemic rats were 77, 50, and 58% of those in healthy control rats, respectively. Normoglycemia was induced by vanadium or insulin therapy or by phlorizin. They concluded that: 1) chronic hyperglycemia largely reduces intracellular G-6-P in all three insulin-responsive tissues; 2) vanadate therapy restores this deficiency, but insulin therapy does not restore G-6-P in muscle tissue; 3) the induction of normoglycemia *per se* (i.e., by phlorizin) restores G-6-P only in the liver; and 4) glucose and vanadate together elevate G-6-P in adipose explants *in vitro* and significantly restore lipogenic capacity above the threshold of G-6-P level. They suggest that hyperglycemia-associated decrease in peripheral G-6-P is a major factor responsible for the peripheral resistance to insulin. The mechanism by which vanadate increases the capacity of peripheral tissues to metabolize glucose and to respond to the hormone involves the elevation of this hexose phosphate metabolite and the resulting cellular consequences of this elevated level of G-6-P. [106]

Geraldes et al. [107] examined the influence of vanadate on glucose metabolism and sodium transport in isolated perfused rat hearts. Hearts were perfused with 10 mM glucose and varying vanadate concentrations (0.7-100 microM) while changes in high energy phosphates (ATP and phosphocreatine (PCr)), intracellular pH, and intracellular sodium were monitored using ^{31}P and ^{23}Na NMR spectroscopy. Tissue lactate, glycogen, and (Na+, K+)-ATPase activity were also measured using biochemical assays. Under baseline conditions, vanadate increased the levels of tissue glycogen two fold and reduced (Na+, K+)-ATPase activity. Significant decreases in ATP and PCr were observed in the presence of vanadate, with little change in intracellular pH. These changes under baseline conditions were less severe when the hearts were perfused with glucose, palmitate and beta-hydroxybutyrate. During ischemia, vanadate did not limit the rise in intracellular sodium, but slowed sodium recovery on reperfusion. The presence of vanadate during ischemia resulted in attenuation of acidosis, and reduced lactate accumulation. Reperfusion in the presence of vanadate resulted in a slower ATP recovery, while intracellular pH and PCr recovery was not affected. These results indicate that vanadate alters glucose utilization and (Na+, K+)-ATPase activity and thereby influences the response of the myocardium to an ischemic insult [107].

Noda et al. [108] studied the effects of vanadium on cardiac performance and energy metabolism in diabetic rat hearts, and concluded that vanadium restored the production of high energy phosphates in the myocardium and improved myocardial dysfunction by regulating metabolic processes [108].

The effect of V^{5+} (as ammonium metavanadate) on the expression of the aryl hydrocarbon receptor (AhR)-regulated gene: cytochrome P450 1a1 (*Cyp1a1*) at each step of the AhR signal transduction was examined. This study provided the first evidence that V^{5+} down-regulates the expression of Cyp1a1 at the transcriptional level through an ATP-dependent mechanism [109].

REFERENCES

[1] Felcman J, Lopes DCN, Cattapan R A, Soto CAT, *Synth. React. Inorg. Met-Org. Chem.*, 2001, 31 (3), 507.

[2] Manning G, Whyte DB, Martinez R, Hunter T, Sudarsanam S, *Science,* 2002, 298, 1912.

[3] Mann M, Ong SE, Gronborg M, Steen H, Jensen ON, Pandey A, *Trends Biotechnol.,* 2002, 20(6), 261.

[4] Cohen P, *Trends in Biochem. Sci.*, 2000, 25, 596.

[5] Semenovski ML, Shumakov VI, Sharov VG, Mogilewvsky GM, Asmolovsky AV, Makhotina LA, Saks VA, *J. Thorac. Cardiov. Surg.*, 94, 1987, 762.

[6] Eldar-Finkelman H, Eisenstein M, *Curr. Pharm. Des*. 2009,15(21),2463.

[7] Schumacher M, Camp S, Maulet Y, Newton M, MacPhee-Quigley K, Taylor SS, Friedmann T, Taylor P, *Nature*, 1986, 319 (6052), 407.

[8] Scolari MJ, Acosta GB, *Amino acids,* 2007, 33 (4), 563.

[9] van Hoof NALM, Koevoets PLM, Hakvoort HWJ, Bookum WMT, Schat H, Verkleij JAC, Ernst WHO, *Physiologia Plantarum*, 2001, 113 (2), 225.

[10] Wanderwerf SM, Cooper MJ, Stetsenko IV, Lutsenko S, *J. Biol. Chem.,* 2001, 276 (39), 36289.

[11] Li C, Peoples RW, Weight FG, *Neurosci. Lett,* 1996, 219 (1), 45.

[12] Hernández-Esquivel L, Marín-Hernández A, Pavón N, Carvajal K, Moreno-Sánchez R, *Toxicol. Appl. Pharmacol.,* 2006, 212 (1), 79.

[13] Kopp SJ, Klevay LM, Feliksik JM, *Am. J. Physiol.*, 1983, 245(5 Pt 1), H855.

[14] Chao JC, Medeiros DM, Altschuld RA, Hohl CM, *Comp. Biochem. Physiol. Comp. Physiol.*, 1993, 104(1), 163.

[15] Philips RSJ, J. Annu. Rev., 1996, 501.

[16] Alves da Silva J, Felcman J, Mercê ALR, Mangrich AS, Lopes RSC, Lopes CC, *Inorg. Chim. Acta*, 2003, 356, 155.

[17] Lomozik L, Gasowska A, *J. Inorg. Biochem.*, 1998, 72, 37.

[18] Silva AM, Mercê ALR, Mangrich AS, Téllez Soto CA, Felcman J, *Polyhedron*, 2006, 25, 1319.

[19] Abbot EH, Mohan MS, *Inorg. Chem.*, 1978, 17, 2203.

[20] Lopes DCN, Doctoral Thesis, Department of Chemistry, PUC-Rio, Brazil, 2001.

[21] IUPAC Stability Constants Database, Release 3, Version 3.02, 1998; compiled by Petit LD, Powell HKJ, Academic Software, Timble, Otley, West Yorkshire, UK.

[22] Silva IG, Master Thesis, Department of Chemistry, PUC-Rio, Brazil, 2003.

[23] Almeida BL, Ramos JM, Versiane O, Sousa M, Téllez Soto CA, Mercê ALR, Mangrich AS, Felcman J, *Polyhedron,* 2008, 27, 3662.

[24] Szyfman NW, Loureiro NP, Felcman J, 32nd Meeting of the Brazilian Chemical Society, published in the Book of Abstracts, 2009.

[25] Lomozick L, Metal Complexes with Polyamines. In the Handbook of Metal-Ligand Interactions in Biological Fluids; Berthon G, Ed.Toulouse, France, 1995, Vol 1, Chapter 2 pp. 686-697.

[26] Seiler N, Delcros JG, Moulinoux JP, *Int. J. Biochem. Cell Biol.,* 1996, 28, 43.

[27] Fang Y, Hoh JH, *J. Am. Chem. Soc.,* 1998, 120, 8903.

[28] Ropars C, Viovy R, *J. Biol. Chem.,* 1958, 230, 797.
[29] Janne J, Poso H, Raina A, *Biochim. Biophys. Acta*, 1978, 473, 407.
[30] Xiao L, Rao JN, Zou T, Liu L, Marasa BS, Chen J, Turner DJ, Zhou H, Gorospe M, Wang JY, *Mol Biol Cell*, 2007, 11, 4579.
[31] Bolognesi ML, Banzi R, Bartolini M, Cavalli A, Tarozzi A, Adrisano V, Minarini A, Rosini M, Tumiatti V, Bergamini C, Fato R, Lenaz G, Hrelia P, Cattaneo A, Recanatini M, Melchiore C, *J Med Chem*, 2007, 50 (20), 4882.
[32] Houen G, *APMIS Suppl.*, 1999, 96, 1.
[33] 33 Tsou KC, Lo KW, Selzer M, Weinstein J, Bunder D, *Excerpta Med. Int. Congr. Ser.*, 1973, 275, 300.
[34] Lomozik L, Gasowska A, Bolewski L, *J. Inorg. Biochem*, 1996, 63, 191.
[35] Bunce S, Kong ESW, *Biophys. Chem.,* 1978, 8, 357.
[36] Kopf-Maier P, Kopf H, *Chem. Rev.,* 1987, 87, 1137.
[37] Scherman SE, Lippard SJ, *Chem. Rev.,*1987, 87, 1153.
[38] Bachrach U, Function of Naturally Occurring Polyamines, Academic Press, N. Y. and London, 1973.
[39] Israel M, Modest EJ, *J. Med. Chem.,* 1971, 14, 1042.
[40] Weinberg ED, *Antimicrob. Agents Chemother.*, 1963, 573.
[41] Almeida BL, Versiane O, Souza, Mercê ALR, Mangrich AS, Felcman J, *Inorg. Chim. Acta*, 2009, 362, 2447.
[42] Somasundara I, Palaniandavar M, *J. Inorg. Biochem.,* 1994, 53, 95.
[43] Gasowska A, Lomozick L, Jastrzab R, *J. Inorg. Biochem.* , 2000, 78, 139.
[44] Gasowska A, Lomozick L, *Polyhedron*, 2002, 21, 745.
[45] Gasowska A, *J. Inorg. Biochem.,* 2003, 96, 346.
[46] Johnson S, *Med. Hypotheses*, 1999, 53(5), 407.
[47] Zhao CR, Shang L, Wang W, Jacobs DO, *J. Surg. Res.,* 2002, 105 (1), 10.
[48] Itani SI, Saha AK, Kurowski TG, Coffin HR, Tomheim K, Ruderman NB, *Diabetes*, 2003, 52 (7), 1635.
[49] Elliot WH, Elliot DC, Biochemistry and Molecular Biology, 2nd. Ed., Oxford, University Press, p.296, 2002.
[50] Hhirabayashi Y, Furuya S, Progress in Lipid Research, 2008, 47 (3), 188.
[51] Nelson DL, Cox MM, Lehninger Principles of Biochemistry, 3rd. Ed., p.510, Worth Publishers, 2004.
[52] Kiss T, Biocoordination Chemistry. Burger K (Ed.). Nova York, Ellis Horwood, 1990.
[53] Hegedüs K, Feketi I, Molnar L, Journal of Neurology (J. Neurol.), German Society of Neurology. Annual meeting N°67, INCONNU (28/09/1994) Swiss Neurological Society. Meeting N°154, INCONNU (28/09/1994) 1995, vol. 242, n° 4, pp. 257-287 (38 ref.), pp. 243-248.
[54] Kass IS, Lipton P, *J. Physiol.*, 1986, 378, 313.
[55] Ho CK, Pei Y, Shuman S, *J. Biol. Chem.*, 1998, 273 (51), 34151.
[56] Wells JA, Werber MM, and Yount RG, *J. Biol. Chem.*, 1980, 255, (16), 7552.
[57] Pecoraro VL, Rawlings J, Cleland WW, *Biochemistry,* 1984, 23(1), 153.
[58] Parsons JF, Lim K, Tempczyk A, Krajewski W, Eisenstein E, Herzberg O, *Proteins*, 2002, 46(4), 393.
[59] Tsuzuki J, Luftig RB, *Virus Res.*, 1985, 2(2), 95.
[60] Li HC, Chan WW, *Eur. J. Biochem.*, 1984, 144(3), 447.

[61] Gavel OY, Bursakov SA, Rocco GDi, Trincao J, Pickering IJ, George GN, Calvete JJ, Brondino C, Pereira AS, Lampreia J, Taveres P, Moura JJG, Moura I, *J. Inorg. Biochem.*, 2009, 102 (5-6), 1380.
[62] Kladova AV, Gavel OY, Mukhopaadhyay A, Boer DR, Teixeira S, Shnyrov VL, Moura I, Moura JJG, Romão MJ, Trincão J, Bursakov SA, *Acta Cryst.,* 2009, F65, 926.
[63] Grasowska A, *Zeitschrift füanorganische und allgemeine Chemie*, 2006, 632 (14), 2281.
[64] Sorenson JRJ, Campbell IR, Tepper LB, Lingg D, *Environ. Health Perspect.*, 1974, 8, 3.
[65] Perl DP, Brody AR, *Science*, 1980, 208, 297.
[66] Mc Laughlin AIG, Kazantzis G, King E, Teare D, Porter RJ, Owen R, *Br. J. Ind. Med.*, 1962, 19, 253.
[67] Hellström HO, Michaëlsson K, Mallmin H, Mjöberg B, *Osteoporos. Int.,* 2005, 12, 1982.
[68] Davis K, Aslam, Pejović-Milić A, Chettle DR, *Med. Phys.*, 2008, 35 (11), 5115.
[69] Kruck TPA, Mc Lachlan DRC, in: Siegel H (Ed.), Metal Ions in Biological Systems, 24, Marcel Dekker, NY, 1988, pp 285-313.
[70] Kohila T, Parkkonen E, Tähti H, *Arch. Toxicol.*, 2004, 78(5), 276.
[71] Matsuzaki S, Manabe T, Katayama T, Nishikawa A, Yanagita T, Okuda H, Yasud Miyata S, Meshitsuka S, Tohyama M, *J. Neurochem.*, 88 (6) (2004) 1345.
[72] Cowburn JD, Farrar G, Blair JA, *Chem. Britain*, 1990, 26 (12), 1169.
[73] Atkári K, Kiss T, Bertani R, Martin RB, *Inorg. Chem.*, 1996, 35(24), 7089.
[74] Riedel G, *Cell. Mol. Life Sci.,* 1999, 55(4) 549.
[75] Platt B, Fiddler G, Riedel G, Henderson Z, *Brain Res. Bull.* , 2001, 55(2), 257.
[76] Kawahara M, *J. Alzheimers Dis*. 2005, 8(2), 171 and 209.
[77] Feng Q, Waki H, *Polyhedron*, 1991, 10, 1527.
[78] Kiss T, Sovago I, Martin RB, *Inorg. Chem.*, 1991, 30, 2130.
[79] Wang XG, Simpson JH, Nelson DJ, *J. Inorg. Biochem.,* 1995, 58, 29.
[80] Martin R, in: Siegel H (Ed.), Metal Ions in Biological Systems, 24, Marcel Dekker, NY, 1988, pp 2-57.
[81] Orvig C, Berthon G, Handbook of Metal-Ligand Interactions in Biological Fluids, Vol 2, Part Five, Chapter 3, Section G, Marcel Dekker,
[82] Finnegan MM, Rettig SJ, Orvig C, *J. Am. Chem. Soc.*, 1986, 108, 5033.
[83] Martell AE, Motekaites RJ, Smith RM, *Polyhedron*, 1990, 9, 171.
[84] Kiss T, Zatta P, Corain B, *Coord. Chem. Rev.*, 1996, 149, 329.
[85] Akitt JW, *Progr. NMR Spectroscopy*, 1989, 21, 49.
[86] Kiss E, Lakatos A, Bányai I, Kiss, T. *J. Inorg. Biochem.*, 1998, 69, 145.
[87] Silva AM, Doctorate dissertation, Department of Chemistry, Pontifícia Universidade Católica do Rio de Janeiro, 2003.
[88] Akitt JW, Multinuclear NMR, Chapter 9, pp259-289, Joan Mason Ed., Plenum Press, New York, 1989.
[89] Alves TVB, Master Thesis, Department of Chemistry, Pontifícia Universidade Católica do Rio de Janeiro, 2010.
[90] Jackson GE, Voyi KV, *Polyhedron*, 1987, 6 (12), 2095.
[91] Mao X, Huang H, Liu M, *Spectrochim. Acta Part A*. 1998, 54, 999.
[92] Baran EJ, *J. Inorg. Biochem*. 2000, 80, 1.

[93] Cantley LC Jr, Josephson L, Warner R, Yanagisawa M, Lechene C, Guidotty G, *J. Biol. Chem.* 1977, 252 (21), 7421 and references cited therein.

[94] Nechay BR, *Annu. Rev. Pharmacol. Toxicol.,* 1984, 24, 501.

[95] Etcheverry SB, Apella MC, Baran EJ, *J. Inorg. Biochem.*, 1984, 20(4), 269.

[96] Monteiro HP, Winterbourn CC, Stern A, *Free Radic. Res. Commun.* 1991,12-13 Pt 1, 125.

[97] Dick DA, Naylor GJ, Dick EG, *Psychol. Med.*, 1982, 12(3),533.

[98] Naylor GJ, Smith AH, Bryce-Smith D, Ward NI, *Biol. Psychiatry*, 1984, 19(5), 759.

[99] Naylor GJ, *Nutr. Health*, 1984, 3(1-2), 79.

[100] Bencherif M, Lukas RJ, *Neurosci. Lett.*, 1992, 134(2), 157.

[101] Liochev S, Fridovich I, *Biochim. Biophys. Acta*, 1987, 924(2), 319.

[102] Liochev S, Ivancheva E, Russanov E, *Free Radic. Res. Commun.* 1988, 4(5), 317.

[103] Liochev SI, Ivancheva EA, *Free Radic. Biol. Med.* 1988, 5(5-6), 349.

[104] Liochev S, Ivancheva E, Fridovich I, *Arch. Biochem. Biophys.* 1989, 269(1), 188.

[105] Tahiliani AG, McNeill JH, *Life Sci.* 1986, 38(11), 959.

[106] Sun Q, Sekar N, Goldwaser I, Gershonov E, Fridkin M, Shechter Y, *Am. J. Physiol. Endocrinol. Metab.* 2000 Aug, 279(2), E403.

[107] Geraldes CF, Castro MM, Sherry AD, Ramasamy R, *Mol. Cell. Biochem.* 1997, 170(1-2), 53.

[108] Noda C, Masuda T, Sato K, Ikeda K, Shimohama T, Matsuyama N, Izumi T, *Jpn Heart J.* 2003, 44(5), 745.

[109] Anwar-Mohamed A, El-Kadi AO, *Drug Metab. Dispos.* 2008, 36(9), 1819.

In: Molecular and Supramolecular Bioinorganic Chemistry ISBN: 978-1-61728-785-5
Ed: A. L. R. Mercê, J. Felcman and M. A. L. Recio

Chapter 4

ALUMINIUM IN WATERS. SOURCES, SPECIATION AND REMOVAL TECHNIQUES

María Ángeles Lobo-Recio[1*], Ana Lucia Ramalho Mercê[2], Maria Eliza Nagel-Hassemer[1] and Flávio Rubens Lapolli[1]

[1]Departamento de Engenharia Sanitária e Ambiental, Laboratório de Reuso da Água – LaRA, Universidade Federal de Santa Catarina, Campus Trindade, Florianópolis – SC, Brazil

[2]Universidade Federal do Paraná – UFPR - CP 19081 – LEQ – DQ Centro Politécnico, Jardim das Américas, Curitiba – Paraná - Brazil

ABSTRACT

Over the past years there has been a growing concern in the scientific community regarding the effects that an excessive ingestion of aluminium may cause on human health. The acute toxicity of aluminium is low; however, a prolonged exposure may provoke serious damages to health. Many researches are directed to the study of the biochemistry of aluminium, as well as to the determination of the effects of its accumulation in the brain, bones, blood stream, etc. One of the most important sources of aluminium ingestion is constituted by drinking water, which contains appreciable amounts of dissolved aluminium derived from the treatment carried out in order to remove the colloidal material present in the water that enters a treatment station. On the other hand, acid rain has also contributed to the increased concentration of aluminium contained in natural waters coming from the dissolution of rocks. Thus, frequently, the water treated and distributed to the population contains relatively high aluminium levels, which can cause operational problems at the station and accumulation in the distribution network, strongly affect people's health and reduce the population of fish when the served waters go back to the receiving body.

In the face of this scenery, this chapter discusses the need for the use of aluminium salts in water treatment and the techniques that can be used to remove the excess of aluminium contained in the treated water. The conventional technique of filtration is generally not sufficient to remove the residual aluminium, so that more and more studies for its remotion are developed through advanced techniques, such as membrane technology and adsorption which can then be followed by flotation. Aluminium removal techniques through the formation of complexes with organic ligands, which in many cases reach almost quantitative results, are of special interest. Among the techniques for removal via complexation, the use of functionalized biopolymers as aluminium ligands has become of growing interest, due to its bioavailability, relatively low cost and no toxicity.

1. Introduction

Aluminium (Al) is the third constituent element of the Earth´s crust and the most abundant metallic element. In the environment, it is present in the form of salts and oxides with diverse applications in the industry: transportation, construction, packaging, agri-food, utensils, pharmacy, surgery, cosmetics, and supply water treatment, among others [1].

The main sources of aluminium are obtained from bauxite, nepheline, cryolite, alunite and leucite. The mineral bauxite is the main commercial source of aluminium. In 2001 the world's reserves totaled 31.3 billion tons, located in tropical and subtropical regions. According to information from the International Aluminium Institute (IAI), bauxite occurs in three types of climates: tropical, subtropical and Mediterranean. World production of bauxite in 2001 was 137.1 million tons: Jamaica at 17%, Suriname, 11.1%, Brazil 10.1%, India 8.6%, Venezuela 4.8% and China, 2.2%. Figures 1 and 2 show the reserves and the world production of aluminium in 2001, respectively.

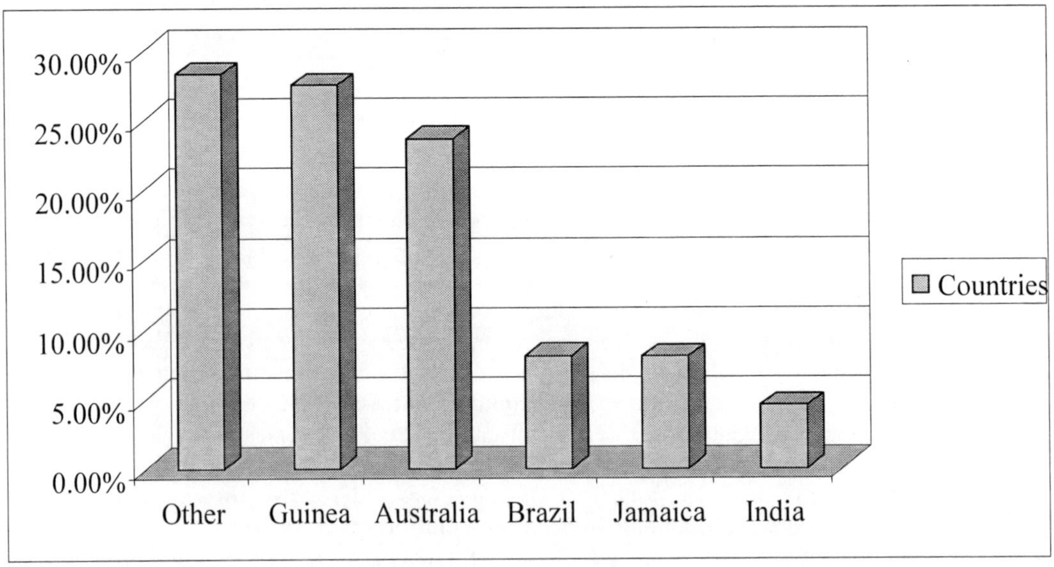

Figure 1. Aluminium worldwide stock (%) in 2001 – ($31.3x10^9$ tons). Adapted from reference [2].

Component of rocks and soil, aluminium is a natural constituent of groundwater and surface water. The highest concentrations of aluminium are found in waters exposed to acid rain, which have a great ability to dissolve the aluminium from the rocks to the point where the aluminium salts are solubilized and transported, being able to be absorbed by plants and animals. The man is therefore exposed to aluminium of natural origin, by direct contact with soil, dust in the air, in the form of particles coming from the soil or, yet by the ingestion of food and water. The amounts of aluminium absorbed by the body vary depending on the levels of aluminium, on the means of contact and also on the habits of each individual consumer. Regardless of the form and route of contact of aluminium, it is poorly absorbed by the body, in which more than 95% is rapidly eliminated.

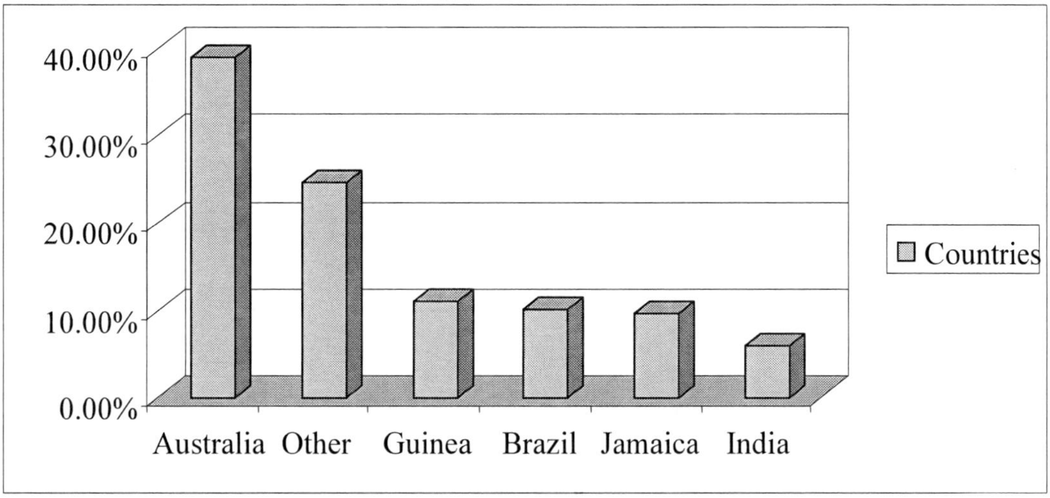

Figure 2. Aluminium worldwide production (%) in 2001 – (137.1x10^6 tons). Adapted from reference [2].

Living in the *"Age of Aluminium"* [3] an increase of exposure and consequently an overload of this element on men are inevitable consequences. Accumulation of aluminium in the human body will directly or indirectly define the impact on human health, but still requires serious considerations and investigations in medicine.

Aluminium toxicity is low. The soluble forms of aluminium ($AlCl_3$, AlF_3 $Al_2(SO_4)_3$ and aluminium citrate) have a greater toxic potential than the insoluble forms such as aluminium hydroxide. The toxic effects of aluminium on humans are due to strong chronic exposure and may cause neurological diseases such as encephalitis, as well as psychomotor, bone and hematologic disturbances (anemia and bone diseases). Aluminium is also responsible for immune and allergic reactions and is also associated to Alzheimer's disease [1].

2. Aluminium in Natural and Drinking Waters

Water is directly responsible for the maintenance of life and well being of mankind, apart from being the most important natural resource for the use of other resources, whether vegetables, animals or minerals. Without water, there would be no conditions for the development of life on Earth. It is the liquid that attends all of man's primary needs and that is present in all living beings.

The presence of Al in drinking water can come from the sources of water supply or from chemicals used in the treatment process.

2.1. Aluminium in Water Treatment

In the complete treatment process for the obtainment of drinking water, many chemicals are used. For example, lime can be used early in the process to assist in the coagulation and in the end of the treatment process to adjust the pH of the water and make it less corrosive. Chlorine is used as a disinfectant and fluoride is used as a preventive in the fight against dental cavities. Iron or, especially, aluminium salts are used in the coagulation and flocculation stages, which are part of the treatment destined to remove the colloidal matter contained in natural waters. The treatment of surface waters with these coagulants has already been used for more than a hundred years throughout the world [4]. The water coming from the washing of filters and the cleaning of the decantation tanks of water treatment plants may become an undesirable contribution of aluminium when improperly released into bodies of fresh water.

In the process of treatment of water supply the colloidal particles are small in size and negatively charged, thus one repelling each other, therefore making their removal not possible by simple processes like decantation and filtration. It is then necessary to destabilize the colloids in order to obtain particle aggregates of higher density and size, which then enables removal through sedimentation and/or filtration, thus eliminating the turbidity of the water. In this colloid destabilization process, called coagulation, aluminium salts are extensively used, since the small size and high positive charge of the aluminium cation attract the colloidal particles, neutralizing their charge and forming aggregates. The size of these aggregates will be increased in the next flocculation stage with the aid of flocculants, generally polyelectrolytes, to form flakes large enough to be removed by decantation and filtration.

When the destabilization of particles is considered, coagulation comprises two primary clotting mechanisms: neutralization of negatively charged particles through their interaction with the positive charge of the dissolved aluminium and colloid scan in the precipitate of $Al(OH)_3$ in balance with the dissolved aluminium which proportion depends greatly on the pH of the water, as it will be seen next.

The aluminium sulfate, as a coagulant, is one of the most used products due to its good efficiency, low cost, ease of application and commercial availability. It can be purchased as powder or granulate and has the chemical formula of $Al_2(SO_4)_3$ x nH_2O where n is equal to 14.3 or 18. However, the aluminium sulfate has some disadvantages which have raised concerns regarding extensive use. Its action as a coagulant agent depends on the pH of the water, with the possibility, after its use, of increasing the concentration of residual aluminium.

Aluminium sulfate is soluble in water, dissociating completely in Al^{3+} cation and sulfate anions. In water, the cation Al^{3+} is found in the form of the aqueous complex $[Al(H_2O)_6]^{3+}$, that experiences hydrolysis processes resulting in the formation of different cationic species more or less electrically charged which are responsible for coagulant properties:

$$[Al(H_2O)_6]^{3+} + H_2O \Leftrightarrow [Al\,(H_2O)_5(OH)]^{2+} + H_3O^+$$

$$[Al(H_2O)_5(OH)]^{2+} + H_2O \Leftrightarrow [Al(H_2O)_4(OH)_2]^+ + H_3O^+$$

$$[Al(H_2O)_4(OH)_2]^+ + H_2O \Leftrightarrow [Al(H_2O)_3(OH)_3] \text{ (insoluble)}$$

$$[Al(H_2O)_3(OH)_3] + H_2O \Leftrightarrow [Al(H_2O)_2(OH)_4]^-$$

When the aluminium sulfate is added to the water in the rapid mixture tank of a water treatment plant, Al^{3+} quickly suffers these hydrolysis reactions to form dissolved aluminium species and/or aluminium hydroxide precipitates. The four main soluble species expected are the ones previously indicated (Al^{+3}, $Al(OH)^{2+}$, $Al(OH)_2^{+}$ and $Al(OH)_4^{-}$) along with the amorphous precipitate of $Al(OH)_3$. The formation of binuclear complexes with hydroxy bridge groups and polynuclear species is also possible.

Aluminium (III) is an amphoteric cation, and its cationic forms are highly soluble in acidic conditions (pH < 6). The anionic form is soluble in alkaline conditions (pH > 8.5) and the neutral form, the aluminium hydroxide ($Al(OH)_3$ or $[Al(H_2O)_3(OH)_3]$), is insoluble in values close to neutral pH (6.5 to 7.5). Consequently, the solubility in water of aluminium (III) strongly depends on the pH. Therefore, the pH of the water should be carefully monitored during the coagulation and flocculation treatment in order to avoid the presence of high levels of dissolved aluminium in the treated water.

Table 1 presents the solubility constants of the different hydrolyzed species of aluminium in equilibrium with the precipitate of $Al(OH)_3$. With the equations related to these equilibriums it is possible to construct a diagram of hydrolyzed aluminium species as a function of the pH, as shown in Figure 3. The total aluminium concentration is given by the equation:

$$C_T(Al^{3+}) = [Al(OH)^{2+}]+[Al^{3+}]+[Al_6(OH)_{15}^{3+}]+[Al_8(OH)_{20}^{4+}]+[Al_2(OH)_2^{4+}]+[Al(OH)_4^{-}]$$

Table 1. Aqueous equilibria between insoluble $Al(OH)_3$ and different aluminium soluble forms. Data from reference [4].

Equilibrium	pKs (25°C)	Related Equation
$Al(OH)^{2+} + 2\,H_2O \Leftrightarrow Al(OH)_3 + 2H^+$	4.64	$\log[Al(OH)^{2+}] = 4.64 - 2pH$
$Al^{3+} + 3\,H_2O \Leftrightarrow Al(OH)_3 + 3H^+$	9.66	$\log[Al^{3+}] = 9.66 - 3pH$
$Al_6(OH)_{15}^{3+} + 3\,H_2O \Leftrightarrow 6\,Al(OH)_3 + 3H^+$	10.96	$\log[Al_6(OH)_{15}^{3+}] = 10.96 - 3pH$
$Al_8(OH)_{20}^{4+} + 4\,H_2O \Leftrightarrow 8\,Al(OH)_3 + 4H^+$	8.58	$\log[Al_8(OH)_{20}^{4+}] = 8.58 - 4pH$
$Al_2(OH)_2^{4+} + 4\,H_2O \Leftrightarrow 2\,Al(OH)_3 + 4H^+$	13.05	$\log[Al_2(OH)_2^{4+}] = 13.05 - 4pH$
$Al(OH)_4^{-} + H^+ \Leftrightarrow Al(OH)_3 + H_2O$	13.91	$\log[Al(OH)_4^{-}] = pH - 13.91$

pKs = - logarithm of the solubility constant

Thus, depending on the dosage of the product used and the pH the amounts of precipitate and hydrolyzed species may be determined.

Several studies address the fractionation of aluminium in drinking water at room temperature. These studies recognized that aluminium in drinking water can exist, apart from its mononuclear form as free aluminium (Al^{3+}) and products of its hydrolysis, complexed by other dissolved inorganic constituents, such as F^- and SO_4^{2-}, or can also be chemically bound to organic ligands or inorganic colloids [6].

Coagulation and flocculation are important stages in the conventional treatment of supply waters, because they have direct relation to the sedimentation and filtration performances. The efficiency of the coagulation and flocculation is influenced by factors such as pH, turbidity, alkalinity, coagulant dosage, agitation intensity, point of application of the coagulant, size and distribution of particles, total dissolved solids, color, etc [4, 6].

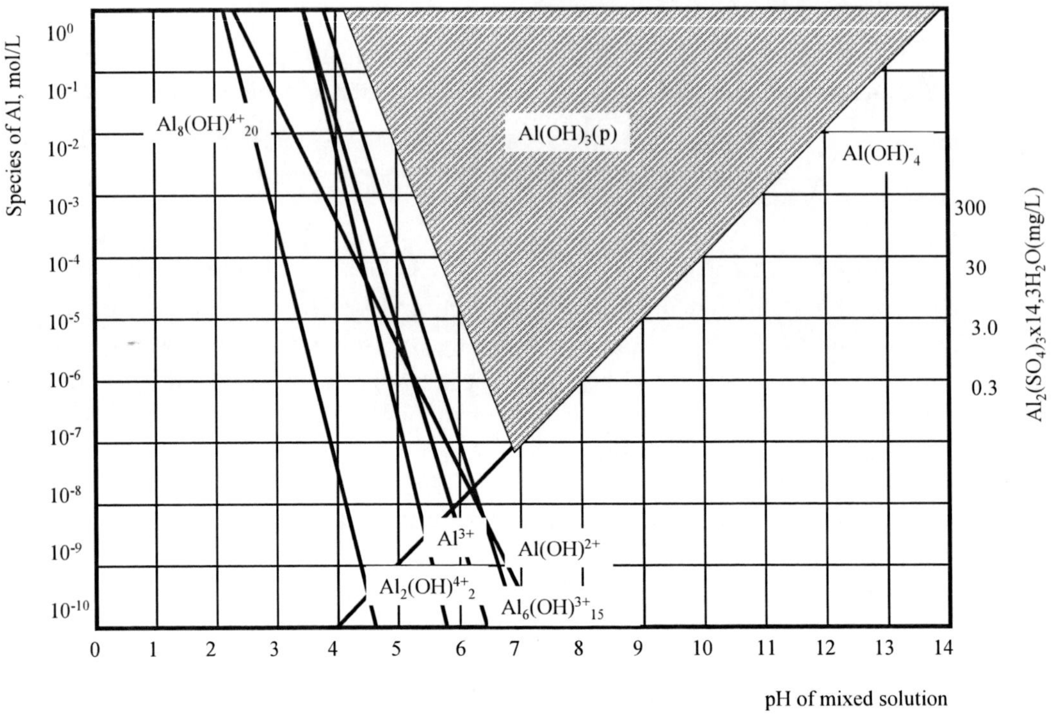

Figure 3. Solubility diagram for aluminium species related with aluminium hydroxide. Adapted from references [4, 5].

The presence of natural organic materials, especially humic aquatic substances and fulvic acids, strongly affect the chemistry of the coagulation. These humic substances are organic ligands that form complexes with dissolved aluminium species.

The coagulation of water containing particles and natural organic matter (NOM) becomes more complicated. Figure 4 shows the factors that affect the coagulation of these particles. Coagulants remove natural organic matter dissolved by complexation reactions followed by a phase change, that is, the natural organic matter is removed from the solution by forming a solid or by adsorption in a solid.

The clotting mechanisms attributed to the removal of natural organic matter are: complexation of the dissolved aluminium, leading to direct precipitation of a solid aluminium - organic matter complex; complexation of natural organic matter with dissolved aluminium leading to adsorption of this complexed material on the solid precipitate, $Al(OH)_3$, and direct adsorption of the natural organic matter on the surface of the solid precipitate of $Al(OH)_3$. It is also important to emphasize that the natural organic matter may cover inorganic particles, changing their behavior in the coagulation [7].

As noted earlier, the aluminium content in water may be presented in insoluble, colloidal and soluble forms. A review of the literature related to studies of aggregate forms of Al in drinking water indicates that there is no uniform definition of terms relating to the various fractions of Al. Nevertheless, there is a general agreement on the following definitions to differentiate among the various fractions. Total aluminium is the sum of the suspended, colloidal and monomeric forms of aluminium. The particle aluminium is the sum of the suspended and colloidal aluminium. The monomeric Al can be divided into two forms: non-labile and labile. Non-labile aluminium is associated to the dissolved organic carbon (DOC), while the labile aluminium includes Al^{3+} in the form of hydroxide and complexed species with fluoride and sulfate [8]. Figure 5 shows a block diagram of the various forms of aluminium.

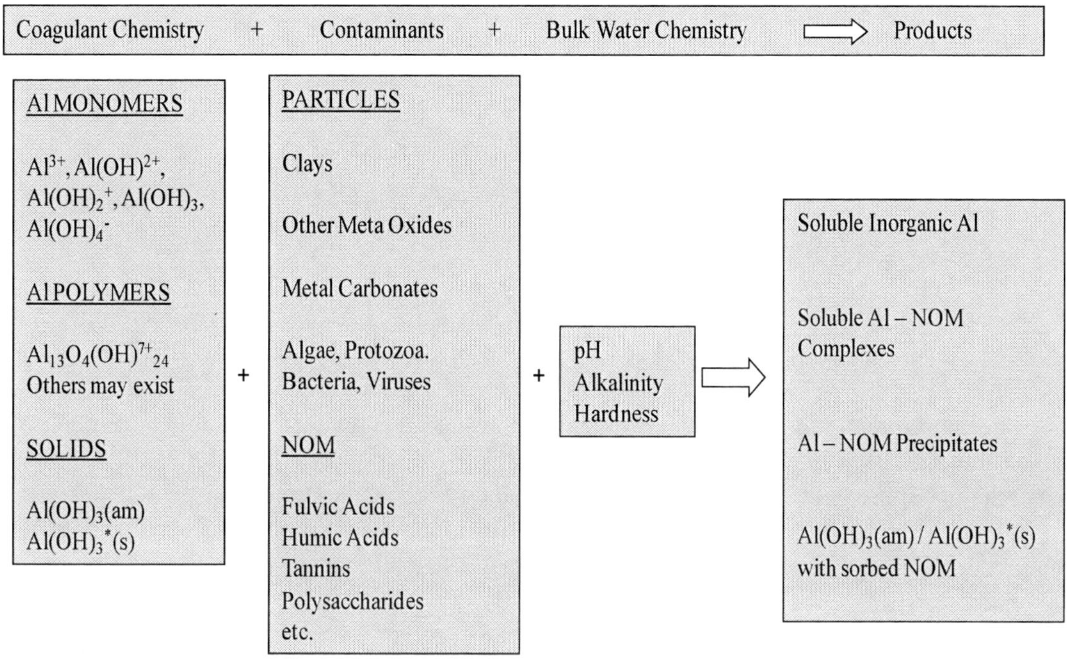

Figure 4. Important factors affecting coagulation of particles and natural organic matter. Adapted from reference [7].

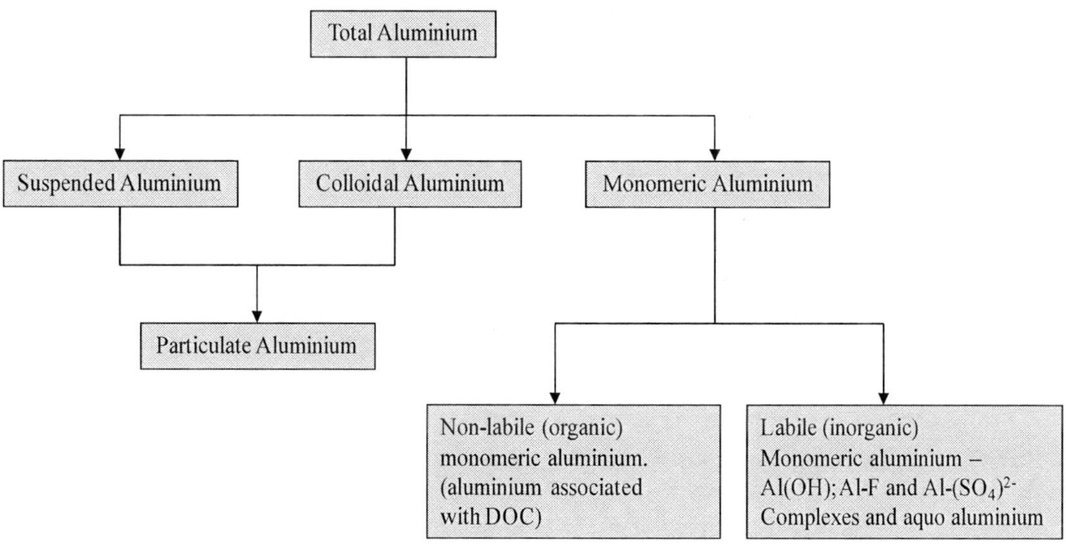

Figure 5. Block diagram showing various fractions of Al. Adapted from reference [8].

The fractions of aluminium are very different in raw water, filtered and treated water. A great part of the Al in raw water is associated to inorganic colloids or is linked to large organic molecules; this water contains only a small mononuclear fraction. The water treatment changes this distribution due to its emphasis on the aggregation of particles and filtration, producing treated water that is proportionally greater in mononuclear species [8, 9]. The aluminium that is present in the aluminium sulfate and the aluminium that is naturally present in raw water are, therefore, processed in various forms during the water treatment. Studies report that when the aluminium sulfate is used as a coagulant for the water treatment an increase of up to 50% in Al concentrations can occur in treated water compared to raw water [6].

2.2. Residual Aluminium

Residual aluminium is considered the Al that leaves the treated water treatment plant and enters the distribution system and that is consisted of dissolved aluminium particles and its species.

The aluminium salts used as coagulants in water treatment can result in residuals of which concentration depends on the characteristics of the source, the amount of coagulant used, the pH of the treated water and the performance of the water treatment plant [4]. A good coagulation performance is essential for the clarification, good filtration and for the control of disinfection by-products. Inappropriate clotting may cause the increase of residual aluminium in treated water.

The presence of residual aluminium in water at concentrations above 0.2 mg/L is associated to risks for the human health, causing gastrointestinal problems, skin rashes and ulcers. Moreover, an excess of aluminium in water can cause operational problems in the treatment plant and distribution system of the treated water. As a result, several regulatory agencies establish the limits of residual aluminium in supply water: EPA - United States Environmental Protection Agency of 0.05-0.2 mg/L Al; Canadian Operational Guidance value, 0.1-0.2 mg/L Al; World Health Organization, maximum Al of 0.2 mg/L, the Health Ministry of Brazil, a maximum of 0.2 mg/L [10, 11].

Aluminium particles can be easily removed with an efficient operation of solid-liquid separation systems, such as clarifiers and filters [8].

As mentioned earlier, temperature, pH and turbidity are important factors in determining the solubility of aluminium and consequently its residual concentration. The solubility of Al^{3+} is minimal in pH values near neutrality due to the precipitation of $Al(OH)_3$. At low temperature (4° C), the pH of minimum solubility increases, thus resulting in high levels of residual aluminium as aluminates [8].

Srinivasan et al. [8] report that there is a correlation between Al and the residual turbidity of the effluent. Specifically, the residual Al concentrations were less than 0.1 mg/L when the turbidity of the effluent was less than 0.15 NTU. This shows that the low turbidity of treated water has a low amount of particles of Al, with a consequent reduction in the residual Al.

2.3. Problems in Water Treatment and Distribution Derived from the Presence of Aluminium

For years, the occurrence of aluminium in treated water was considered an undesirable aspect of good treatment practices. There is considerable concern throughout the world regarding the levels of aluminium found in supply sources and in the drinking water treatment, not only in respect of the health of people but also of operational and distribution problems.

From the engineering point of view, there is concern about the use of Al salts as coagulants in the treatment of drinking water, since high concentrations of aluminium in treated water can result in a series of problems in water supply. Thus, the formation of a precipitate of aluminium hydrate in the distribution system can increase the turbidity of the water and the number of complaints regarding the clarification conditions of treated water.

The presence of aluminium flakes in the output of the water treatment system, derived from a disabled decantation and filtration, may interfere in the disinfection process because the solid particles contained in water constitute a type of refuge for microorganisms, where they can be guarded from the action of the disinfectant. Another problem attributed to the increase in the concentration of aluminium is the deposition of Al hydrolysis products in the walls of the pipes, reducing their flowage capacity [8].

2.4. Health Effects Due to the Presence of Aluminium in Water

Connections were established between the Al present in the water for human consumption and neurological disorders such as dialysis encephalopathy. In addition, links to Alzheimer's disease have been established [6, 12]. Less deleterious effects, compromising psychomotor functions, are also verified in patients undergoing dialysis. These patients may present attention, motor coordination and memory problems and also vision disturbances. Other effects caused by exposure to aluminium are: effects on bones, as well as hematological, respiratory, immune, allergic and carcinogenic effects [1].

Flaten [13] states that, despite the inconsistency of epidemiological, toxicological and mechanistic data, one cannot discard the relation between drinking water containing aluminium and Alzheimer's disease. Walton [14] raises hypotheses in which aluminium has an important role and in which it must be considered as a causative factor of Alzheimer's disease.

The presence of aluminium in drinking water has originated discussions concerning possible effects on health due to the connection of aluminium to Alzheimer's disease or dialysis encephalopathy [6].

Srinivasan et al. [8] reported several studies in which animals that were exposed to Al under controlled conditions demonstrated a correlation between neuropathological disorders and aluminium ingestion. Aluminium has low acute toxicity. Chronic exposure data are limited, but indicate that Al probably interferes with the absorption of phosphorus and causes weakness, bone pain and anorexia [6].

It is reported that the OH and F complexes of aluminium are highly labile and may be more bioavailable and harmful than organic forms or particles of Al [8].

Based on studies on Al toxicity it was confirmed that the positive charges of aluminium hydroxide species are more toxic for fish than organic complexes of aluminium [8]. Petterson et al. [15] confirmed the importance of water quality regarding the acidification, particularly for the general toxic effects of aluminium on fish ectoparasites in freshwater environments.

The many studies carried out have indicated that aluminium is related to the increase of diseases in humans. The researches on aluminium and its effects should be thoroughly examined through epidemiological studies for a better comprehension of the problem and also to provide solutions that seek the prevention and preservation of the human health.

3. Treatment Processes for the Removal of Aluminium from Waters

During the last few years, increasing attention has been focused on the separation and/or determination of trace-metal ions in the environment. This field of interest is becoming a real challenge due to the specificity, accuracy and sensitivity required by increasingly stringent regulations.

Traditional technologies for the treatment of drinking water (coagulation, flocculation, decantation, filtration and disinfection) are often not efficient in the removal of water-soluble metal traces. Normally the soluble metallic ions are removed through precipitation. However, the removal of metals in the form of oxides, hydroxides, carbonates or sulfates, which is obtained through the addition of coagulants, has some disadvantages, which we can mention: inefficient precipitation if the metals are complexed or as anions; the lowest metal concentration achieved is limited by the solubility product; a great amount of sludge is generated after the decantation; and the need for further filtration for the removal of suspended solids and clarification of treated water.

Aluminium is one of the inorganic metals traces present in drinking water. Al-based coagulants are commonly used in drinking water treatment to enhance the removal of particulate, colloidal and dissolved substances via coagulation processes. A portion of the alum added to the raw water is not removed during treatment and remains as residual aluminium in the treated water [8].

The conventional treatment for removal of dissolved metals is chemical precipitation followed by decantation and filtration. However, this technique is not, sometimes, effective enough, what requires the application of advanced techniques to remove these species, such as adsorption, adsorption / flotation (adsorptive flotation), membrane filtration, complexation, ionic exchange and extraction [5, 16, 17]. In addition to the conventional filtration technique, the most used advanced techniques to remove aluminium from waters are described next.

3.1. Filtration

The filtration unit is indispensable in the traditional water treatment system, which is important for ensuring the safety and quality of water. With the development of water treatment, many kinds of filtration tanks have emerged. The normal rapid filtration tank has the longest application history and is still being used prevalently. To enhance its ability in holding filtration tank impurities, the single layer filter, double layer filter, and homogeneous filter are used, and these are the major filter forms in most water plants [18].

Aluminium salts are predominantly employed in the water treatment system. Although they have a good coagulation effect, their application is the main factor causing high residual aluminium in most drinking water.

In several works regarding water, in order to increase coagulation performance and to control the coagulant process conveniently, the actual coagulant dosage is generally considerably higher than the theoretical dosage. Under this condition, partial aluminium will react with the pollutants and deposit in the settlement tank, while the excess part will still exist in effluent water as residual aluminium and will enter the filtration unit. Considering that turbidity in the filtration effluent is correlated to the residual Al concentration, the removal of turbidity can reduce the residual aluminium concentration including suspended and colloidal aluminium.

Zouboulis et al. [19] studied the operation of mono and dual media filter beds in the drinking water treatment. A comparison was attempted between a sand filter bed and a dual media filter bed, consisting of sand and anthracite layers. Two alternative treatment schemes were examined: the conventional coagulation-flocculation/sedimentation, followed by gravity filtration and the direct filtration (coagulation-flocculation and filtration, but without sedimentation). The coagulant used was aluminium sulfate. The results obtained indicated that the two filter beds produced water with the same constant (high) quality, with final turbidity below 0.1 NTU, and low residual aluminium concentration (30 μg/L) at the outlet of filter beds (approximately 93% removal for both).

Small variations of pH (from 7 to 7.3) showed no significant effect, regarding the filtration cycles and the removal of turbidity. However, at higher pH values (7.5 – 8) the residual aluminium concentrations were more likely to increase, over the highest permissible limit for drinking water, which is of 200 μg/L. It can be stated that pH values over 7.3 cannot be used safely, because in connection with a slight increase of the filtered water turbidity, these values may result in higher residual aluminium concentration. It is known from the literature that the residual aluminium concentration is also increased by elevated turbidity values [19].

According to Lettermann and Driscoll [20], the residual aluminium in the filtration effluent has relation to Al_T in raw water and residual turbidity in the filtration effluent, what can be expressed by the following equation:

$$Al_e = 0.1 Al_r^{0.36}\ T_e^{0.32}$$

where, Al_e represents the residual aluminium concentrations in the filtration effluent water, Al_r is the total aluminium concentration in raw water, and T_e is the residual turbidity in effluent water. Through the experiential equation, we can obtain the total residual aluminium concentration in the filtration effluent.

Wen-Dong et al. [18] studied the removal rules of different aluminium species in the single layer filter (sand) and double layer filter (anthracite and silica). The species referred in this research include total aluminium (Al_T), suspended aluminium (Al_{SUS}), colloidal aluminium (Al_C), soluble aluminium (Al_S), inorganic aluminium (Al_I), organic aluminium (Al_O), and soluble-colloidal aluminium ($Al_{(C+S)}$). The water sources include Kaolin soil water (waterl), PAC water (water2), and HA/$AlCl_3$ water (water3), and the aluminium concentration was controlled between 0.5 and 1.5 mg/L, maintaining similar water quality as the settlement effluent (PAC - Poly Aluminium Chloride; HA - Humic Acid).

Table 2 shows the results for aluminium species removal of water1, water2 and water3 in rapid filtration.

The residual aluminium concentration derived from water1, water2 and water3, after direct rapid filtration was still above the national standard, 0.2 mg/L and the removal percentage of the soluble aluminium was very low in all cases.

Table 2. Aluminium removal rate of water1, water2 and water3 in rapid filtration

Sample	Rapid filtration	Removal rate (%)						
		Al_T (c)	Al_{SUS} (d)	$Al_{(C+S)}$ (e)	Al_C (f)	Al_S (g)	Al_O (h)	Al_I (i)
Water1	SF(a)	3	4	2	-	-	-	-
	DF(b)	8	10	5	-	-	-	-
Water2	SF(a)	33	40	19	20	7	-	-
	DF(b)	40	45	30	32	6	-	-
Water3	SF(a)	43	47	34	36	7	37	22
	DF(b)	52	56	38	42	7	41	37

(a) single layer filter; (b) double layer filter, (c) total aluminium, (d) suspended aluminium, (e) soluble-colloidal aluminium, (f) colloidal aluminium, (g) soluble aluminium, (h) organic aluminium, (i) inorganic aluminium.

Generally, most of the small natural particles were nonleachable in sand filtration. In this case, the suspended aluminium in water1 originated from Kaolin, which contained several small natural particles, was difficult to be removed without a coagulant agent. The particulate aluminium in water2 and water3 originated from fresh-forming aluminium flakes were easier to be removed in the filtration process. Thus, the removal rate of total aluminium in the water1 effluent was lower than in water2 and water3. Water3 had a higher total aluminium removal rate than water2: the HA in water3 could react with aluminium ions and its hydrolysis species forming colloid matter and suspended flakes, which were easier to be removed in the filtration process. The aluminium removal rate (R) was not only connected to the filtration media's chemical character, but was also related to the particle diameter of the different aluminium species: $RAl_{SUS} > RAl_C > RAl_S$.

The double layer filter (DF) and single layer filter (SF) could not remove small particles or soluble aluminium; it is necessary to maintain certain turbidity in raw water to accelerate the formation of large settleable or leachable particles in the treatment process.

Different water quality will lead to different aluminium residual and species distribution. Natural small particles that adsorbed some amount of soluble or small particle aluminium on their surface were difficult to be removed by direct fast filtration. The experiment results showed that coagulation and settlement will accelerate the removal of aluminium in the filtration process.

3.2. Membrane Filtration

Micro filtration (MF) and nano filtration (NF) are new techniques in the drinking water treatment system. Micro-membranes, with a pore size of 0.02 - 1.0 μm, remove pollutants by mechanical selection and diffusion function. Nano filtration membranes are more effective in removing organic matter and inorganic ions because of its considerably smaller average pore size. Therefore, membrane filtration will be an effective technology in aluminium removal.

Wen-Dong et al. [18] studied the removal rules of different aluminium species in the membrane filtration units, in different types of water, after rapid filtration. The water sources include Kaolin soil water (water1), PAC water (water2), and HA/$AlCl_3$ water (water3). In the experiment, soluble-colloidal aluminium ($Al_{(C+S)}$) was defined as the part of aluminium that can pass through a 0.45 μm membrane filter, therefore, only soluble aluminium (Al_S) and colloidal aluminium (Al_C) were analyzed.

Residual aluminium could not be detected in NF effluent water. Nano filtration, with small membrane pore size, could remove most multi valence salts. Thus, NF achieved a nearly 100% aluminium removal rate [18].

Table 3 shows the membrane micro filtration experiments results on water1, water2 and water3, after rapid filtration.

Table 3. Micro filtration results for water1, water2 and water3 after rapid filtration

Sample	Removal rate (%)								
	Water1	Water2			Water3				
	$Al_{(C+S)}^{(a)}$	$Al_{(C+S)}^{(a)}$	$Al_C^{(b)}$	$Al_S^{(c)}$	$Al_{(C+S)}^{(a)}$	$Al_C^{(b)}$	$Al_S^{(c)}$	$Al_O^{(d)}$	$Al_I^{(e)}$
Initial time	100	43	49	15	2	3	2	4	1
3h later	-	91	100	100	72	100	100	74	71

(a) soluble-colloidal aluminium, (b) colloidal aluminium, (c) soluble aluminium, (d) organic aluminium, (e) inorganic aluminium.

Micro filtration could remove the particles between 0.05 - 10 μm. The average diameter of Kaolin soil's used was about 1.69 μm. Therefore, most aluminium species in water1 could be removed.

In water2, after 3h, the $Al_{(C+S)}$ concentration in the filtration effluent water decreased with time and the final aluminium concentration was stable at a considerably lower level with a 91% aluminium removal rate. This can be explained by the fact that the micro membrane pore size can become smaller as the filtration continues. Finally, a layer of gel will form on the membrane surface, which will enhance the filtration performance. The same occurred with the $Al_{(C+S)}$ from water3. Al_C and Al_S could be removed effectively in water1 and water2.

According to the authors [18], NF was able to remove almost all aluminium species, while MF with a larger membrane pore diameter allowed partial $Al_{(C+S)}$ passing through the system. Also, when the residual aluminium concentration in coagulant effluent water is higher than 1 mg/L, MF or NF may be required to ensure the water quality safety.

3.3. Adsorption

Adsorption is the process of accumulation of substances that are in solution on an adequate solid surface. In adsorption, a porous solid at microscopic level is capable of retaining particles of a given solute when it comes into contact with a solution. Adsorbate is the substance that is being removed from the solution; Adsorbent or support is the solid on which the adsorbate is accumulated.

Different process conditions such as pH, initial adsorbent/adsorbate concentrations, temperature and the percentage of aluminium removal are important to ascertain the mechanism for aluminium adsorption [21].

The pH value has a very significant role on the adsorption of aluminium as it affects the solubility of aluminium to a great extent and consequently it becomes necessary to determine the pH at which maximum adsorption takes place.

Activated carbons are unique and versatile adsorbents because of their extended surface area, micro-porous structure, high adsorption capacity and high degree of surface reactivity. They find their wide applications in the removal of color, odor, taste and other undesirable organic and inorganic impurities from different aqueous media [16].

Suitable low-cost adsorbents which have comparable sorption efficiency can be very effective for the removal of aluminium from drinking water. In this context, one area being explored is the use of naturally occurring materials which have the potential for adsorbing pollutants, such as rice husk carbon, commercial granular activated carbon, powdered marble wastes and powdered activated carbon prepared from olive stones [21].

Singh et al. [21] researched the sorption of aluminium from drinking water using low-cost adsorbents. The adsorbents tested were rice husk carbon (RHC), commercial granular activated carbon (GAC) and activated rice husk carbon (ARHC). Among the various sorbents studied, rice husk carbon has shown the maximum adsorption capacity for aluminium compared to activated rice husk char and activated carbon. The adsorption capacity of rice husk carbon decreased with the increase of aluminium concentration. For commercial activated carbon the results indicated that the adsorption is not very favorable in relation to the adsorbent.

The value of adsorption parameters also gives an indication that rice husk char (RHC) is the most suitable adsorbent for the removal of aluminium from water. More than 97% aluminium can be removed with RHC.

Adsorption technology is also used along with techniques such as flotation, in order to optimize the separation of the saturated adsorbent from the solution.

Flotation is a solid/liquid or liquid/solid (or both) separation process which has recently received considerable interest due to its simplicity, rapidity, good separation yield for small impurity agent concentrations (10^{-6} - 10^{-2} mol dm^{-3}), large possibility of applications for species having different nature and structure, flexibility and friability of equipment and processing for recovery. For the aforementioned reasons a combination of adsorption and flotation into a unified operation termed sorptive-flotation could be considered as a vital process.

Ghazy et al. [16, 17] conducted experiments in the laboratory to remove aluminium from water using low-cost adsorbents and flotation agents. In studies carried out in 2003, the authors established a selective, rapid and simple procedure for the removal of aluminium ions from aqueous solutions using powdered marble wastes (PMW) as a sorbent (which may be inexpensive, readily available and an alternative to existing commercial adsorbents) and oleic acid (HOL) as a surfactant (combined process of sorption and flotation). Good results were obtained, for which nearly 100% of aluminium was removed at pH 7 and after 5 min of shaking, and at room temperature (~25°C). The aluminium ions were essentially held by a PMW sorbent, and would not leach out by acids due to the solubility of the sorbent. In further experiments (2006), the authors conducted tests based on the use of powdered activated carbon (PAC), which was prepared from olive stones (organic sorbent) generated as plant wastes and modified with an aqueous oxidizing agent, with HNO_3 as an effective sorbent and oleic acid (HOL) as a surfactant. Good results were obtained, for which nearly 100% of aluminium was removed at pH 7 and at room temperature (~25°C), and an adsorption capacity of 1.6 mg Al/g adsorbent was determined. In the tests carried out with conventional commercial activated carbon, there were not good results in aluminium removal.

3.4. Removal of Aluminium from Waters via Complexation

The aluminium removal from water methods described above have some disadvantages as seen before. To summarize, precipitation-filtration processes require a large consumption of precipitant and sometimes the filtration of sludge is difficult; adsorption processes require an often difficult stage, directed to separate adsorbent and water; adsorption-flotation processes require organic agents which residually remain in the treated water; and problems with fouling and high cost are associated to membrane technologies [22].

Due to this situation, other alternatives become necessary in order to achieve acceptable aluminium levels according to the ranges established by the guidelines of different countries [16]. In Brazil, for example, the maximum value admitted for aluminium in drinking water is 0.2 mg/L [11]. However, it is important to emphasize that these established limits are referred to levels which do not cause damage to water supply systems, but are not referred to the safe limits for the health.

Aluminium(III) is a small (ionic radius 0.51 Å) and highly polarizing cation which easily forms coordination compounds with N-, O- and P-donor ligands. In general, the organic compounds used as ligands are synthetic and of high costs. For this reason, the removal of aluminium via complexation requires the formation of complexes which have to be stable enough to be removed from water but also capable of being easily decomplexed further for the reuse of the ligand.

The removal of aluminium via complexation has been used to purify both drinking water and dialysis water. Aluminium complexation has also been successfully used in the selective and quantitative determination of aluminium from diluted solutions through the formation of coordination compounds with specific ligands [23, 24, 25].

3.4.1. Aluminium Complexation Techniques

The different complexation techniques applied in order to remove aluminium from aqueous solutions can be classified in three groups: (i) forming insoluble complexes which are removed by filtration, (ii) forming soluble complexes which are removed by solvent extraction, (iii) chelating solid phase extraction. In all cases, the success of the technique employed depends not only on achieving a very high aluminium removal yield but also on recovering the complexing agent easily and completely for its posterior utilization. Further, in this work, these three techniques are described and illustrated with some examples. Table 4 below shows some of the ligands and materials of common use in techniques of aluminium removal by complexation.

i. Formation of insoluble complexes

Rodil et al. [22] used sodium di-(n-octyl) phosphinate [$(C_8H_{17})_2POONa$] as complexing agent for aluminium (III) as part of a systematic study of removal of different metals via complexation with the ligand mentioned above [26, 27]. The authors worked with diluted and acidic aluminium synthetic solutions and studied the influence of the pH of the initial solution, the pH in the complexation reaction, the ratio ligand/Al on the complexation, as well as the effect of ions commonly present in natural waters and effluents. The best results were obtained for an initial pH of 2.75, a complexation pH of 4 and a ratio ligand/Al of 3/1, with the formation of an insoluble complex, AlL_3, and an essentially complete aluminium removal. The white insoluble complex was removed by filtration on 0.2 μm filters. The presence of chloride, nitrate or calcium did not affect the removal of aluminium. The method showed two important advantages: first, it is easily applicable, and second, the ligand can be easily recovered and the aluminium concentrated in aqueous solution by treatment with 3 M NaOH. Aluminium remains in solution as aluminate and the ligand is extracted with diethyl ether.

Table 4. Ligands frequently used in the removal of aluminium via complexation

Ligand	Complexation Technique	Support [Organic extraction agent]	Optimum complexation pH	Ref.
Sodium di-(n-octyl) phosphinate $(C_8H_{17})_2POONa$	Formation of insoluble complexes	---	4	22
8-hydroxyquinoline 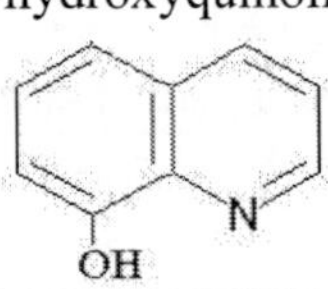	Formation and solvent extraction of soluble complexes	[Methyl isobutyl ketone]	5	28, 29
	Chelating solid phase extraction	Iontosorb oxin (IO)	2.5 - 6	29
Salicylic Acid 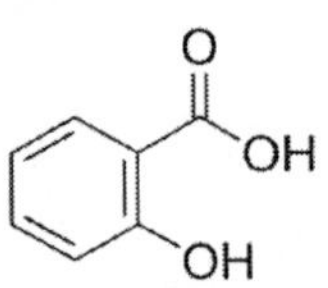	Chelating solid phase extraction	Amberlite XAD-4	2.5	30, 31
	Chelating solid phase extraction	Iontosorb Salicyl (IS)	2.5 - 6	28
Alizarin yellow 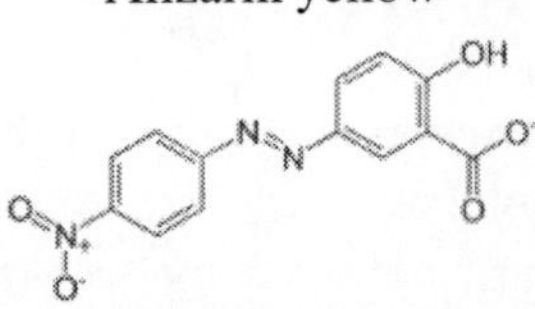	Chelating solid phase extraction	poly(2-hidroxyethyl methacrylate) (mPHEMA)	5	32, 33

ii. Formation and solvent extraction of soluble complexes

8-hydroxyquinoline is a much employed ligand in this technique. The so-called "8-hydroxyquinoline method" was applied by Boudot et al. [28] which consists of a flash reaction of 1% 8-hydroxyquinoline in 2% acetic acid solution with an Al(III) solution sample followed by extraction of the complex product with methyl isobutyl ketone. The method quantitatively separates the soluble inorganic monomeric species of Al (aqua, hydroxo, sulfate and phosphate Al complexes), with the exception of fluoride complexes which are not completely recovered. The dimeric Al-Si species are also separated but the dimeric Al-PO_4 ones are not recovered. Al-organic ligand complexes are only partially separated by this method.

The phytotoxicity of aluminium is mainly attributed to its labile aqua, hydroxo and probably sulfate Al cationic monomer and some polymer (e.g., Al_{13}) complexes. The Al complexes with organic and some inorganic (fluoride, phosphate, carbonate, silicate) ligands seem to be non toxic species. Matúš [29] applied the 8-hydroxyquinoline method to soil solutions and surface waters in an attempt to separate the phytoavailable and phytotoxic aluminium species fractions. For this reason, very short reaction times (5s) and a work pH of 5 were used. The fraction of labile Al species was separated by this method, i.e., inorganic aqua, hydroxo and sulfato Al cationic monomers (which are responsible for aluminium toxicity) but, in a small extension, some complexes with organic ligands of low molecular weight and fluoride complexes (non toxic) were also removed. Nevertheless, the 8-hydroxyquinoline method was shown to be the most selective towards the fraction of labile Al species among the five methods tested in this study.

iii. Chelating solid phase extraction

Chelating solid phase extraction consists in covalently attaching the ligand to support materials, such as inorganic adsorbents, polymeric resins, magnetic beads, etc., in order to obtain modified matrixes which provide a higher effectiveness and selectivity in the removal of aluminium.

Boudenne et al. [30] modified a commercial resin (Amberlite® XAD-4) on which salicylic acid was chemically bonded as a chelating agent. With the objective of selectively trapping aluminium from samples of synthetic and real treated drinking waters in order to determine on-line the aluminium contents of treated and finished water samples, the same research group used this modified resin into a microcolumn and studied the factors which affect the extraction-elution steps of aluminium onto this support [31]. The goal of these authors was to carry out a very interesting and promising attempt to develop a multi-criteria approach for systematically determining the best conditions to affect the complexation/decomplexation in order to achieve an effective aluminium recovery (95% minimum) from water. They applied an empirical mathematical quadratic model in which six variables were considered: three of them directly related to the extraction stage via complexation (flow-through sample volume, sample percolation rate and sample metal concentration) and the other three ones related to the elution stage (eluent volume, elution flow-rate and concentration of eluent).

The results proved that the elution stage is largely predominant over the extraction stage and that volume and concentration of the eluent (HCl) was the most influential parameter in aluminium recovery of the initial water samples. The parameter values for the optimized extraction procedure were established as: flow-through sample volume of 1.5 mL, a sample percolation (extraction) flow-rate of 2 mL/min, an HCl volume of 2 mL, an elution flow-rate of 3 mL/min and an HCl concentration of 0.10 mol/L.

The maximum tolerable amounts of foreign ions that do not interfere in aluminium determination were also determined for an aluminium concentration of 100 μg/L. The detection limit of this procedure was 1.1 μg/L of aluminium in 1.5 mL of sample solution.

Another very interesting study regarding aluminium removal by chelating solid phase extraction is constituted by the research of Denizli et al. [32] and Yavuz et al. [33], who used, as aluminium complexing ligand, alizarin yellow covalently supported on magnetic beads of poly(2-hidroxyethyl methacrylate) (mPHEMA), a highly hydrophilic polymer. The mPHEMA beads, in the size fraction of 80-120 μm, were produced by a dispersion polymerization technique in presence of magnetite particles (Fe_3O_4, diameter < 5 μm). According to these authors, the magnetic adsorbents combine the best characteristics of the fluidized beds (low pressure drop and high feed-stream solid tolerances) and of the fixed beds (absence of particle mixing, high mass transfer rates and good fluid-solid contact). The mPHEMA beads show a rough surface, due to the abrasion of the magnetite particles, and inner macropores. These two factors contribute to providing a high attachment ability for the alizarin yellow, increasing the ability to complex aluminium(III) and, consequently, enhancing the removal capacity of this adsorbent in relation to the non-magnetic PHEMA.

Alizarin yellow contains phenolic hydroxyl, carboxylic, nitro and azine groups (Table 4). The complexation with aluminium cations occurs through oxygen and nitrogen atoms. The attachment of the alizarin yellow to the mPHEMA beads was carried out through the addition of alizarin yellow in absolute ethanol to dry mPHEMA beads in $SOCl_2$ (Fig. 6). After 24 hours under reflux and subsequent purification, alizarin yellow-attached beads were obtained containing 208 μmol of dye per g of polymer.

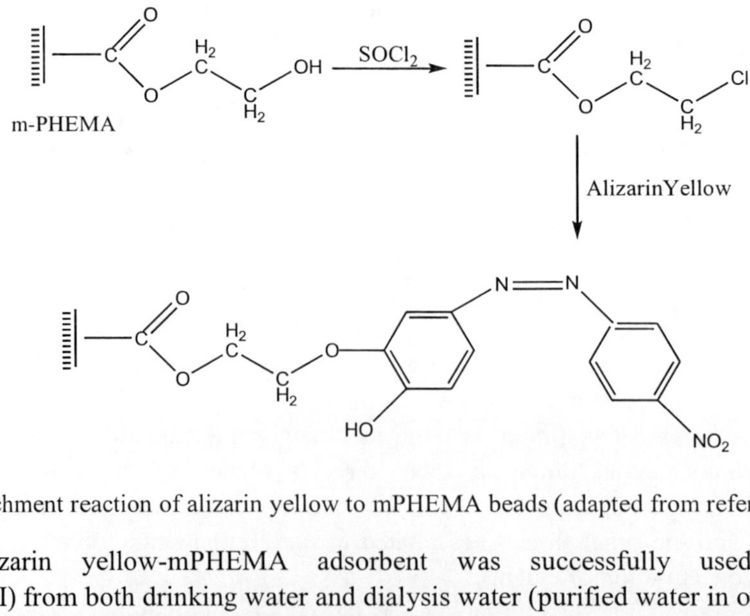

Figure 6. Attachment reaction of alizarin yellow to mPHEMA beads (adapted from reference 32).

The alizarin yellow-mPHEMA adsorbent was successfully used to remove aluminium(III) from both drinking water and dialysis water (purified water in order to prepare solutions for dialysis), showing to be a very effective aluminium adsorbent. Aluminium removal experiments were carried out in both batch [32] and magnetically stabilized fluidized bed systems [33]. In both cases, the optimum working pH was of 5±0.1.

In the batch system, the maximum adsorption capacity was 420 μmol Al(III)/g Polymer (11.8 mg/g) for water samples with Al(III) concentration ≥ 30 mg/L. It seems, on the basis of mass-stoichiometry, that each attached molecule of alizarin yellow complexed two Al(III) ions via phenolic, carboxylic and nitro groups. The vacant coordination sites were in turn occupied by water molecules. No more extensive studies about the structure of the aluminium complex formed were carried out.

The magnetically stabilized columns require faster processing times and high flow-rates with a much lower operating pressure than a packed-bed column [33]. In the magnetically stabilized fluidized bed experiments, a significant decrease in the adsorption capacity (from 17.2 to 6.9 mg Al(III)/g Polymer) was observed with the increase of the flow-rate (from 0.5 to 3.0 mL/min). The maximum adsorption capacity was 647 μmol Al(III)/g Polymer (17.5 mg/g) for water samples with Al(III) concentration ≥ 50 mg/L, a value 65% higher than the batch system. According to the authors, this fact can be explained in two ways: a) the ligand-Al(III) dissociation rate in the batch system is higher than the association rate in the fluidized bed system, and/or b) in the presence of flow, the Al(III) solution is forced to penetrate into the mPHEMA macropores which also contain attached alizarin yellow, thus increasing the adsorption capacity of the system.

Competitive adsorption of other metal ions was also investigated. The alizarin yellow-mPHEMA adsorbent can also complex Cu(II), Fe(III) and Zn(II) ions, but in these cases its adsorption capacity is much lower than the Al(III) case, showing a relatively high specificity towards this last one, probably due to its smaller size.

Aluminium(III) is easily decomplexed from the alizarin yellow-mPHEMA adsorbent through a treatment with HNO_3, which allows the repeated reuse of the adsorbent without noticeable losses in its adsorption capacity.

Other chelating solid phase extractants include Iontosorb as support ion-exchanger cellulose resins and 8-hydroxyquinoline and salicylic acid (Iontosorb Oxin (IO) and Iontosorb Salicyl (IS), respectively) as covalently attached ligands [29]. These extractants were applied to filtered water samples and H_2O extracts from soil and sediments using the batch technique, by stirring together sample and extractant for 1 min of extraction time followed by centrifugation. This method is less selective towards the fraction of labile Al species than the above described "8-hydroxyquinoline method", probably because of the slower kinetics of the reaction between labile Al species and these agent functional ligands and the influence of many sorption and diffusion effects on the reaction rate between liquid and solid phases. However, these chelating solid phase extraction methods via IO or IS can be more convenient in order to carry out experiments to assess aluminium environmental risk because they are simple and of low cost, what, in this case, dispenses the use of the organic extraction agent (methyl isobutyl ketone).

3.4.2. Aluminium Complexation by Functionalized Biopolymers

In recent years, water remediation through the complexation of metals with biopolymers has been an object of growing interest. Although the complexation techniques employed with biopolymeric ligands can be any of the ones described above, the aluminium complexation via biopolymeric ligands exhibits some advantages and differences and for this reason it is separately considered in this chapter.

Biopolymers are substances of natural occurrence and, consequently, abundant, renewable, non toxic and of low cost. These characteristics turn either natural or modified biopolymeric materials into very interesting complexing agents in the remediation of large quantities of water and effluents.

In order to determine the ability of a biopolymer as a remediation agent, it is important to determine the speciation of the complexes as a function of the pH, their stability in solution and in solid state and the ease of decomplexing and recovering the biopolymeric ligand due to its further reuse [34].

In particular, regarding the removal of aluminium, the most employed biopolymers are those with O-donor or N-donor functional groups, due to the hard Lewis acid characteristic of the Al^{3+} ion. Thus, studies with galactomannan (Gal), carboxymethylcellulose (CMC), chitin and chitosan, among others, have been reported. The structures of these biopolymers are represented in Figure 7.

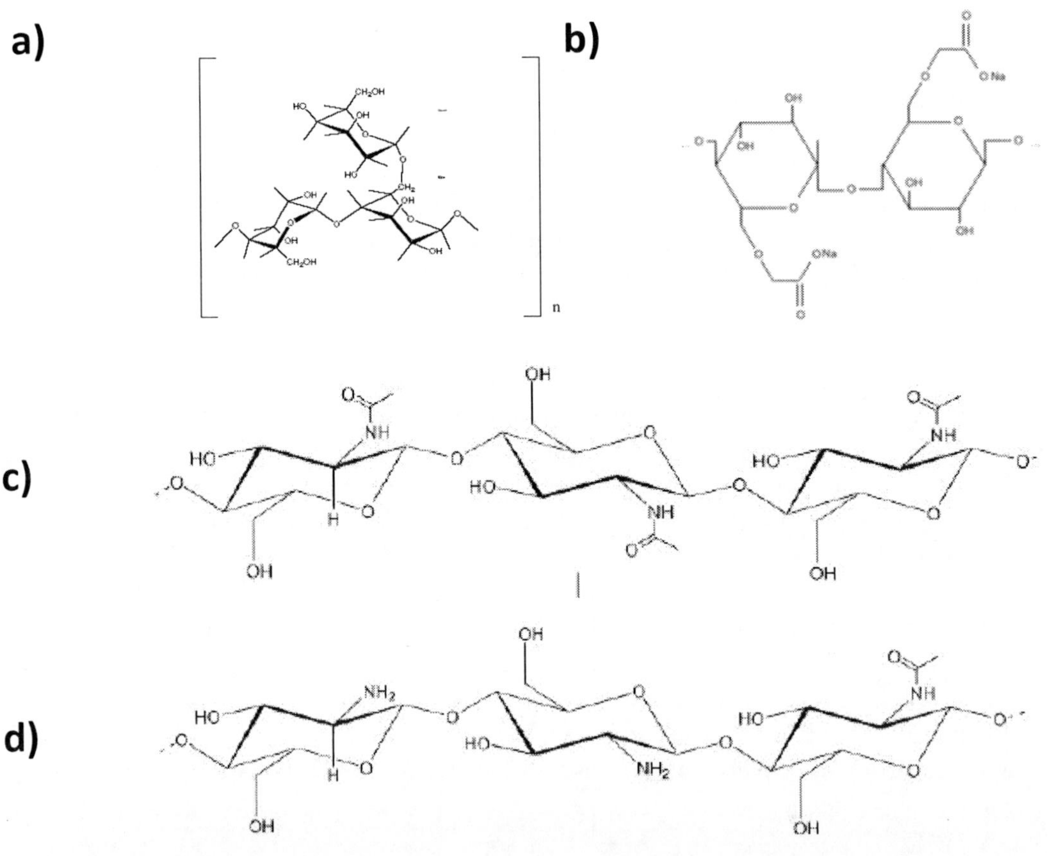

Figure 7. Partial structures of: a) galactomannan mannose-galactose ratio of 2:1; b) sodium carboxymethylcellulose; c) chitin; d) chitosan.

Galactomannan is a polysaccharide constituted by monomeric moieties of mannose and galactose whose coordination sites are constituted by hydroxyl groups. It can be extracted, for instance, from seeds of *L. leucocephala* [35] among other *Leguminosae*.

Carboxymethylcellulose is also a polysaccharide which consists in repeated unities of β-1,4 linked anhydroglucose residues substituted by sodium carboxymethyl groups [34]. Chitin is one of the most abundant natural homopolymers of β(1-4) bonded N-acetyl-D-glucosamine. It can be extracted, for instance, from crustacean shells. Chitin in different degrees of deacetylation produces different mixtures of of β(1–4)-2-acetamide-2-deoxy-D-glucose and β(1–4)-2-amino-2-deoxy-D-glucose called chitosans, whose properties (water solubility and ability of complexation among them) are dependent of such degree of deacetylation. Therefore, both hydroxyl and amine groups constitute the main coordination sites of chitosan [36].

Mercê et al. [34-35] have researched the physico-chemical properties of Al-O-donor biopolymer complexes in solution and in solid state and studied the complexing capability of these biopolymers in the removal of aluminium from polluted river water. These studies also employed potentiometric, spectroscopic and thermal analysis of Al-galactomannan and Al-carboxymethyl cellulose complexes allowing their identification and the evaluation of their chemical and physical properties.

Potentiometric essays along with the use of mathematical models [37, 38] allowed the identification of three different Al-Gal complexes in solution, whose binding constants and speciation diagram are respectively showed in Table 5 and Figure 8. Analogous studies allowed the identification of a sole Al-CMC complex in solution (Table 5), which precipitated at low pH values. In the formation equilibria of the complexes, L stands for the monomeric moiety of the biopolymer coordinated with the Al^{3+} ion. The stability constants of these Al-complexes were higher than those of the Gal- and CMC-complexes with other metallic ions, such as Co(II), Cu(II), Pb(II), Ni(II) and Mo(VI), also determined by the same authors with the same analytical techniques [34,35,39]. The values of the binding constants are an indication of a high affinity between these biopolymers and the Al(III) ion, probably due to its high charge and small size.

Table 5. Logarithms for the binding constants of the complexes between Al^{3+} and galactomannan (Gal) and carboxymethylcellulose (CMC).

Ligand	Complex Equilibrium	log K
Gal	[Al-L]/[Al][L]	16.53±0.09
	$[Al\text{-}L_2]/[Al\text{-}L][L]$	16.37±0.09
	$[Al\text{-}L_3]/[Al\text{-}L_2][L]$	11.20±0.09
CMC	[Al-L]/[Al][L]	19.5±0.2

Thermal analysis and IR spectroscopy demonstrated the maintenance of the complexes in solid state and their stability even in relatively high temperatures.

UV-VIS spectroscopy and IR studies, in solution and solid state respectively, showed a different behavior of these biopolymers in relation to the coordination sites. Thus, it seems that in the Al-Gal complexes each coordinated monosaccharide moiety is bonded to the Al(III) via primary deprotonated C6-OH and possibly secondary deprotonated C3-OH hydroxy basic sites and so, the six coordination positions of the Al^{3+} in the $Al\text{-}L_3$ Al-Gal complex would be occupied by oxygen atoms of the galactomannan, yielding a less stable complex. In contrast, in the CMC-Al complex the coordination occurs via deprotonated C6-OH and/or carboxylate groups, this last one, preferentially, and only two coordination positions are occupied as maximum by oxygen atoms of the carboxymethylcellulose. This different coordination behavior would explain the different solubility characteristics observed among the complexes.

The high values of the binding constants and the formation of complexes at low pH values make these biopolymers potentially pronc for the remediation of acid effluents derived from aluminium mining or industrial processing, in addition to making its application possible in the municipal water and wastewater treatment, which may contain a high charge of residual aluminium.

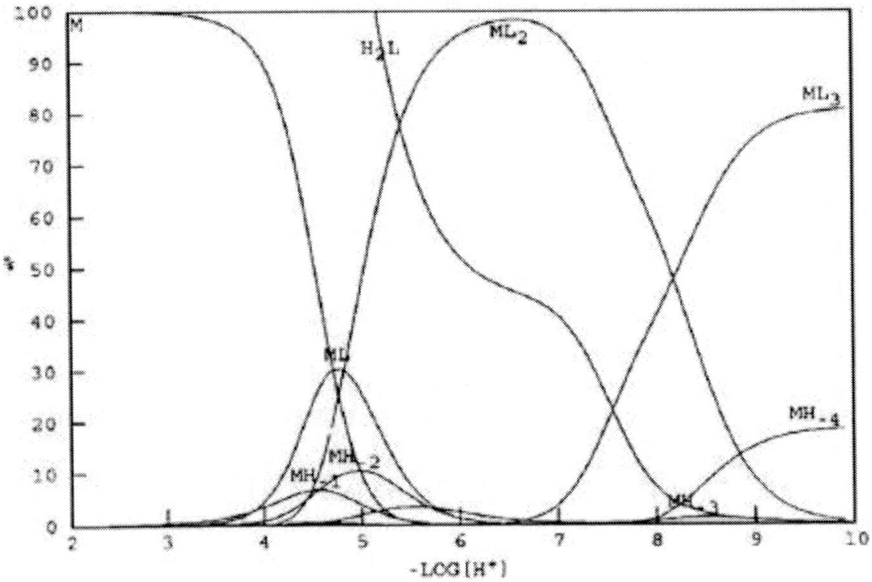

Figure 8. Species distribution diagram of the 0.09 mmol solution of galactomannan (L) with Al^{3+} (M, 0.03 mmol) of pH values from 2.0 to 10.0. Percentage of a species present with the metal ions set at 100%. H_{-x} represents (OH^-). (From reference 35).

However, the high solubility of the Gal-Al complexes implies the necessity of implementing a precipitation, extraction or immobilization technology that would yield a complicate remediation procedure. For this reason, the authors studied the aluminium contaminated water remediation based in the formation of the Al-CMC complex, which is water-insoluble in a large pH range and consequently may be easily removed by mechanical procedures.

Aluminium removal essays by CMC were carried out on Iraí river water samples (the river from which the city of Curitiba, Paraná – Brazil is supplied), which were artificially contaminated by Al(III) in an attempt to simulate a possible environmental accident. Competitive aluminium removal studies in the presence of other metallic contaminants were also carried out.

In order to evaluate the effect on aluminium complexation by CMC of the natural organic matter content present in the river water, a parallel study was performed with distilled and deionized water. All essays were carried out in triplicate with 1 g CMC per liter of sample. The results obtained are shown in Table 6.

As it can be seen from the results, CMC showed to be a very good remediation agent, capable of removing very high percentages of metals especially after 150 min of contact time. This period is very adequate for water treatment plants, where the water is kept between 2.5 and 3 hours.

Aluminium removal was high, only exceeded by the virtually quantitative cobalt removal. The presence of humic substances in the river water was not very influent on metal removal with exception of the copper case, whose ability to complex humic substances is well known [40].

The removal of aluminium via CMC complexation has the advantage of a higher efficiency regarding the conventional method of precipitation as aluminium hydroxide. Although being more expensive than the currently employed precipitant agents, CMC can be easily decomplexed from its Al complexes through treatments, for instance, with mineral acids and can also be recovered for later reuse.

Table 6. Percentages of metal removal by CMC (1 g/L) in the contaminated Iraí River and in distilled and deionized waters.

t(min)	Metal M	Iraí River water		Distilled and deionized water	
		$[M]_o$ (mol/L)	Reduction (%)	$[M]_o$ (mol/L)	Reduction (%)
90	Al	1.34×10^{-3}	70	1.62×10^{-3}	95
120			70		95
150			92		99
90	Cu	1.4×10^{-4}	31	8.0×10^{-5}	19
120			32		50
150			36		100
90	Co	1.2×10^{-4}	83	6.0×10^{-5}	100
120			85		100
150			100		100
90	VO	6.9×10^{-4}	62	5.2×10^{-4}	66
120			86		83
150			86		83

Chitin is the second most abundant biopolymer on Earth after cellulose. [41] Chitin is an excellent physical sorbent, especially at low pH, and has been shown to remove metals like aluminium, arsenic, chromium, copper, iron, manganese, nickel, and zinc from aqueous solutions [42, 43, 44, 45].

Daubert and Brennan [46], for example, used crab shell chitin in the anaerobic passive bioremediation of acid mine drainage (AMD) due to its nitrogen content (6-7%), which is very adequate to act as a substrate of the sulfate-reducing bacteria, the key organisms in biological AMD treatment. The authors used the least refined grade of crab shell chitin, SC-20 (20% pure chitin), not only because it is very cheap but also because it contains approximately 40% calcium carbonate by weight, which can serve as a buffer and increase the alkalinity of the treated water. Aluminium contents in samples that came in contact with chitin were dropping as the pH values were increasing, and after five days their concentrations were lower than the detection limit of the ICP technique employed (0.05 ppm). In this system, the removal of aluminium was attributed to a physical sorption on chitin at a low pH early in the experiment, followed by a precipitation as aluminium hydroxide at a higher pH later in the experiment.

The same aluminium removal mechanisms were observed by Simionato et al. [47] in the treatment with chitin, extracted from silkworm chrysalides, of textile wastewater.

Therefore, chitin acts mainly through an adsorption mechanism, but at pH values higher than 6 the $Al(OH)_3$ precipitation is the dominant process. So, it is expected that chitosan, the deacetylated derivative of chitin, is more effective as a remediation agent at pH values near neutrality due to its ability to remove metal ions not only *via* adsorption but also *via* complexation, specially through the amine groups, not available in chitin.

The high ability of chitosan to strongly coordinate metallic ions via N and/or O is well known. This ability is mainly due to three factors: i) high hydrophilicity given the great number of hydroxyl groups that chitosan possesses, (ii) a large number of chemically active amino groups and (iii) a flexible structure of polymer chains that makes it compatible with the adsorption of metal ions [36].

The high capacity of chitosan as a complexing agent has been widely studied [39,36,48]. However, in water and wastewater treatment researches, chitosan has been more extensively studied as a successful coagulant and as a flocculant agent in substitution of aluminium sulfate or other inorganic and polymeric coagulants, due to chitosan's non toxic property, in order to contribute to the removal of colloidal matter [49].

Works studying the removal of residual aluminium after the use of aluminium sulfate as coagulant in water and wastewater treatment are scarce. Simionato et al. [47] employed chitosan to remove aluminium and other pollutants from a textile effluent after a photodegradative pre-treatment and optimum results were obtained according to the authors. The maximum adsorption capacity was 21.35 mg Al per gram of chitosan at pH ~5. The authors proposed an adsorption mechanism that follows a Freundlich isotherm model, i.e., a multi-layer and reversible adsorption compatible with a physisorptive process.

In contrast, Septhum et al. [50] proposed a Langmuir isotherm model for the Al(III) adsorption from synthetic solutions on chitosan. This model assumes a mono-layer adsorption, in accordance to the kinetic studies effected by the same authors, which showed that the Al(III) adsorption on chitosan follows a pseudo second-order kinetic, consistent with a chemisorptive rate-limiting step and indicative of covalent bonds between Al(III) and chitosan $-NH_2$ groups. These authors obtained a maximum adsorption capacity for chitosan of 45 mg Al per g chitosan at pH 4, pH value in which the dominant mechanism for the removal of aluminium is probably a chemisorptive complexation process.

Chitosan microspheres have been successively employed in the treatment of coal mining effluents, achieving not only a virtually quantitative aluminium, iron and copper removal, but also increasing the pH close to neutrality [51]. Protonation of chitosan $-NH_2$ groups are in this case responsible by the acidity decrease and the process follows, as in the case above, a Langmuir isotherm model. Chitosan microspheres could be recovered, and therefore reused, by treatment with sulfuric acid and then sodium hydroxide.

It seems, therefore, that the adsorptive behavior of Al(III) on chitosan is strongly dependent of the process conditions, such as pH, temperature and Al(III) and chitosan initial contents. Thus, at very low pH some $-NH_2$ groups can be protonated, and so decreasing the complexing capacity of chitosan but allowing the pH increase of very acidic effluents, whereas pH values higher than 5 implicate two competitive processes, adsorption/complexation and $Al(OH)_3$ precipitation.

Final Considerations

The excellent coagulant properties and the low cost of aluminium salts are the main reasons for its extensive use in the water treatment plants worldwide. Nevertheless, the increasing concerns about aluminium effects on health have motivated researches about different processes to remove it from treated water. Conventional aluminium removal techniques, decantation and filtration, are not sufficient to remove the residual dissolved aluminium, whose concentration is strongly dependent on the pH, and hence the need for the use of advanced technologies. Nanofiltration and adsorption-flotation are techniques virtually capable of removing aluminium quantitatively, but they potentially have the problem of the high cost in the first case and the risk of contaminating the treated water with surfactant residues in the second. Aluminium removal via complexation has demonstrated to be an adequate technique when the complexes formed are insoluble and can be easily removed by conventional procedures or can be attached to an inert matrix. Special attention is directed towards the use of functionalized biopolymers as aluminium complexing agents, due to their availability and no toxicity.

References

[1] C. Gourier – Fléry, N. Fléry, Aluminium, *EMC – Toxicologie Pathologie* (2004) 79-95.

[2] P. M. Branco, Dicionário de Mineralogia – Sumário Mineral – DNPM, 2002. *http://www.dnpm-pe.gov.br/detalhes/alumínio.htm,* 2010.

[3] C. Exley, Aluminium and Medicine. In: *Molecular and Supramolecular Bioinorganic Chemistry – Applications in Medical Sciences*, Ed. A.L.R. Mercê et al, Nova Science Publishers Inc., New York, 2009, 246 p.

[4] L. di Bernardo, L. P. Sabogal Paz, *Seleção de Tecnologias de Tratamento de Água, EDiBe Ed.*, São Carlos – SP, 2008, V.1, 878 p. and V.2, 682 p.

[5] [5] Metcalf & Eddy, *Wastewater Engineering: Treatment and Reuse*, MgGraw-Hill ed., New York, 4th ed. rev 2003, 1819 p.

[6] P. T. Srinivasan, T. Viraraghavan, Characterization and concentration profile of aluminium during drinking-water treatment, *Water SA* 28(1) (2002) 99-106.

[7] D. J. Pernitsky, J. K. Edzwald, Selection of alum and polyaluminium coagulants: principles and applications. *Journal of Water Supply: Research and Technology—AQUA* (2006) 121-141.

[8] P. T. Srinivasan, T. Viraraghavan, K. S. Subramaian, Aluminium in drinking water: an overview, *Water SA* 25 (1999) 47-55.

[9] S. X. Campos, L. di Bernardo, E. M. Vieira, Influência das características húmicas na eficiência da coagulação com sulfato de alumínio, *Engenharia Sanitária e Ambiental* 10(3) (2005) 194-199.

[10] S. Kvech, M. Edwards, Solubility controls on aluminium in drinking water at relatively low and high pH. *Water Research* 36 (2002) 4356-4368.

[11] BRASIL, Portaria 518, 2004. *Padrões de Potabilidade* - Ministério da Saúde, Brasil.

[12] D. Ribes, M. T. Colomina, P. Vicens, J. L. Domingo, Effects of oral aluminium exposure on behavior and neurogenesis in a transgenic mouse model of Alzheimer's disease, *Experimental Neurology* 214 (2008) 293-300.

[13] T. P. Flaten, Aluminium as a risk factor in Alzheimer's disease, with emphasis on drinking water. *Brain Research Bulletin* 55(2) (2001) 187-196.

[14] J. R. Walton, Aluminium in hippocampal neurons from humans with Alzheimer's disease, *Neurotoxicology* 27 (2006) 385-394.

[15] R. A. Pettersen, L. A. Vollestad, L. E. W. Flodmark, A. B. S. Poléo, Effects of aqueous aluminium on four fish ectoparasites, *Science of the Total Environment* 369 (2006) 129-138.

[16] S. E. Ghazy, S. E. Samra, A. E. M. Mahdy, S. M. El-Morsy, Removal of aluminium from some water samples by sorptive-flotation using powdered modified activated carbon as a sorbent and oleic acid as a surfactant, *Analytical Sciences* 22 (2006) 377-382.

[17] S. E. Ghazy, S. E. Samra, A. E. M. Mahdy, S. M. El-Morsy, Flotation-separation of aluminium from some water samples using powdered marble waste and oleic acid, *Analytical Sciences* 19 (2003) 1401-1406.

[18] W. Wen-Dong, Y. Hong-Wei, Z. Hua-Zhang, J. Zhan-Peng, Transfer and transport of aluminium in filtration unit, *Journal of Environmental Sciences* 19 (2007) 897-901.

[19] A. Zouboulis, G. Traskas, P. Samaras, Comparison of single and dual media filtration in a full-scale drinking water treatment plant, *Desalination* 213 (2007) 334- 342.

[20] R. D. Letterman and C. T. Driscoll, Survey of residual aluminium in finished water, *Journal of American Water Works Association* 80 (4) (1988) 154–158.

[21] T. S. Singh, B. Parikh, K. Pant, Investigation on the sorption of aluminium in drinking water by low-cost adsorbents, *Water SA*. 32 (2006) 49-54.

[22] E. Rodil, R. Dumortier, J. H. Vera, Removal of aluminium from aqueous solutions using sodium di-(*n*-octyl) phosphinate, *Chemical Engineering Journal* 97 (2004) 225-232.

[23] P. Vanloot, J.-L. Boudenne, L. Vassalo, M. Sergent, B. Coulomb, Experimental design approach for the solid-phase extraction of residual aluminium coagulants in treated waters, *Talanta* 73 (2007) 237-245.

[24] C. Brach-Papa, B. Coulomb, C. Branger, A. Margaillan, F. Théraulaz, P. Van Loot, J.-L. Boudenne, Fluorimetric determination of aluminium in water by sequential injection through column extraction, *Analytical and Bioanalytical Chemistry* 378 (2004) 1652-1658.

[25] V. Arancibia, C. Muñoz, Determination of aluminium in water samples by adsorptive cathodic stripping voltammetry in the presence of pyrogallol red and a quaternary ammonium salt, *Talanta* 73 (2007) 546-552.

[26] J. O. Esalah, M. E. Weber, J. H. Vera, Removal of lead from aqueous solutions by precipitation with sodium di-(*n*-octyl) phosphinate, *Separation and Purification Technology*. 18 (2000) 25-36.

[27] J. O. Esalah, M. E. Weber, J. H. Vera, Removal of lead, cadmium and zinc from aqueous solutions by precipitation with sodium di-(*n*-octyl) phosphinate, *Canadian Journal of Chemical Engineering* 78 (2000) 948-954.

[28] J. P. Boudot, D. Merlet, J. Rouiller, O. Maitat, Validation of an operational procedure for aluminium speciation in soil solutions and surface waters, *Science of the Total Environment* 158 (1994) 237-252.

[29] P. Matúš; Evaluation of separation and determination of phytoavailable and phytotoxic aluminium species fractions in soil, sediment and water samples by five different methods, *Journal of Inorganic Biochemistry* 101 (2007) 1214-1223.

[30] J.-L. Boudenne, S. Boussetta, C. Brach-Papa, C. Branger, A. Margailan, F. Théraulaz, Modification of poly(styrene-*co*-divinylbenzene) resin by grafting on an aluminium selective ligand, *Polymer International* 51 (2002) 1050-1057.

[31] P. Vanloot, J.-L. Boudenne, L. Vassalo, M. Sergent, B. Coulomb, Experimental design approach for the solid-phase extraction of residual aluminium coagulants in treated waters, *Talanta* 73 (2007) 237-245.

[32] A. Denizli, R. Say, E. Piskin, Removal of aluminium by alizarin yellow-attached magnetic poly(2-hydroxyethyl methacrylate) beads, *Reactive and Functional Polymers* 55 (2003) 99-107.

[33] H. Yavuz, R. Say, M. Andaç, N. Bayraktar, A. Denizli, Performance of dye-affinity beads for aluminium removal in magnetically stabilized fluidized bed, *Biomagnetic Research and Technology* 2(5) (2004) (open access on line).

[34] A. P. Franco, M. A. Lobo-Recio, B. Szpoganicz, A. López-Delgado, J. Felcman, A. L. R. Mercê, Complexes of carboxymethylcellulose in water. Part 2. Co^{2+} and Al^{3+} remediation studies of wastewaters with Co^{2+}, Al^{3+}, Cu^{2+}, VO^{2+} and Mo^{6+}, *Hydrometallurgy* 87 (2007) 178-189.

[35] S. C. Lombardi, A. L. R. Mercê, Properties of complexes of galactomannan of *Leucaena leucocephala* and Al^{3+}, Cu^{2+} and Pb^{2+}, *Bioresource Technology* 89 (2003) 63-73.

[36] R. B. Hernández, A. P. Franco, O. R. Yola, A. López-Delgado, J. Felcman, M. A. Lobo-Recio, A. L. R. Mercê, Coordination study of chitosan and Fe^{3+}, *Journal of Molecular Structure* 877 (2008) 89-99.

[37] A. Martell, R. Motekaitis, The Determination and Use of Stability Constants, VCH, New York, 1992.

[38] A. E. Martell, R. M. Smith, *NIST Critical Stability Constants of Metal Complexes*, version 6.0; NIST Database 46, Gaithersburg, MD, USA, 2001.

[39] A. L. R. Mercê, A. P. Franco, A. López-Delgado, R. B. Hernández, M. A. Lobo-Recio, J. Felcman; Metal Ions and Biopolymers Complexes for Protecting and Improving the Environment. Complexation, Speciation and Remediation Studies of Ni^{2+} and Mo^{6+} with Chitosan and Carboxymethylcellulose, in: *Reactive and Functional Polymers Research Advances*, M. I. Barroso (Ed.), Chapter 5, Nova Science Publishers Inc., New York, 149-173, 2007.

[40] P. Zhou, H. Yan, B. Gu, Competitive complexation of metal ions with humic substances, *Chemosphere* 58 (2005) 1327-1337.

[41] P. Beaney, J. Lizardi-Mendoza, M. Healy, Comparison of chitins produced by chemical and bioprocessing methods, *Journal of Chemical Technology & Biotechnology* 80 (2005) 145-150.

[42] D. J. Hawke, S. Sotolongo, F. J. Millero, Uptake of Fe(II) and Mn(II) on chitin as a model organic phase, *Marine Chemistry* 33 (1991) 201-212.

[43] B. J. Mcafee, W. D. Gould, J. C. Nadeau, A. C. A. da Costa, Biosorption of metal ions using chitosan, chitin, and biomass of *Rhizopus oryzae*, *Separation Science and Technology* 36 (2001) 3207-3222.

[44] L. D. Franco, R. D. C. Maia, A. L. F. Porto, A. S. Messias, K. Fukushima, G. M. de Campos-Takaki, Heavy metal biosorption by chitin and chitosan isolated from *Cunninghamella elegans* (IFM 46109), *Brazilian Journal of Microbiology* 35 (2004) 243-247.

[45] K. Vijayaraghavan, J.R. Jegan, K. Palanivelu, M. Velan, Nickel recovery from aqueous solution using crab shell particles, *Adsorption Science & Technology* 23 (2005) 303-312.

[46] L. N. Daubert, R. A. Brennan, Passive Remediation of Acid Mine Drainage Using Crab Shell Chitin, *Environmental Engineering Science* 24(10) (2007) 1475-1480.

[47] J. I. Simionato, A. T. Paulino, J. C. García, J. Nozaki, Adsorption of aluminium from wastewater by chitin and chitosan produced from silkworm chrysalides, *Polymer International* 55 (2006) 1243-1248.

[48] A. Gamage, F. Shahidi, Use of chitosan for the removal of metal ion contaminants and proteins from water, *Food Chemistry* 104 (2007) 989-996.

[49] L. C. K. Moraes, R. Bergamasco, C. G. Tavares, D. Hennig, M. C. Bongiovani, Utilization of the Coagulation Diagram in the Evaluation of the Natural Organic Matter (NOM) Removal for Obtaining Potable Water, *International Journal of Chemical Reactor Engineering* 6 (2008) A87. http://www.bepress.com/ijcre/vol6/A87.

[50] C. Septhum, S. Rattanaphani, J.B. Bremner, V. Rattanaphani, An adsorption study of Al(III) ions onto chitosan, *Journal of Hazardous Materials* 148 (2007) 185-191.

[51] R. Laus, R. Geremias, H. L. Vasconcelos, M. C. M. Laranjeira, V. T. Fávere, Reduction of acidity and removal of metal ions from coal mining effluents using chitosan microspheres, *Journal of Hazardous Materials* 149 (2007) 471-474.

In: Molecular and Supramolecular Bioinorganic Chemistry ISBN: 978-1-61728-785-5
Ed: A. L. R. Mercê, J. Felcman and M. A. L. Recio

Chapter 5

ADVANCES IN METALS SEPARATIONS TECHNOLOGY FROM LIQUID EFFLUENTS

F.J.Alguacil, M.Alonso, F.A. López, A.López-Delgado*, and I.Padilla
Centro Nacional de Investigaciones Metalúrgicas (C.S.I.C.), Madrid, Spain

1. OVERVIEW

There is a growing demand to avoid the production and disposal of liquid effluents containing toxic metals, also some of these streams contained valuable metals, thus, a higher emphasis to recover and recycle (or sale) the valuable products and the safe disposal of the toxic products contained within these effluents are continuously demanded either by regulatory and cost pressures. Several technologies can be used to remove, separate and/or concentrate these toxic and valuable elements from liquid effluents, thus in this chapter, three separations technologies are considered which can be used to improve the selective removal and concentration of metals from liquid effluents, wastewaters or contaminated waters prior to its final recycle, sale or disposal.

Liquid-liquid extraction is a proven technology for the recovery of heavy metals and other pollutants and is useful in medium and large scale operations and when the solute concentration is high. Extractants are becoming increasingly selective, allowing specific compound to be separated from the aqueous phase while others are retained in it. The capital outlay for such equipment can be expensive, large volumes of organic phases are required and performance would be limited by hydrodynamic constrains such as flooding and entrainment. Sometimes the potential cross-contamination of the aqueous stream with the organic solution must be considered. In the operation, a water insoluble extractant dissolved in a suitable diluent is contacted with the aqueous phase and the metal or solute transfers into the organic phase. This loaded organic phase is then further contacted with a second or strip aqueous

* Centro Nacional de Investigaciones Metalúrgicas (C.S.I.C.), Avda. Gregorio del Amo 8, 28040 Madrid (Spain).
*E-mail corresponding author: alopezdelgado@cenim.csic.es
*E-mail corresponding author: malobo@ens.ufsc.br

phase. The extracted species transfer back into the strip phase, from which is finally processed and also the organic phase is regenerated and is returned to a further extraction operation. Since metals can be present in solution as cations, anions or neutral compounds, three types of extractants are considered: i) acidic and chelating extractants, both generally known as cationic reagents, ii) basic or anionic and (iii) neutral or solvating reagents.

Liquid membranes technologies using organic phases extended the range of conditions under which liquid-liquid extraction processes are viable. The organic phase used in these technologies resembled very closely those used in liquid-liquid extraction operation. The term liquid membrane is given to a system in which the membrane that separates the feed and stripping phases consists of a thin film or organic solution. Two types of liquid membranes are considered here. The first, a surfactant or emulsion liquid membrane, consisting of a water-in-oil emulsion formed from droplets of the aqueous strip solution contained in the organic phase. The latter consists of an organic diluent, an organic extractant or carrier to combine to the species to be transported and a surfactant to stabilize the emulsion. Once formed, this emulsion is put into contact with the aqueous feed solution in a suitable contactor. The solute of interest in the feed, reacts with the carrier phase at the feed-organic interface and migrate across the organic solution to the inner aqueous strip interface where it is stripped. The organic extractant which is regenerated, then migrates back to the feed interface and the extraction process continues. When the extraction is complete, the loaded emulsion phase is separated from the aqueous raffinate and it is broken to separate the aqueous strip and the organic phase. Then, all the above process is repeated. In the second type of liquid membranes considered here, the organic phase, now without the surfactant, is immobilized into the pores of a porous solid polymer, resulting in the so-called supported liquid membranes. The polymeric support, which is generally of a hydrophobic nature, consists of microporous membranes in the form of a sheet or hollow fiber. The polymer is wet by the organic solution which migrates (and fill) into the pores to produce the liquid membrane. This then forms a physical barrier between the aqueous feed and strip solutions. The mechanism of transport is somewhat the same as in the surfactant liquid membranes. Recent modifications or improvements to the supported liquid membranes operation mode included: non-dispersive solvent extraction (NDSX), pseudo-emulsion hollow fiber strip dispersion (PEHFSD) and hollow fiber renewal liquid membrane (HFRLM).

Adsorption and ion exchange processing are useful to remove species from non-clarified aqueous solutions and when the concentration of the pollutant/valuable metal is extremely low. In these technologies, a solid matrix replaces the extractant or carrier as the medium for mass transfer. The apparent lack of selectivity of various adsorbents, i.e. activated carbon, can be overcome by the use of byosorbents, i.e. chitosan, algaes, fungi, etc. Ion exchange technology consists in the exchange of ions in solution with the counterions of the same charge contained in insoluble synthetic polymers. Coordination and chelating ion exchange resins are polymers with covalently bound side chains containing donor atoms that are able to form complexes with selected metal ions or with their complexes. Since some of the properties of such exchangers are not good enough, specially when refereed to selectivity, the concept of solid polymeric extractants (TVEX and Levextrel resins) and solvent impregnated resins (SIRs) have been developed.

2. METALS: ESSENTIALS AND TOXICS FOR HUMANS

A number of metals are of concern because of their toxic properties, equally some metals are also essential for both survival and health of humans. Thus, metals have been classified as essential, beneficial or detrimental. Some of these metals are either essential or beneficial under the correct intake, whereas an excess of their intake can result in severe (cancer and non-cancer) diseases to humans. Sometimes, this toxicity comes *via* the presence of the metal in the terrestrial and aquatic biota and the interaction of the population with them. The next Tables 1 and 2 summarized some peculiarities of such behaviour.

Table 1. Essentiality and toxicity of selected metals.

Metal	Essential	Toxic
Selenium	Constituent of enzymes, to decrease the risk of Keshan disease	Selenosis, eurological symptons
Iron	Essential nutrient	Potential interaction with other elements, hepatomegaly, cardiac disease, liver cirrhosis
Manganese	Essential nutrient	Neurological disordes
Zinc	To avoid the risk of: growth retardation, the delay in sexual and skeletal maturation, diarrhea and alopecia	Supression of immune response, decreased hig-density lipoprotein, anemia from copper deficiency
Copper	To avoid the risk of: hypochronic anemia, neutropenia, hypopigmentation of hair and skin, abnormal bone development, vascular abnormalities	Gastric irritation
Molybdenum	Required for growth and reproduction	Strong interaction with copper: antagonistic or beneficial?

Table 2. Target sites of metals accepted as carcinogens.

Metal	Target site
Arsenic and compounds	Lung, urinary bladder, kidney, liver, skin
Beryllium and compounds	Lung
Cadmium and compounds	Lung, liver, prostate, kidney
Hexavalent chromium compounds	Lung, sinonasal cavity, liver
Nickel compounds	Lung, sinonasal cavity

Not all the arsenic compounds are equally toxic, generally, trivalent arsenic ones are more toxic than their As(V) counterparts and inorganic arsenic compounds are more toxic than organoarsenicals.

Mercury vapour can cause acute damage to the lungs and death from pneumonia when inhaled at extremely high concentrations. On the other hand, methyl mercury has the property of inflicting damage almost exclusively on the nervous system, mainly the central nervous system.

Aluminium is highly abundant in the environment and represents almost 8% of the earth´s crust, only being exceeded by oxygen and silicon. Despite this abundance, this element (and compounds) is present in relatively small amount in healthy living systems. Its toxicity is based in chronic renal failure, though other problems associated with Al *ingesta* can occur, i.e. gastrointestinal failure and neurodegenerative diseases.

The toxic properties of lead have been known for centuries. This element is found in food, air and water, thus humans are exposed to it in varying quantities. Once lead is adsorbed into the organism, it is stored in the bone, kidney and liver. Among other effects, neurotoxicity of lead has been demonstrated. Investigations about the risk of lead-induced cancer in humans is somewhat inconsistent in their findings.

3. The Needed of Metals Removal from Liquid Effluents

Besides the concerns about the relationship between metals and metals compounds and cancer and non-cancer diseases, industries are being forces either by regulatory, cost pressures and social pressures to minimize the amount and environmental sensitivity of the effluents they produced.

Natural and anthropogenic activities are responsible for the presence of these metals in the environment, where they are accumulating at a quick pace, due to the growing demands of the modern society.

The presence of metals in water is the result of mineral-weathering processes, though soluble metal compounds and the mobilization of metal species compounds from sediments also contributed to the presence of these elements, in whatever form, in waters or liquid effluents. Also, waters can be contaminated with liquid wastes from various industries. Additionally, solid wastes from these industrial processes or municipal incineration can find its way and contaminated waters if not properly disposed of.

Liquid effluents have traditionally been treated by a combination of physico-chemical processes:

- flocculation,
- precipitation,
- filtration,

and biological processes:

- activated sludge,
- biofilm,

which are usually effective in the removal of organic pollutants but fail to reduce the concentration of heavy metals to levels suitable for their safe disposal in whatever form. Also, these technologies usually do not give the selectivity to built-up streams with a value for their recycle or reuse, even in the same industry.

Three separations technologies (liquid-liquid extraction, liquid membranes and ion exchange and adsorption) can provide the means to split a given liquid effluent into two or more liquid phases that are either recyclable, saleable (i.e. to recover the metal or metal compound) or waste (Figure 1). Once the effluent is generated, the downstream of the operation requires treatment of the solution by some of the separation technologies for the recovery of the metal (or whatever metal compound). Sometimes, the liquid effluent may be treated directly to recover the metal species, but, very often, they may first have purified and concentrated before the final product can be obtained effectively.

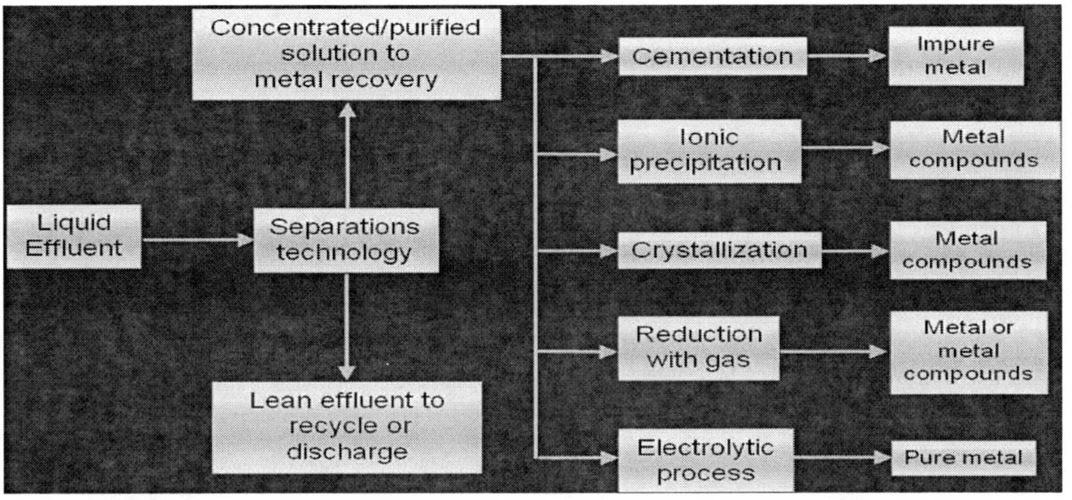

Figure 1. Separations technology: Where do they stand?.

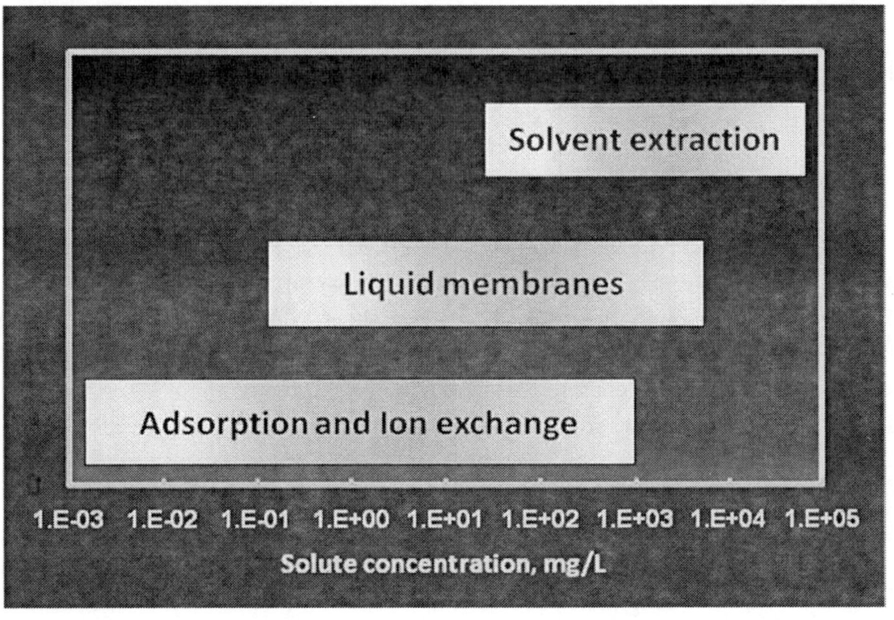

Figure 2. Approximate metal concentrations range for separations technology.

By a first consideration of the metal concentration in the liquid solution, each one of the above technologies is useful in a determined range of solute concentrations (Figure 2).

Liquid-liquid extraction can be used where the concentration of the solute is high, whereas ion exchange and adsorption processes are useful at very dilute solute concentrations. Liquid membrane technologies work appropriately between the both above.

4. Liquid-Liquid Extraction

The large amount of continuous research and development work, carried out over past decades, on both extractants and operational systems have resulted in the adoption of liquid-liquid extraction (LLE onwards) as a unit process in a number of hydrometallurgical operations for the recovery of heavy metals. As an extension of its applicability, LLE is considered the most established separations technology for liquid effluents recycling.

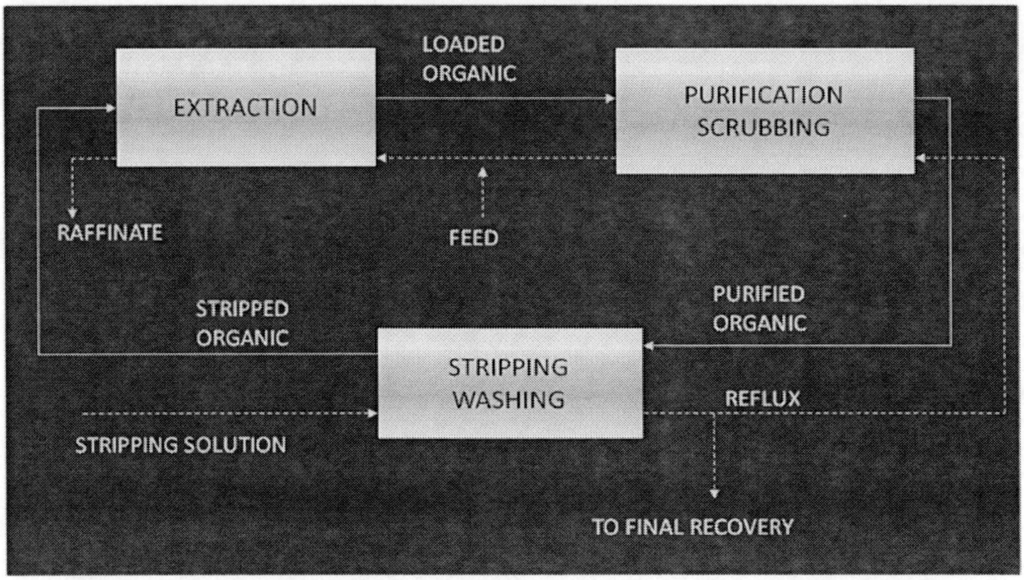

Figure 3. The general process of liquid-liquid extraction.

A scheme of a typical LLE circuit is shown in Figure 3. The metal-bearing liquid solution (i.e. a liquid effluent) and an organic solution are fed, usually in a counter-current mode, into a suitable contactor in which the two phases are mixed. As a consequence, the metal of interest is transferred from the aqueous to the organic solution. After convenient settling, the aqueous phase or raffinate is separated and treated for the recovery of other metals, recycled to some point upstream or downstream of the general process treatment of the given effluent or waste.

The loaded organic solution goes to another stage in which is scrubbed, with suitable aqueous solution, to remove some co-extracted impurities from the above step, and then, the loaded organic solution passes to a further stage in which the metal is stripped from the organic solution by a suitable aqueous (stripping) solution producing a concentrated and purified solution of the metal, which is processed to metal (or metal compound) recovery, dumping, etc.

The now free-metal organic solution is back recycled to another extraction stage, since in the stripping stage is conveniently regenerated, if not, the organic solution is further treated in another stage (not shown in the figure) prior entering the extraction operation.

For the successful of the whole operation, very often each one of the above steps may involve several contactors in each step.

The key of the operation is evidently the organic solution, this solution is normally composed of an extractant or mixed extractants and a diluent; sometimes a third component or modifier is also needed.

During all these years, extractants (an organic compound in itself) are becoming increasingly selective, allowing the specific separation of some solutes from another. In LLE practice three main types of extractants are considered:

- Acidic and chelating reagents.
- Basic or anionic reagents.
- Neutral or solvation reagents.

Acidic and chelating reagents extracted metals by a cation- related exchange reaction:

$$nM^{+}_{aq} + nHR_{org} \Leftrightarrow MR_{n_{org}} + nH^{+}_{aq} \tag{1}$$

where M represents the metal and HR the extractant, and the subscripts aq and org to the aqueous and organic phases, respectively. Non-chelating acidic reagents are organic derivatives of phosphorous acids and monocarboxylic acids, whereas chelating acidic extractants are usually oxime derivatives. In both cases of extractants, and accordingly to eq.(1), extraction occurs shifting the equilibrium to right, and stripping is carried out by the use of a highly acidic aqueous solution allowing to shift the equilibrium to the left.

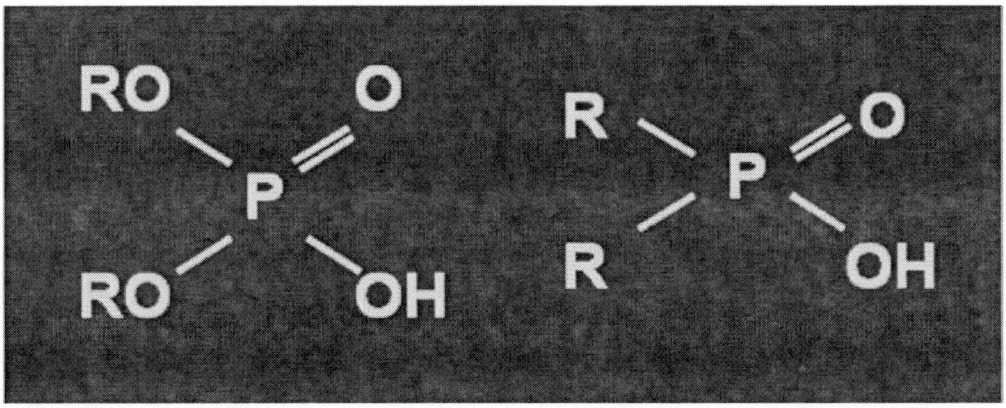

Figure 4. Active substance of DEHPA (left) and Cyanex 272 (right). (R represented alkyl chains).

Figure 4 represented the general structure of some of these, and most use, derivatives of phosphorous acids. DEHPA or di(2-ethylhexyl) phosphoric acid is probably one of the most versatile of all the extractants ever used, though this versatility is the cause of one of its major drawbacks: certain lack of selectivity. This extractant is commercialized under various names and manufacturers such as DEHPA, D2HEPA, DP-8R, etc. Cyanex 272 (manufactured by CYTEC) is the commercial name of a phosphonic acid derivative, being the chemical composition of its active substance di(2,4,4-tri-methylpentyl) phosphinic acid.

These extractants extracted metals at pH values ranging 1-7, though metal extractability can vary depending upon the aqueous media (sulphate, chloride, etc.) of the stream. In general, the lowest the metal charge the lower the metal extraction, and smaller ions are extracted better than large ions having the same charge. Sometimes, sodium or ammonium salts of these extractants are used to change the extraction performance of a given reagent.

Included in the group of carboxylic acids derivatives are the various Versatic acids (manufactured by Shell) and Naphtenic acids (various manufacturers) obtained from the distillation of crude petroleum.

The active substance of these reagents is expressed as R-COOH, where R are alkyl chains for the Versatic acids and cyclic aliphatic chains for the Naphtenic acids. Generally, these reagents extracted the same metal ion at less acidic pH value than the phosphorous acids derivatives do.

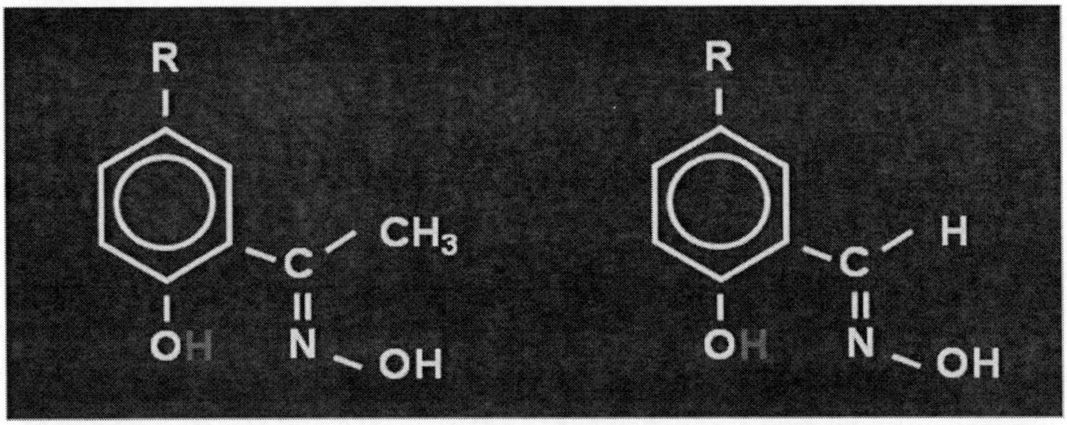

Figure 5. Oxime derivatives used for LLE operations.Ketoximes (left) and salicylaldoximes (right).

Chelating extractants are those which contain a group with an easily dissociated proton, and near to it a donor atom with a free electron pair. The most used extractans of this type are oxime derivatives (Figure 5), being the nitrogen atom the responsible for donating the electron pair to complex the metal ion. On the other hand, the hydrogen of the phenolic group is the responsible for the cationic exchange with the metal ion.

Commercially available reagents of this type are named under their manufacturers, i.e., LIX reagents (Cognis) and Acorga (CYTEC). Also, some of these reagents are mixtures of ketoximes and salicylaldoximes to improve the extraction characteristic for a given metal.

These extractants had been mainly developed to extract copper (II) at pH values of 1-3, whereas for the extraction of this element, and also zinc(II), nickel (II) and cobalt (II), at alkaline pH values, β-diketones (R'-CO-CH_2-CO-R'') are the extractants to be used. Examples of this last type of reagents are LIX 54 (Cognis) and Hostarex DK16 (Hoechst).

Basic or anionic extractants, extracted anionic metal complexes by an anion exchange reaction, that in the particular case of the extraction of cadmium(II) from highly chloride medium is:

$$CdCl^{2-}_{4_{aq}} + 2R_3NH^+X^-_{org} \Leftrightarrow (R_3NH^+)_2CdCl^{2-}_{4_{org}} + 2X^-_{aq} \quad (2)$$

From the above, in the extraction of the metal compound the extraction reaction is shifted to the right. (aq and org subscripts refereed to the aqueous and organic solutions, respectively).

In order to achieve this exchange, the amine (R_3N) must be first converted to an appropriate amine salt to provide the anion (X^-) to exchange with the metal compound:

$$HX_{aq} + R_3N_{org} \Leftrightarrow R_3NH^+X^-_{org} \tag{3}$$

that is, the amine reacts with a i.e. mineral acid (HX) to form the amine salt in the organic solution. This reaction can occur in the same extraction step of the metal itself, if suitable acidity to form the amine salt is available in the aqueous solution, or in a step prior to metal extraction. The above reaction means that amines can be used to remove inorganic acids, i.e. mineral acids, from a given effluent.

In any case, the amine salt should be considered as being the active extractant and not the free amine (though there are some exceptions to this general rule).

Basic reagents comprised amines and quaternary ammonium salts, some of these industrial available extractants are shown in Table 3.

Table 3. Basic extractants.

Type	Formula	Name	Manufacturer
Primary amine	RNH_2	Primene 81R	Rohm and Haas
		Primene JMT	Rohm and Hass
Secondary amine	R_2NH	Amberlite LA2	Rohm and Haas
Tertiary amine	R_3N	Alamine 304	Cognis
		Alamine 336	Cognis
		Hostarex A324	Hoechst
		Hostarex A327	Hoechst
Quaternary amine salt	$R_3NCH_3^+Cl^-$	Aliquat 336	Cognis

R represents alkyl chains

It is clear that, if a quaternary ammonium salt is used, the reaction of the formation of the amine salt is not needed to extract the metal, thus these compounds can be directly used to extract metals even at alkaline pH values.

The stripping reaction can involve a variety of chemical reactions which basically depends upon the metals to be stripped.

The third type of extractants, neutral or solvating reagents, are based on the solvation of neutral inorganic molecules or complexes by electron-donor extractants. By means of this solvation, the solubility of inorganic species in the organic solution is increased.

There are two main groups of extractants belonging to this type of reagents; extractants containing oxygen bonded to carbon, such as ethers, esters, alcohols and ketones, and those containing oxygen or sulphur bonded to phosphorous.

Alcohols are used for the extraction of mineral acids, i.e. phosphoric and nitric acids, whereas only one ketone: methylisobutyl ketone, has gained some interest for its practical use as is the separation of zirconium and hafnium, and the separation of niobium and tantalum, also it seems to be a good extractant for Au(III). The extraction of this element can be also accomplished with one commercially available ether such dibutyl carbitol (2,2′-dibutoxy diethyl eter.)

Some examples of extractants with P=O bond are shown in Table 4.

Table 4. Neutral derivatives of organophosphorous acids and phosphine.

Type	Active substance	Name	Manufacturer
Phosphate	$(RO)_3PO$	TBP	Various
Phosphonate	$(RO)_2RPO$	DBBP	Various
Phosphine oxide	R_3PO	TOPO	Various
		Cyanex 921	Cytec
		Cyanex 923	Cytec

R represents alkyl chains

Though it is very difficult to give an overall or general extraction reaction for these extractants, the extraction mechanism is mainly the same for all of them and it is the oxygen of the phosphoryl group responsible for the coordination bond with the metal species. These reagents can also extract acids from aqueous solutions, i.e:

$$H_2CrO_{4_{aq}} + nL_{org} \Leftrightarrow H_2CrO_4 \cdot L_{n_{org}} \quad (4)$$

where L may represents any of the compounds shown in Table 4, and n is the number of molecules of the solvation reagent involved in the reaction, this number must be determined from experimental results.

In the case of phosphates and phosphonates derivatives, there is the possibility of more than one coordination bonds formation through their other oxygen atoms, influencing the extraction as well as the stripping performance of a given system.

As in the case of amines, stripping of neutral reagents is characteristic of each system, and not a general rule may be given; however, in a number of extraction systems involving these neutral extractants, water is an effective stripping reagent.

One of the features of these solvation reagents is that they can be used without diluent, that is in concentrations up to 100%, and thus improving their extraction abilities. The exception to this rule are TOPO, Cyanex 921 and Cyanex 471X (alkyl thiophosphate derivative) which are solids.

A peculiar case in LLE practice is ionic liquids. They are a new generation of extractants, considered as “green solvents”, which have been proposed for metal separation processes. These extractants presented some specific features such as non-flammability, high thermal stability, etc., which make for their application also in many fields, i.e. catalysts for organic reactions, plasticizers, fungicides, electrochemical applications, etc.

Ionic liquids are ionic media resulting from combinations of organic cations and various anions that may be liquid at room temperature. The adopted upper temperature limit for the classification “ionic liquid” is 100°C and higher melting point named to these systems as “molten salts”. Table 5 showed some applications of these reagents in the extraction of metals.

Table 5. Selected ionic liquids and extracted metals.

Ionic liquid	Acronysm or name	Metal
1-octyl-3-methylpyridinium tetrafluoroborate	$[3MOPYR]^+[BF_4]^-$	Hg(II)
1-methyl-1-butyronitrilepiperidinium bis(trifluoromethyl)sulfonylimide	$[MPIPCN]^+[Tf_2N]^-$	Pd(II)
1-alkyl-3-methylimidazolium tetrafluoroborate*	$[C_nmim]^+[BF_4]^-$	Ce(III),Eu(III),Y(III)
Tri-hexyl(tetradecyl)phosphonium chloride	CYPHOS IL101	Hg(II)

*Mixed with (octyl(phenyl)-N,N-diisobutylcarbamethylphosphine oxide (CMPO)

However, some undesiderable performances of these extractants had led to develop the so-called task-specific ionic liquids (TSILs), which are ionic liquids with functional groups covalently tethered to the cationic or anionic part of the salt. These new branch of reagents can apparently extract cations and anions without ion exchange. As a rule, these TSILs are not used as a only component ionic liquid, but they are mixed with conventional ionic liquids. Table 6 showed some of the TSILs and their selected targets:

Table 6. Selected TSILs and extracted metals.

TSIL generic cation	Metal
imidazolium	Hg(II), Cd(II)
piperidinium	Hg(II), Cu(II), Sn(IV)
pyrrolidinium	Hg(II), Cu(II)

Either of the extractants considered for LLE practice should have the following characteristics:

- be relatively inexpensive,
- have a low solubility in water (aqueous solutions) and high solubility in aliphatic or aromatic diluents,
- have a good stability,
- will not form stable emulsions with the aqueous solutions, and good coalescing properties when mixed with diluents or modifiers (if necessary),
- have high metal (solute) loading capacity, and the most possible selectivity against other metals (solutes),
- be non-flammable (high flash point), non-volatile and non-toxic,
- have good kinetics of extraction and stripping,
- be easy to be stripped from metal (solute) loaded.

The second most important part of the organic phase in LLE operation is the diluent, which refers to the organic liquid in which the extractant and modifier are dissolved to form the organic solution. Often, the diluent comprises the major component of the organic solution. Diluents are usually relatively inexpensive crude oil fractions, and can be aliphatics, aromatics or mixtures of both types of oils. The major properties of the diluents in LLE practice are basically the same than that of the extractants, and they are used to:

- decrease the viscosity of the extractant,
- decrease emulsion-forming tendencies of extractants, and improving dispersion and coalescence properties of the organic phase,
- provide a suitable concentration of the extractant as required for each particular system.

Diluents are generally considered as inert and do not enter into the extraction mechanism, but this rule is sometimes far from the reality, and in fact the properties of the diluent can highly vary the results and performance of a given organic solution. Broadly speaking, the performance, and thus the choice, of a diluent for a particular process can not be predicted before experimentation.

Figure 6. Third phase formation in liquid-liquid extraction.

A third component of the organic phase is the modifier. Sometimes, a third phase or a secondary organic phase, appeared in the settler either after extraction or stripping operation and after phases settling. The formation of this third phase can visualize in the left-hand side vessel of Figure 6, whereas in the right-hand side the correct appearance of the common two phases (organic and aqueous) can be also seen. The formation of this phase must be avoided, in operation a system with this third phase is unpracticable, and the problem is generally solved by changing the diluent of the organic solution or by adding a modifier. The modifiers are organic compounds, with the same properties of the extractants, and the most used are: alcohols (2-ethylhexanol, n-decanol, p-nonylphenol) and tri-n-butyl phosphate (TBP), some of these in fact act as extractants by themselves. The required amount of modifier to be added to an organic phase to avoid the problem of third phase formation must be determined experimentally.

Figure 7. Mixer-settler units in liquid-liquid extraction operation.

Equipment in LLE extraction practice is related to mixer-settlers (Figure 7, in which two mixer-settler are shown under operation). The smaller vessels are the mixers in which the aqueous and organic phases are intimately mixed, and the emulsion passed continuously to the settlers (the bigger vessels) where emulsion splits into two phases, the upper corresponding to the organic phase, and the lower corresponding to the aqueous phase. Also in the Figure, the principle of continuous counter-current operation can be seen, the organic phase from the left-hand settler unit passes continuously to the mixer of the right-hand unit, whereas the aqueous phase from the right-hand settler unit passes continuously to the mixer of the left-hand unit. The sequence is repeated as many units (or stages) are necessary and connecting the extraction, scrubbing and stripping operations. In continuous co-current operation mode, all the phases flows in the direction.

In industry, most common equipment is box-type mixer-settler units, though columns are not uncommon and also centrifugal separators can be found, especially if mud or a solid product is formed during the operation.

For liquid effluents with a low dilute concentration of the metal to be separated, the technology is limited by the need for high aqueous to organic ratios. Other odds in relation with the use of this technology in the treatment of a number of liquid effluents are summarized as:

- capital outlay for equipment can be expensive,
- relative large volumes of organic solutions are required,
- performance may be limited by flooding and entrainment,
- potential for cross-contamination of the aqueous stream with the organic solution,
- aqueous solutions may be as well clarified as possible.

5. Liquid Membranes

A membrane is a semi-permeable barrier through which selected species may diffuse. Normal membrane technologies, such as nanofiltration, ultrafiltration, etc, operate primarily through size exclusion, thus, the selectivity for specific solutes is somewhat limited.

Thus, combining the physically selective permeability of the above membrane operations with the chemically selective separation provided by the LLE process, greatly enhanced the performance of the membrane technology, and then the concept of liquid membranes (LMs) arises.

Two important phenomena must be immediately considered in LMs practice: the facilitated transport and the coupled transport.

The LM is located between a liquid effluent and a stripping solution, then the carrier phase, similar to the organic phase used in LLE operation, binds to the diffusing species on one side of the membrane, carrying it across to the other side and then, owing to the alternate chemical environment, releases it in the strip phase. This is the so-called facilitated transport.

The phenomena of coupled transport in LMs are due to that species are pumped "uphill" against its own concentration gradient as a consequence of an existing concentration gradient of a second species.

In LMs both types of transport occur simultaneously resulting in the facilitated coupled transport. Two types of facilitated coupled transport must be considered in LMs operations:

- Countertransport. Metal (solute) and counter ion diffuse in opposite directions. This type of transport is typical of cationic/chelating extractants.
- Cotransport. Metal (solute) and counter ion diffuse in same direction. This is typical of basic and solvation extractants.

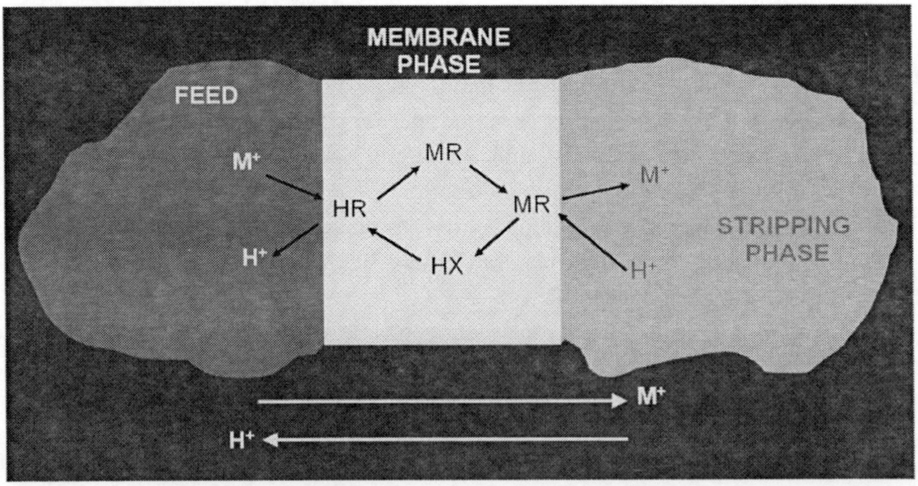

Figure 8. Mechanism of countertransport.

In countertransport mode (Figure 8), the target species and the counter ion (usually protons) flow in opposite direction across the membrane.

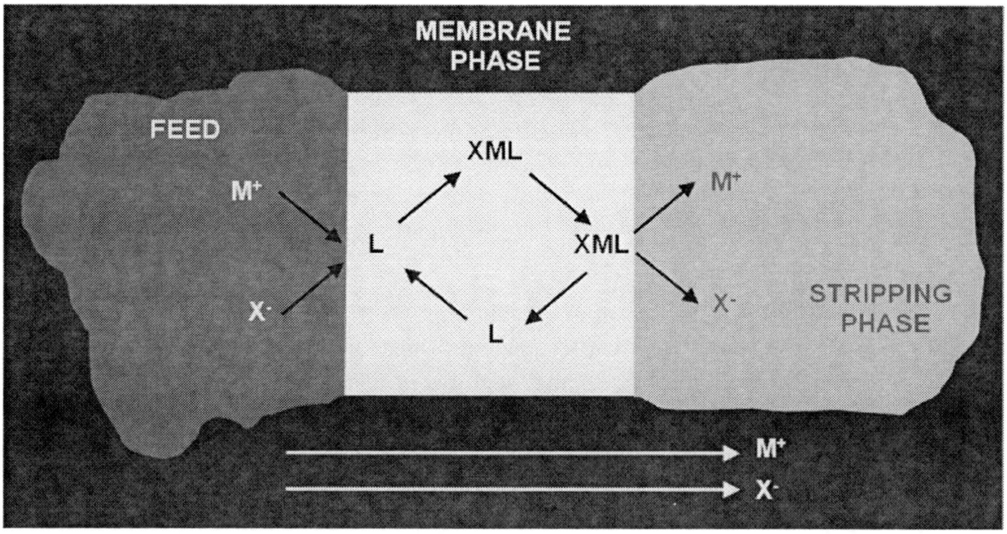

Figure 9. Mechanism of cotransport.

Figure 9 shows the principle of the cotransport mode using a solvation extractant (L), it can be seen that a neutral species is formed in the membrane phase, and both the metal (M^+) and the counter ion (X^-) flows in the same direction. Very often the metal forms an anionic complex in the aqueous feed solution and the counter ion is a cation.

Two types of LMs can be considered, unsupported and supported liquid membranes. In the first case, only emulsion liquid membranes will be described, since bulk liquid membranes has only a limited utility in lab and batch scale.

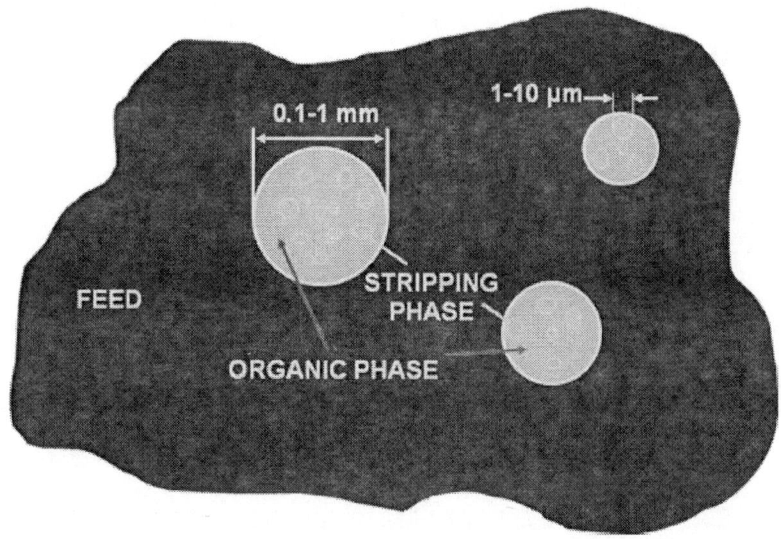

Figure 10. Schematic view of emulsion liquid membranes.

Emulsion liquid membranes (ELMs) are based in the formation of a double emulsion, being the concept shown in Figure 10. These are a three phase system in which the liquid membrane, containing the organic phase (extractant and diluent) and a surfactant, forms relatively large droplets which encapsulated the smaller drops of the strip solution. The emulsion so formed is stable due to the convenient size of the drops and the content of the emulsifier or surfactant. Then, this emulsion is dispersed into the aqueous feed solution containing the metal (solute) to be extracted, and the process of pertraction or mass transfer takes placc. Once the process is ended, the feed solution and the liquid membrane/strip solution emulsion are separated, and this last emulsion is broken-up to recover the loaded strip phase, which is further processed, and the liquid membrane is back-recycled to form a new first emulsion with new strip solution.

The process is quick due to the high interfacial areas produced by the emulsion, making ELMs attractive for the treatment of low concentrations of the permeating species. On the other hand, problems encountered with the operation are related to emulsion stability and liquid membrane rupture.

In the case of emulsion stability, this is refereed to the stability of the liquid membrane/strip solution emulsion, which is not always broken-up easily. Sometimes this can be improved by the use of bi-functional surfactants, acting as extractant as well as emulsifiers.

The problem of liquid membrane rupture is closely linked to membrane elasticity or rigidity, depending of the point of view. Sometimes this elasticity is a function of the temperature, thus, an increase in the temperature will help in the rupture of the emulsion. Also, long chain polymers can be added to the liquid membrane to decrease the rigidity of the membrane film.

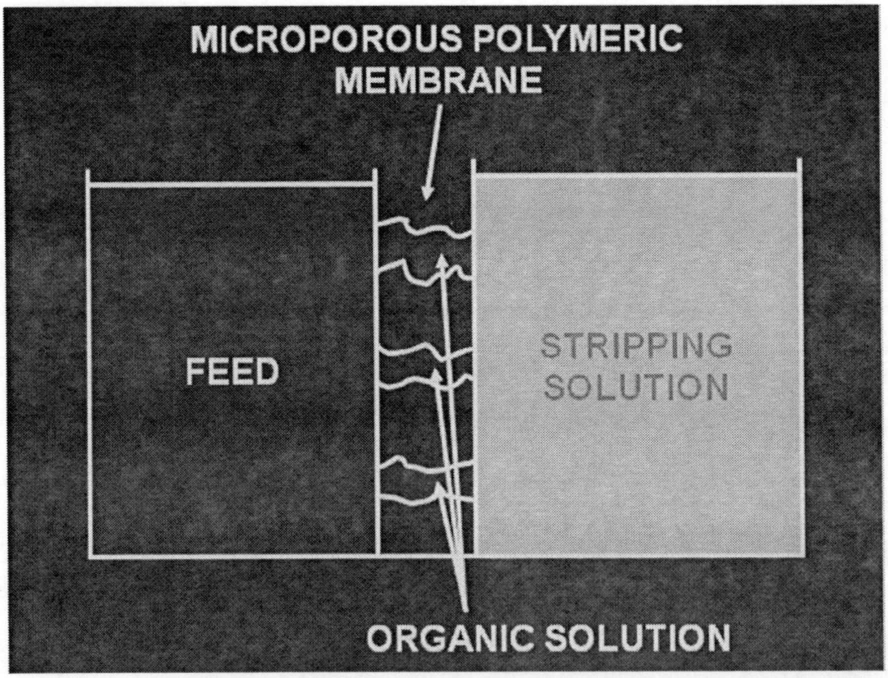

Figure 11. Schematic view of supported liquid membranes (flat-sheet configuration).

Supported liquid membranes (SLMs) are based on the use of a porous solid membrane (polymer or ceramic) which supports or holds the organic phase and separates the feed and stripping solutions (Figure 11). The pores of the thin solid membrane are completely filled with the organic solution or carrier phase and this impregnation produces relatively stable heterogeneous solid/liquid membranes. Often, this solid support is hydrophobic in its nature, which facilitated wetting by the organic solution and the reject of the aqueous phases.

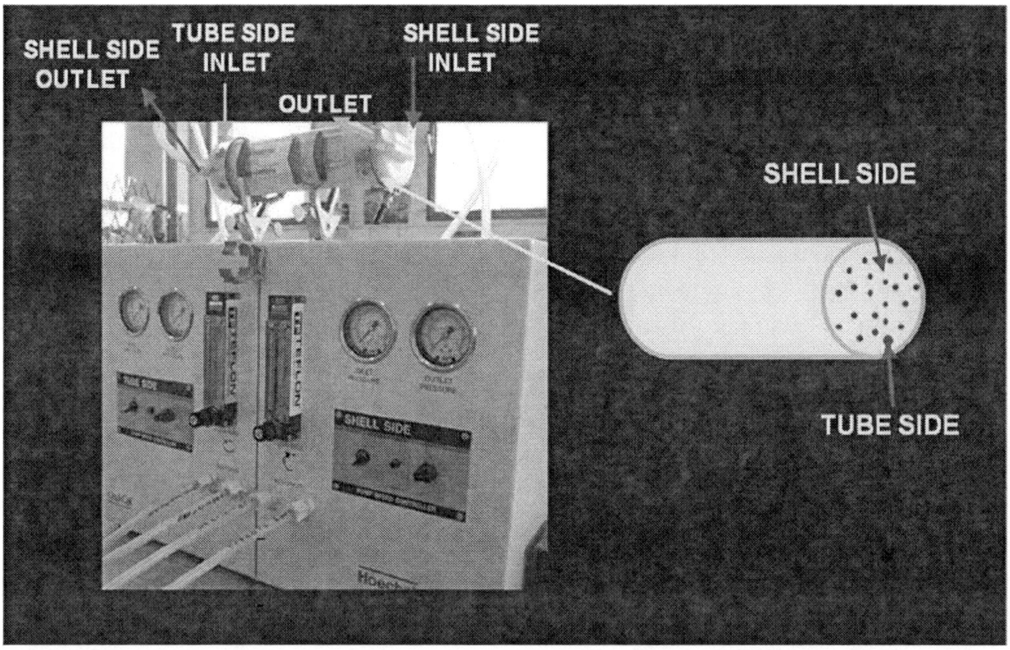

Figure 12. Hollow fiber module equipment.

Though flat-sheet supported liquid membranes are useful to obtain first data about the permeation of a given solute, for practical purposes this configuration is not very effective and the hollow fiber and spiral modules are then used since they provide high surface area to volume ration and allow to work continuously and not only in batch conditions.

In hollow fiber modules (Figure 12), the organic phase is confined, by capillarity, within the pores of the hollow fibers, and the aqueous feed and strip solutions circulate by each side of the fiber, thus, the solid membrane serves both as a support for the organic phase and as a uniform barrier between the aqueous solutions, which usually circulate within the inner (feed solution) and the outer (strip solution) of the fiber. This result in two aqueous/organic interfaces with well defined transfer areas.

Spiral wound modules consist of a sandwich of flat-sheet membranes alternate by spacers wrapped around a central axe.

From a practical point of view, one of the claimed advantages of supported liquid membranes is that the extraction, stripping and regeneration of the organic phase operations are combined and processed simultaneously in a single step. Also the small amount of organic phase used in operations makes feasible the use of expensive and more specific extractants. Lack of membrane stability under long-term operations is the major drawback associated with the use of SLMs because long-lived membranes are needed in order to minimize secondary wastes.

The instability of SLMs may be due either to the organic phase, the aqueous phases and the solid support itself.

Several attempts to improve the stability of these membranes include:

- devices for the refilling, with fresh organic solution, of the solid support to compensate for the loss of the organic phase,
- sandwich type membranes, in which a small reservoir of the organic solution is placed between two or more supports,
- gelation of SLMs, a gel is formed to avoid the flowing of the organic phase away from the gelled membrane,
- polymer inclusion membranes, in which the solid support is created with the carrier phase included within,
- other advanced technologies, though not essentially supported liquid membranes in themselves, used a solid microporous barrier to separate the aqueous phases from the organic solution. Including in this concept are: non-dispersive solvent extraction (NDSX), pseudo-emulsion hollow fiber strip dispersion (PEHFSD) and hollow fiber renewal liquid membrane (HFRLM) technologies.

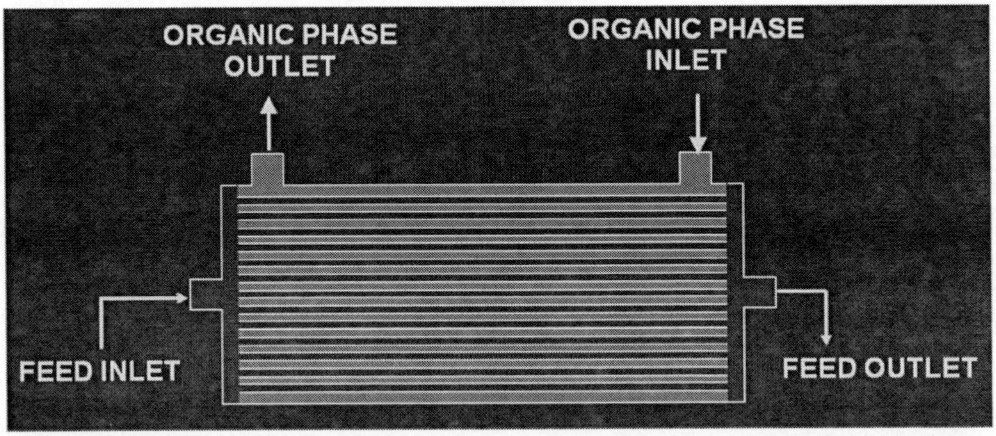

Figure 13. Single mode configuration in NDSX (extraction module).

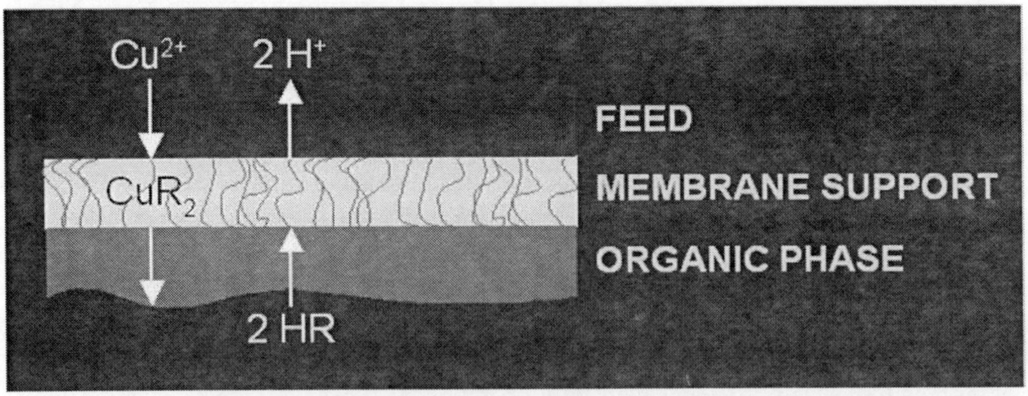

Figure 14. Mass transfer in single mode configuration (extraction module).

In NDSX operation mode two hollow fibers modules, similar to those used in supported liquid membranes with hollow fiber modules, are needed to complete the operation; in the first module (Figures 13 and 14), the organic phase circulates by one side of the fibers, whereas the feed solution circulates by the other side of the fiber bundle, the loaded organic phase then passed to the second module where the stripping operation is carrying out. Then, the organic solution is recirculated to the first module.

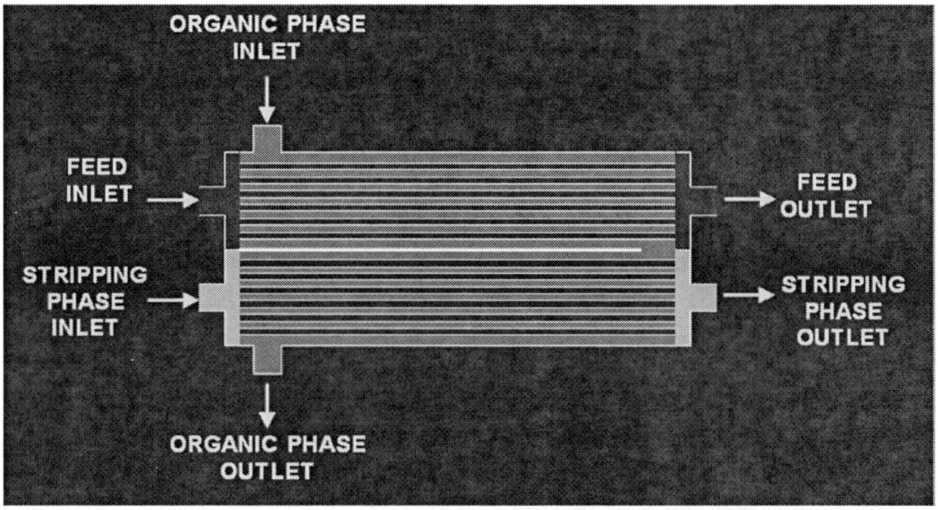

Figure 15. Dual mode configuration in NDSX.

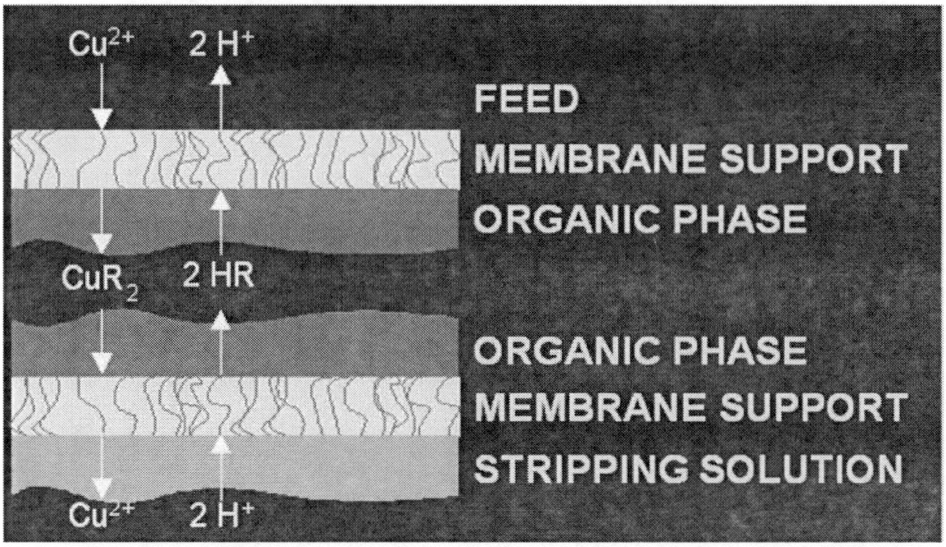

Figure 16. Mass transfer in dual mode configuration.

By the use of dual function modules (Figures 15 and 16), the extraction and stripping of the solute are carrying out simultaneously in a unique hollow fiber module containing one set of fibers for extraction and another set for stripping.

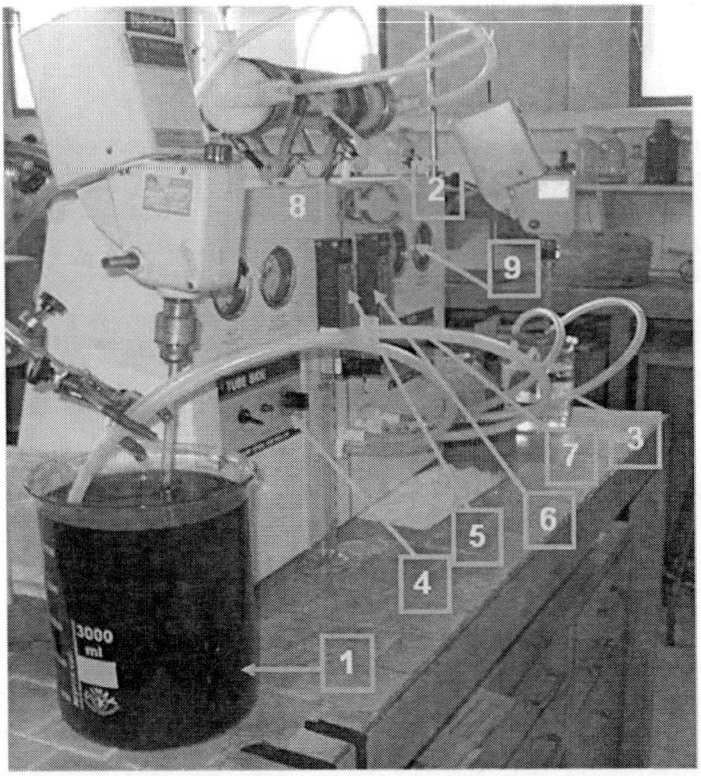

Figure 17. Pseudo-emulsion based hollow fiber with strip dispersion before operation. 1) Feed phase reservoir tank. 2) Hollow fiber module. 3) Pseudo-emulsion phase reservoir tank. 4) Pump control of feed phase. 5) Feed phase flowmeter. 6) Pseudo-emulsion phase flowmeter. 7) Pump control of pseudo-emulsion phase. 8) Pressure gauges of feed phase. 9) Pressure gauges of pseudo-emulsion phase.

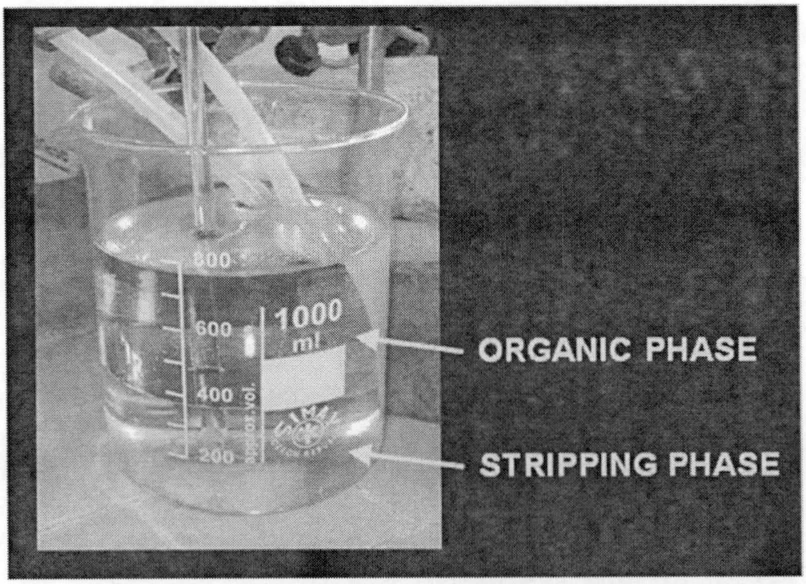

Figure 18. Detail of the organic/stripping phases reservoir tank before operation

Figure 19. View of pseudo-emulsion based hollow fiber with strip dispersion under operation.

PEHFSD process comprises a unique membrane module for extraction and stripping, one stirred tank for feed phase homogenization and one stirred tank for preparation of a pseudo-emulsion (Figure 17). This pseudo-emulsion consists of an organic phase and a strip solution which is dispersed into the organic solution (Figure 18). Both phases are pumping to the module by two gear pumps capable of variable flows. When the process begins (Figure 19), the organic phase of the pseudo-emulsion wet the microporous wall of the fiber because of its hydrophobic nature (Figure 20). When the operation ends (Figure 21), the characteristics of the pseudo-emulsion should be such that it should have clear and fast phase separation when mixing is stopped (Figure 22). The recovery of the extracted metal (solute) from pseudo-emulsion can be accomplished, pseudo-emulsion breaks down after the mixing of the organic and receiving phases stopped, and strip and organic phases separate (typically very few minutes) automatically.

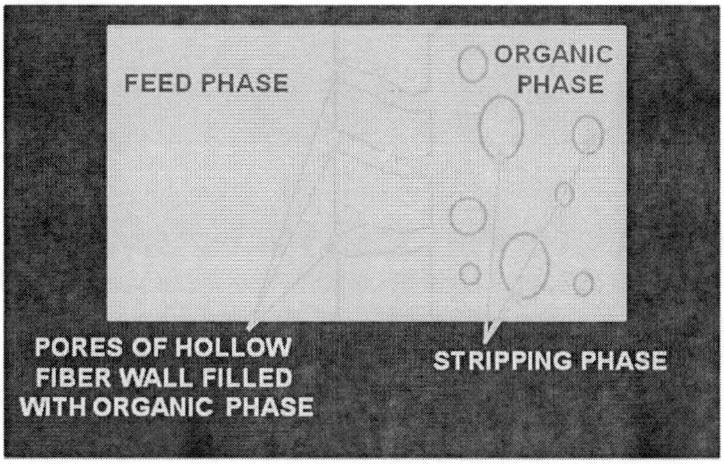

Figure 20. Enlarger view of pseudo-emulsion based hollow fiber with strip dispersion under operation and inside the module.

Figure 21. View of pseudo-emulsion based hollow fiber with strip dispersion at the end of the operation.

Figure 22. Detail of the organic/stripping phases reservoir tank after operation.

Typically, in NDSX and PEHFSD technologies the feed solution flows through the tube side of the fiber and the organic solution (NDSX) or the pseudo-emulsion phase (PEHFSD) flow through the shell side of the fiber.

In HFRLM operation mode the same device as above is used, but the organic phase is dispersed into the above and the pseudo-emulsion flows through the tube side of the fibers.

In the three above advanced membrane technologies, the interface between the aqueous/organic phases was maintaining at the pore by applying a higher pressure to the aqueous phase than to the organic or pseudo-emulsion phases. The differential pressure was always kept below the breakthrough pressure.

Also, these technologies can work either in countercurrent or cocurrent flows of the respective phases, and in the recirculation or one through modes of the feed solution.

As it can be understand, the extractants used within these membranes technologies are the same that used in LLE operation, and the organic solutions presented the same composition of extractant and diluent and modifier when needed. However, the characteristics of liquid membranes allowed the use of more expensive extractants i.e. crown ethers, which are not worth to be used in LLE due to its high costs.

6. Adsorption and Ion-Exchange

The major difference between these technologies and LLE and LMs technologies are that in adsorption and ion-exchange, the active substance for mass transfer is contained in a solid matrix or is the solid matrix itself, and they can be used in non-clarified solutions. As it was pointed out before (Figure 2), these technologies are useful when the target species is in the ppm or less concentration range, at higher metal (solute) concentrations the size of the equipment should be such that make such technologies unviable.

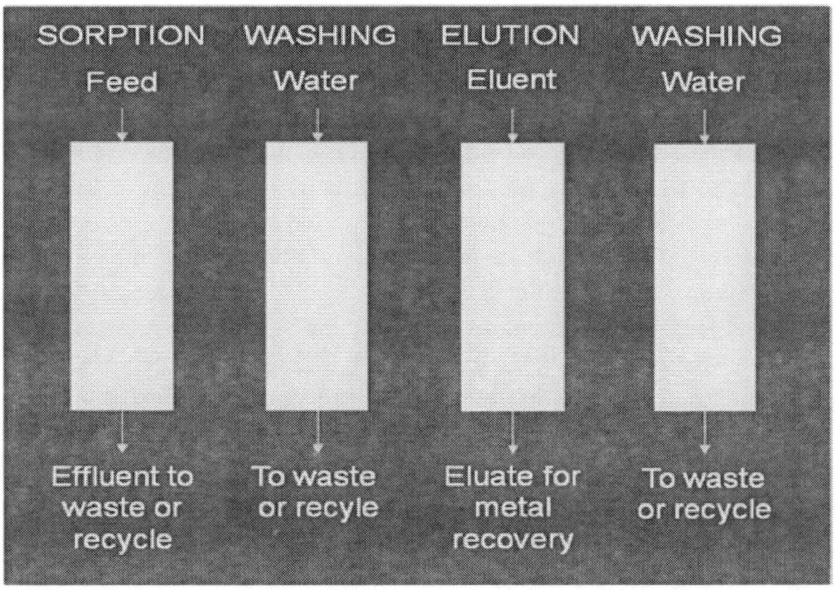

Figure 23. Sequence of operations in adsorption and ion-exchange processes.

As it can be seen from Figure 23, operationally both processes involve the scheme of sorption or loading, washing, elution and finally another washing, though this last step is not always needed. In columns, the adsorptive material packed the equipment, flowing the aqueous solution downstream (more often) or upstream.

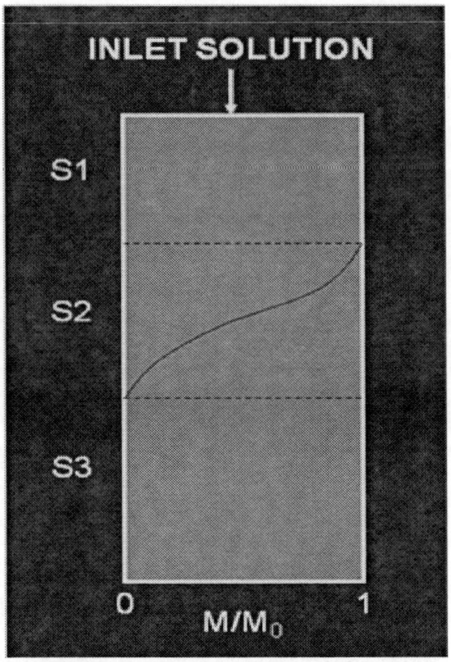

Figure 24. Mass transfer in columns. Downstream mode. M= metal concentration in the outlet solution. M_0= metal concentration on the initial solution.

The concept of mass transfer in adsorption/ion-exchange columns is illustrated in Figure 24, which represented an infinitely long column through which a metal-bearing solution is being passed. The concentration of the metal in the barren solution (as a fraction of the initial influent concentration) at any point in the column is plotted horizontally. The portion S1 where the concentration of the outlet solution is equal to the concentration in the initial solution (M/M_0= 1) represents the length of the column saturated with metal. In the portion S2, called mass transfer zone, the value of the M/M_0 ratio lies between 0 and 1, because the material packed into the column is in the process of retaining the metal species. The bottom section or S3, is completely discharged from the metal, and is normally named as length of unused bed. In any column, S2 travels down at a fixed rate and has a fixed length which determines the height of column required to treat a given solution. Though this is more of laboratory experimentation and calculation, generally for larger-scale plants, the optimum bed depth is about 1.8 m.

Typical material for adsorption processes is activated carbon. They are prepared by heating (600° C), in absence of any gas, solid carbonaceous materials: coal, wood, nut shells, etc. The as-obtained material is further heated (400-800° C), in a gas environment, to render the final product. Most of them provided large surface area and high porosity and they can be wetted by water.

Other alternatives to the use of activated carbon to remove metals or solutes are: tailored bentonite clays, blast furnace sludges and rolling mill scale. Various other inorganic materials had also been investigated for metals removal from different aqueous solutions (Table 7).

Table 7. Selected inorganic materials for metals removal from aqueous solutions.

Material	Target element
Titanium oxide	Cr(VI)
Calcine hydrotalcite	Cu(II), Zn(II), Cd(II)
Modified silica	Cu(II), Ni(II), Co(II), Zn(II)
Metal sludge	V(V)
Rolling mill scale	Pb(II)
Nanocristalline mackinawite (FeS)	Hg(II)

Biomaterials presented a higher selectivity degree that inorganic adsorbents and even ion exchange resins. Some of the biomaterials used to remove hazardous metals from liquid effluents are: chitosan, algaes, fungi, bacterial biomass, etc. To improve the effectiveness of biosorbents, they can be immobilised on to an inert framework. Also, the application of untreated plant wastes. i.e. rice husk, sawdust, sugarcane bagasse, etc., as adsorbents can also bring a number of problems such a lesser adsorption capacity than treated plant wastes, high chemical oxygen demand and biological chemical demand as well as total organic carbon due to release of soluble organic compounds contained in the plant materials. The increase of the three latter values can cause decrease of oxygen content in water and thus affecting the aquatic life.

Table 8 showed some of biomass species used for metal compounds removal from different aqueous media.

Table 8. Selected biomass materials for metals removal from aqueous solutions.

Material	Target element
Cystoseira indica	Cr(VI)
Fucus serratus	Cu(II)
Gloecapsa sp.	Pb(II)
Spirulina platensis	Au(III)
Neurospora crassa	Ag(I)
Sargassum fluitants	Au(I)
Aspergillus versicolor	Hg(II)
Modified cellulose	Hg(II), Cd(II), Co(II), Al(III)
Chitosan derivatives	Pd(II), Pt(IV)

Besides the above commented usefulness of adsorption techniques, sorbents are relative cheapness, not the case of activated carbon, many of the sorbents are waste products in themselves, and they are possible to use as decontaminant natural barriers.

The main drawback for such materials is their apparent lack of selectivity.

Ion-exchange resins can be classified as natural and synthetic. Among the first, zeolites (analite, chabazite), fluoroapatites are included. Synthetic resins are produced by reaction of styrene with divinyl benzene (which provides of cross-linking capacity to the material). To this matrix a functional group is added. According to this functional group, four main types of resins are used (Table 9).

Table 9. Classification of synthetic resins.

Type	Functional group	pH range of utility
Strong acid	R-SO_3H	0-14
Weak acid	R-COOH	4-14
Strong basic	R-$CH_2N^+(CH_3)_3X^-$	0-14
Weak basic	R-$CH_2N(CH_3)_2$	0-5
R is the inert matrix. X is the counter ion		

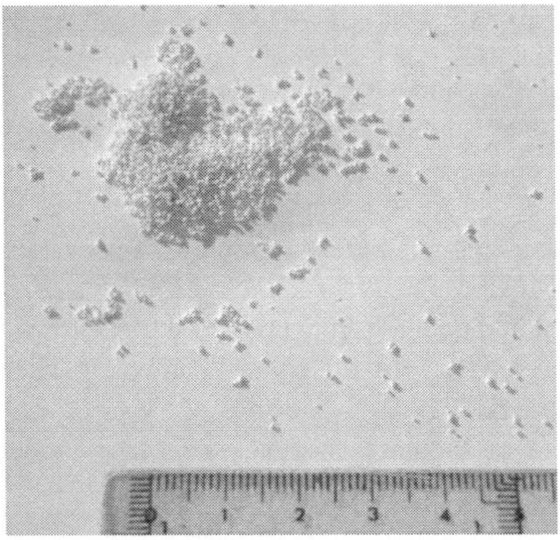

Figure 25. View of a typical ion-exchange resin.

These resins are usually produced as beads, generally in less than 1 mm diameter range and a typical aspect of such resins is shown in Figure 25.

Similarly to the case of LLE extractants, ion-exchange resins are commonly known by the name given by their respective manufacturers rather than other terminology: Lewatit, Dowex, Amberlite, etc.

The mechanism of exchange with a given cation or anion is similar to that showed in the case of the respective type of extractants, i.e.:

$$2R-SO_3H_r + Hg^{2+}_{aq} \Leftrightarrow Hg^{2+}\left(R-SO_3^-\right)_{2_r} + 2H^+_{aq} \quad (5)$$

$$R-CH_2N^+\left(CH_3\right)_3Cl^-_r + HCrO_{4aq} \Leftrightarrow R-CH_2N^+\left(CH_3\right)_3HCrO_{4_r} + X^-_{aq} \quad (6)$$

where r and aq represents the resin and aqueous phases, respectively.

The main drawbacks for the use of resins are: their apparent lack of selectivity, somewhat difficult to eluate (strong basic resins) and proned to swelling, physical breakdown, chemical degradation and solubility (osmotic shock).

Swelling is of considerable practical importance in column operation. When a dry ion-exchange resin is put into contact with water (i.e. a liquid effluent), a certain amount of water enters the solid resin phase, and the resin swells. A resin proned to swelling presented a low degree of cross-linking, a high exchange capacity, complete ionization, and large and strongly solvated counter ions. Often, when a resin swells, high pressures are usually developed inside the packed resin bed, and movements in the bed may take place causing undesirable channelling, and even under extremely conditions, a column may burst.

Improvements to the use of these resins included:

- improving their performance under extreme conditions,
- improving their selectivity: coordination and chelating resins; solvent impregnation resins (SIRs); and solid extractants (Levextrel and TVEX resins).

Coordination and chelating ion exchange resins are polymers with covalently bound side chains containing donor atoms that are able to form complexes with selected metals or their complexes. i.e. Dowex A1, Duolite ES-63, Lewatit TP-207.

SIRs are prepared by adsorption of a liquid extractant on polymer supports.

Levextrel and TEVX resins are macroporous styrene-divinylbenzene copolymers containing a metal (solute) selective extractant entrapped within the polymer matrix.

The main difference between the SIRs and Levextrel and TEVEX resins, is that in SIRS the extractant impregnated the solid matrix, whereas in Levextrel and TEVX resins the extractant is added directly to the mixture of monomers during the bead polimerization process.

CONCLUSIONS

Three inter-related technologies capable of separating a metal-bearing hazardous/valuable product streams from liquid effluents had been discussed.

The probable complexity of such liquid effluents make of interest to explore the usage of hybrid systems: combination of two of the above technologies, to, i.e. increase selectivity.

The target is the consecution of a *zero effluent* or even a *closed cycle process*, whereby all waste streams are recycled and the only output from the process is "at best" saleable product.

ACKNOWLEDGMENT

To the CSIC for support.

SELECTED LITERATURE

Liquid-Liquid Extraction

Agrawal, A., Pal, C., Sahu, K.K. Extractive removal of chromium (VI) from industrial waste solution. *Journal of Hazardous Materials* 159, 458-464 (2008).

Alguacil, F. J., Alonso, M., Navarro, P., Lopez-Delgado, A. Modelling copper(II) liquid-liquid extraction in the system Acorga M5640-Exxsol D100-$CuSO_4$-H_2SO_4. *Journal of Chemical Research* 3, 196-197 (2004).

Alguacil, F. J., Alonso, M. Liquid-liquid extraction of Au(I) by Amberlite LA2 and its application to a polymer-immobilized liquid membrane system. *Gold Bulletin* 38, 68-72 (2005).

Cerpa, A., Alguacil, F. J.Separation of cobalt and nickel from acidic sulphate solutions using the mixture DP-8R and ACORGA M5640. *Journal of Chemical Technology and Biotechnology* 79, 455-460 (2004).

Dietz, M. L. Ionic liquids as extraction solvents: Where do we stand? *Separation Science and Technology* 41, 2047-2060 (2006).

Fabrega, F., M., Mansur, M. B. Liquid-liquid extraction of mercury (II) from hydrochloric acid solutions by Aliquat 336. *Hydrometallurgy* 87, 83-90 (2007).

Lobo-Recio, M. A., Alguacil, F. J., Lopez-Delgado, A. Co-extraction and selective stripping or iron(III), HNO_3 and HF from stainless steel rinse waters. *AIChE Journal* 50, 1150-1155 (2004).

Sastre, A. M., .Kumar, A., Alguacil, F. J. Solvent extraction of silver by LIX 79: experimental equilibrium study. *Journal of Chemical Technology and Biotechnology* 79, 306-310 (2004).

Zhang, W., Liu, J., Ren, Z., Du, C., Ma, J. Solvent extraction of chromium (VI) with tri-n-butyl phosphate from aqueous acidic solutions. *Journal of Chemical Engineering Data* 52, 2220-2223 (2007).

Liquid Membranes

Alguacil, F.J., Lopez-Delgado, A., Alonso, M., Sastre. A.M. The phosphine oxides Cyanex 921 and Cyanex 923 as carriers for facilitated transport of chromium (VI)-chloride aqueous solutions. *Chemosphere* 57, 813-819 (2004).

Alguacil, F. J., Alonso. M. Description of transport mechanism during the elimination of copper (II) from wastewaters using supported liquid membranes and Acorga M5640 as carrier. *Environmental Science & Technology* 39, 2389-2393 (2005).

Alguacil, F. J., Alonso, M., Lopez, F., Lopez-Delgado, A. Uphill permeation of Cr(VI) using Hostarex A327 as ionophore by membrane-solvent extraction processing. *Chemosphere* 72, 684-689 (2008).

Barboiu, M., Cazacu, A., Michau, M., Caraballo, R., Arnal-Herault, C., Pasc-Banu, A. Functional organic-inorganic hybrid membranes. *Chemical Engineering and Processing* 47, 1044-1052 (2008).

Chiha, M., Samar, M. H., Hamdaoui, O. Extraction of Cr(VI) from sulphuric acid aqueous solutions by a liquid surfactant membrane (LSM). *Desalination* 194, 69-80 (2006).

Galan, B., Castañeda, D., Ortiz, I. Integration of ion exchange and non-dispersive solvent extraction processes for the separation and concentration of Cr(VI) from ground waters. *Journal of Hazardous Materials* 152, 795-804 (2008).

Güell, R., Antico, E., Salvado, V., Fontas, C. Efficient hollow fiber supported liquid membrane system for the removal and preconcentration of Cr(VI) trace levels. *Separation and Purification Technology* 62, 389-393 (2008).

Kulkarni, P. S., Kalyani, V., Mahajani, V. V. Removal of hexavalent chromium by membrane-based hybrid processes. *Industrial & Engineering Chemistry Research* 46, 8176-8182 (2007).

Kumar, A., Haddad, R., Alguacil, F. J., Sastre, A. M. Improved kinetics-based gold cyanide extraction with mixture of LIX 79 + TOPO utilizing hollow fiber membrane contactors. *Chemical Engineering Journal* 100, 11-22 (2004).

Kumar, A., Haddad, R., Alguacil, F. J.,Sastre A. M. Comparative performance of non-dispersive solvent extraction using a single module and the integrated membrane process with two hollow fiber contactors. *Journal of Membrane Science* 248, 1-14 (2005).

Kumbasar, R. A. Studies on extraction of chromium (VI) from acidic solutions containing various metal ions by emulsion liquid membrane using Alamine 336 as extractant. Journal of Membrane Science 325, 460-466 (2008).

Ren, Z. Q., Liu, J. T., Zhang, W. D., Du, C. S., Ma, J. N. Treatment of chromium (VI)-containing wastewater with hollow fiber renewal liquid membrane process. *Electroplating Finishing* 11, 20-23 (2006).

Samaniego, H., San Roman, M. F., Ortiz, I. Kinetics of zinc recovery from spent pickling effluents. *Industrial & Engineering Chemistry Research* 46, 907-912 (2007).

Sastre, A. M., Alguacil, F. J., Alonso, M., Lopez, F., Lopez-Delgado, A. On cadmium (II) membrane-based extraction using Cyanex 923 as carrier. *Solvent Extraction and Ion Exchange* 26, 192-207 (2008).

Ulbricht, M. Advanced functional polymer membranes. *Polymer* 47, 2217-2262 (2006).

Adsorption and Ion-Exchange

Ahmady-Asbchin, S., Andres, Y., Gerente, C., Le Cloirec, P. Biosorption of Cu(II) from aqueous solution by *Fucus Serratus*: Surface characterization and sorption mechanisms. *Bioresource Technology* 99, 6150-6155 (2008).

Alguacil, F. J., Alonso, M., Lozano, L. J. Chromium (III) recovery from waste acid solution by ion exchange processing using Amberlite IR-120 resin: batch and continuous ion exchange modelling. *Chemosphere* 57, 789-793 (2004).

Alguacil, F. J., Adeva, P., Alonso, M. Processing of residual gold (III) solutions via ion exchange. *Gold Bulletin* 38, 9-13 (2005).

Anirudhan, T. S., Suchithra, P. S. Synthesis and characterization of tannin-immobilized hydrotalcite as a potential adsorbent of heavy metal ions in effluent treatment. *Applied Clay Science* 42, 214-223 (2008).

Baral, S. S., Das, S. N., Rath, P. Hexavalent chromium removal from aqueous solutions by adsorption on treated sawdust. *Biochemical Engineering Journal* 31, 216-222 (2006).

Basha, S., Murthy, Z. V. P., Jha, B. Biosorption of hexavalent chromium by chemically modified seaweed, *Cystoseira indica*. *Chemical Engineering Journal* 137, 480-488 (2008).

Bhatnagar, A., Minocha, A. K., Pudasainee, D., Chung, H.-K., Kim, S.-H., Kim, H.-S., Lee, G., Min, B., Jeon, B.-H. Vanadium removal from water by waste metal sludge and cement immobilization. *Chemical Engineering Journal* 144, 197-204 (2008).

Bhattacharya, A. K., Naiya, T. K., Mandal, S. N., Das, S. K. Adsorption, kinetics and equilibrium studies on removal of Cr(VI) from aqueous solutions using different low-cost adsorbents. *Chemical Engineering Journal* 137, 529-541 (2008).

Copello, G. J., Varela, F., Martinez Vivot, R., Diaz, L. E. Immobilized chitosan as biosorbent for the removal of Cd(II), Cr(III) and Cr(VI) from aqucous solutions. *Bioresource Technology* 99, 6538-6544 (2008).

Cui, J., Zhang, L. Metallurgical recovery of metals from electronic waste: A review. *Journal of Hazardous Materials* 158, 228-256 (2008).

Das, S. K., Das, A. R., Guha, A. K. A study on the adsorption mechanism of mercury on *Aspergillus versicolor* biomass. *Environmental Science & Technology* 41, 8281-8287 (2007).

Debnath, S., Ghosh, U. C. Kinetics, isotherm and thermodynamics for Cr(III) and Cr(VI) adsorption from aqueous solutions by crystalline hydrous titanium oxide. *The Journal of Chemical Thermodynamics* 40, 67-77 (2008).

Franco, A. P., Lobo-Recio, M. A., Szpoganicz, B., Lopez-Delgado, A., Felcman, J., Mercê, A. L. R. Complexes of carboxymethylcellulose in water. Part 2. Co^{2+} and Al^{3+} remediation studies of wastewaters with Co^{2+}, Al^{3+}, Cu^{2+}, VO^{2+} and Mo^{6+}. *Hydrometallurgy* 87, 178-189 (2007).

Galan, B., Castañeda, D., Ortiz, I. Removal and recovery of Cr(VI) from polluted ground waters: A comparative study of ion-exchange technologies. *Water Research* 39, 4317-4324 (2005).

Gode, F., Moral, E. Column study on the adsorption of Cr(III) and Cr(VI) using Pumice, Yarikkaya brown coal, Chelex-100 and Lewatit MP 62. *Bioresource Technology* 99, 1081-1991 (2008).

Jeong, H. Y., Klaue, B., Blum, J. D., Hayes, K. F. Sorption of mercuric ion by synthetic nanocrystalline mackninawite (FeS). *Environmental Science & Technology* 41, 7699-7705 (2007).

Kabay, N., Solak, O., Arda, M., Topal, Ü., Yuksel, M., Trochimczuk, A., Streat, M. Packed column study of the sorption of hexavalent chromium by novel solvent impregnated resins containing Aliquat 336: Effect of chloride and sulfate ions. *Reactive and Functional Polymers* 64, 75-82 (2005).

Leyva-Ramos, R., Jacobo-Azuara, A., Diaz-Flores, P. E., Guerrero Coronado, R. M., Mendoza-Barron, J., Berber-Mendoza, M. S. Adsorption of chromium (VI) from an aqueous solution on a surfactant-modified zeolite. *Colloids and Surfaces A: Physicochemical and Engineering Aspects* 333, 35-41 (2008).

Martin, M. I., Lopez, F. A., Alguacil, F.J. Possibility of using by-products of the steel making industry for removing lead from aqueous solutions. *Revista de Metalurgia Madrid* 44, 258-264 (2008).

Memon, S., Tabakci, M., Roundhill D. M., Yilmaz, M. Synthesis and evaluation of the Cr(VI) extraction ability of amino/nitrile calyx[4]arenes immobilized onto a polymeric backbone. *Reactive and Functional Polymers* 66, 1342-1349 (2006).

Misaelides, P., Zambouilis, D., Sarridis, Pr., Warchol, J., Godelitsas, A. Chromium (VI) uptake by polyhexamethylene-guanidine-modified natural zeolitic materials. Microporous and Mesoporous Materials 108, 162-167 (2008).

Mureseanu, M., Reiss, A., Stefanescu, I., David, E., Parvulescu, V., Renard, G. Modified SBA-15 mesoporous silica for heavy metal ions remediation. *Chemosphere* 73, 1499-1504 (2008).

Navarro, P., Vargas, C., Alonso, M., Alguacil, F. J. Towards a more environmentally friendly process for gold: models on gold adsorption onto activated carbon from ammoniacal thiosulfate solutions. *Desalination* 211, 58-63 (2007).

Nishihama, S., Nishimura, G., Hirai, T., Komasawa, I. Separation and recovery of Cr(VI) from simulated plating waste using microcapsules containing quaternary ammonium salt extractant and phosphoric acid extractant. *Industrial & Engineering Chemistry Research* 43, 751-757 (2004).

O´Conell, D. W., Birkinshaw, C., O´Dwyer, T. F. Heavy metal adsorbents prepared from the modification of cellulose: A review. *Bioresource Technology* 99, 6709-6724 (2008).

Oliveira, D. Q. L., Gonçalves, M., Oliveira, L. C. A., Guilherme, L. R. G. Removal of As(V) and Cr(VI) from aqueous solutions using solid waste from leather industry. *Journal of Hazardous Materials* 151, 280-284 (2008).

Park, H.-Y., Tavlarides, L. L. Adsorption of chromium (VI) from aqueous solution using a imidazole functionalized adsorbent. *Industrial and Engineering Chemistry Research* 47, 3401-3409 (2008).

Raungsomboon, S., Chidthaisong, A., Bunnag, B., Inthorn, D., Harvey, N. H. Removal of lead (Pb^{2+}) by the Cyanobacterium *Gloeocapsa* sp. *Bioresource Technology* 99, 5650-5658 (2008).

Saha, B., Gill, R. J., Bailey, D. G., Kabay, N., Arda, M. Sorption of Cr(VI) from aqueous solution by Amberlite XAD-7 resin impregnated with Aliquat 336. *Reactive and Functional Polymers* 60, 223-244 (2004).

Sankararamakrishnan, N., Dixit, A., Iyengar, L., Sanghi, R. Removal of hexavalent chromium using a novel cross linked xanthated chitosan. *Bioresource Technology* 97, 2377-2382 (2007).

Schmidt, B., Wolters, R., Kaplin, J., Schneiker, T., Lobo-Recio, M. A., López, F. A., López-Delgado, A., Alguacil, F. J. Rinse water regeneration in stainless steel pickling. *Desalination* 211, 64-71 (2007).

Sirola, K., Laatikainen, M., Lahtinen, M., Paatero, E. Removal of copper and nickel from concentrated $ZnSO_4$ solutions with silica-supported chelating adsorbents. *Separation and Purification Technology* 64, 88-100 (2008).

Trochimczuk, A. W., Kabay, N., Arda, M., Streat, M. Stabilization of solvent impregnated resins (SIRs) by coating with water soluble polymers and chemical crosslinking. *Reactive and Functional Polymers* 59, 1-7 (2004).

Viamajala, S., Peyton, B. M., Gerlach, R., Sivaswamy, V., Apel, W. A., Petersen, J. N. Permeable reactive biobarriers for in situ Cr(VI) reduction: Bench scale tests using *Cellulomonas* sp. Strain ES6. *Biotechnology and Bioengineering* 101, 1150-1162 (2008).

Wu, Y., Ma, X., Feng, M., Liu, M. Behavior of chromium and arsenic on activate carbon. *Journal of Hazardous Materials* 159, 380-384 (2008).

In: Molecular and Supramolecular Bioinorganic Chemistry ISBN: 978-1-61728-785-5
Ed: A. L. R. Mercê, J. Felcman and M. A. L. Recio

Chapter 6

REVIEW: AGING AND NEURODEGENERATION – A BIOCHEMICAL APPROACH

***Luciana Dornnelas*[1], *Ana Lucia Ramalho Mercê*[≠2] *and Judith Felcman*[*1]**

[1] Pontifícia Universidade Católica do Rio de Janeiro – PUC – Rio, Rio de Janeiro – RJ – Brazil

[2] Universidade Federal do Paraná – UFPR, Curitiba – Paraná – Brazil

1. INTRODUCTION

As medicine and other fields of science are rapidly evolving, people are presented with myriads of medications, being constantly developed and improved. As a cause of these new ways of life and technology at hand, life expectancies have raised all over the world and is still increasing year after year.

With an aging population, concerns about neurodegenerative diseases are justifiable as the number of cases being reported are equally soaring.

Statistics foresee that the aging population, above 65 year-old, in 2025 will reach almost 1 billion souls and this number will double by 2050.

Aging normally brings with it more susceptibilities to neurodegenerative diseases as a normal decline in brain activity is expected. The synapses processes causing transmissions between two neurons or nerve-cells are gradually being diminished. The Alzheimer's disease (AD), for instance, flies at people above 65 year-old normally. An average of 1% of above 65 till 69 year-old population presents this neurodegenerative disease, but this percentage raises drastically to 45 to 50% to people above 69 year-old. [1,2]

Besides AD, other neurodegenerative diseases are unfortunately becoming very common like Parkinson disease (PD), Huntington disease (HD), amyotrophic lateral sclerosis (ALS) and the transmissible spongiform encephalopathies which are called Scrapie when developed

≠ Universidade Federal do Paraná – UFPR, CP 19081 – LEQ – DQ - Centro Politécnico, Jardim das Américas, 81531-980 – Curitiba – Paraná – Brazil, anamerce@ufpr.br

in ovines, and bovine spongiform encephalopathy when presented by bovines and popularly known as mad cow disease, and Creutzfeld-Jacob disease when developed by men. All these spongiform encephalitis variations are prion diseases.

Depending on the different neurodegenerative process in different areas of the central nervous system (CNS), a direct consequence is reflected in the way the symptoms are developed and on which brain region is affected. For this reason neurodegenerative diseases (ND) present an heterogeneous group of possible pathologies.

However, although being able to present many pathologic processes, the different ND present some factors in common like malformation of proteins and an imbalance in the metal ion concentrations which lead to free radical formation, oxidative stress which ends up in neural death.

Once damaged, the neural cell not only loses its function and dies but also it can no longer be replenished or replaced, since neural cells are not able to be reconstituted or substituted. Hence, that part of the brain affected most will not have its normal operation back and their specific functions are forever compromised. There is another aggravating factor, the process of degeneration of neural cells is progressive in all types of neurodegenerative diseases occurring sometimes with very high speed leading to death within four years after the onset of the disease as is the case of amyotrophic lateral sclerosis (ALS).

Several factors are related to neurodegeneration: destruction of neural cells, loss of synapses and neural death. Recent studies indicate that methionine is an amino acid that interferes a lot in oxidative stress [3-4] and homocysteine, when produced in excess reduces mental performance and increases the risk of Alzheimer's disease [5]. The combination of folate, vitamin E and acetylcarnitine provides a synergistic protection against oxidative stress [6] and B vitamins have been identified as the best protection against the rise in concentration of homocysteine [7].

Many investments are being made worldwide in order to obtain drugs to combat and / or delay the process of neurodegeneration. Drugs for prevention and that are capable of stopping these diseases are extremely important for health. As the brain is consisted of several segments of peptides and metal ions, the bioinorganic chemistry studies shed some light on the problems posed to a better understanding of brain functions and synapses to drive the development of drugs and dietary supplements which can slow these degenerative processes.

2. Relevance: Aging Population, a Global Concern

With life expectancy of the world population increasing every year, and as aging causes the neural functions to be reduced and synapses compromised, the brain is more susceptible to degeneration. Hence, knowing the changes that occur with aging is essential in order not to mistake the natural physiological processes with pathological ones [8], caused by neurodegenerative diseases. However, it is not entirely clear the way which the various organs, tissues and cells age, mainly because these processes do not occur evenly [9]. Research groups around the world are dedicated to the study of the various processes of aging.

* Pontifícia Universidade Católica do Rio de Janeiro – PUC – Rio, Rua Marquês de São Vicente, 225, Gávea, 22631-180 - Rio de Janeiro – RJ – Brazil, felcman@puc-rio.br

3. BIOCHEMISTRY AND PHYSIOLOGY OF AGING

3.1. Vulnerability in Neurodegenerative Diseases

Aging is characterized by a decline in organic functions that cause a number of changes throughout the body, leading to a reduction of functional capacity. The central nervous system (CNS) is greatly affected by the processes of aging that are characterized by morphological and histological changes, and in the neurotransmitting substances that lead to various modifications in brain physiology [10]. All the physiological changes related to aging are the consequence of shifts in biochemical processes associated with these systems. The biochemical shifts of the CNS lead to physiological changes in either the macroscopic and microscopic level.

3.1.1. Macroscopic Physiological Changes in the CNS

Important alterations in the brain are the changes in cerebral sulci and gyri that can also be found during the aging process [9-11].

The sulci suffer a narrowing, as the gyri become wider and deeper, with more pronounced atrophy in the frontal and temporal lobes, especially in the amygdala-hippocampal complex temporal lobe (role in learning and memory) and prefrontal cortex (motor area), and less intense in the occipital area [8-9].

The cerebral white matter (WM) suffers a reduction in the earlier parts of the corpus callosum (fibers inter-hemispheric frontal and temporal), around the age of 65 in some individuals. On computerized tomography (CT) scans areas of decreased density in the white matter are seen after 70 year-old that may be related to hypoperfusion [8-9].

Due to loss of nerve fibers of the white matter and loss of neuronal and cortical and subcortical loss, the ventricular enlargement evidenced is of the bilateral type, symmetrical and more pronounced in the anterior horn of the lateral ventricles. It was also observed an increase in the volume of fluid and changes in the meninges, especially in the arachnoid, where there is, after 60 year-old, a fibrous thickening in the parasagittal area. Also, due to the reduction in brain volume, both the subdural and subarachnoid space or cavity, swell. The cerebellum also suffer weight loss and atrophy of the three cortical layers. Finally, there is atrophy of the brainstem [8-9, 11].

3.1.2. Microscopic Physiological Alterations of CNS

Microscopically speaking, the physiological changes of the CNS arise from normal and gradual loss of nerve cells and/or from the gradual accumulation of chemical changes that result in disturbances of the function of specific chemical systems [12-14].

Large changes in the biochemistry of the brain with advancing age and those many physiological changes at the microscopic level that are also observed as neuronal atrophy, degeneration of granule - vascular appearance of lipofuscin and changes in neurotransmitter systems - are mainly due to an increase of neurotoxic free radicals that accumulate over the years. These changes are common with aging, normal process that is not always accompanied by neurodegenerative diseases, but leaving all the CNS more vulnerable to neurodegeneration. To a better understanding of why these microscopic physiological changes make the brain more susceptible to various degenerations, biochemical details of these processes are described below.

3.1.2.1. Neuronal Atrophy

The neuronal atrophy and cell death are among the main causes of nervous system disorders associated with aging [8] and are characterized by loss of white matter in the brain. The synthesis of mRNA in the aged brain decreases significantly, especially between 50 and 70 year-old, probably due to a change in the stability of the RNA molecule, thus reducing the protein synthesis [8,11]. The cytoplasmic RNA is reduced especially in the neurons of the frontal cortex, hippocampal gyrus, pyramidal cells of the hippocampus and Purkinje cells of the cerebellum. This reduction occurs with the accumulation of lipofuscin in nerve cells and reducing the size of the nucleus and NISSL body (large granular bodies of proteins that permeate dyes forming the cytoplasmic reticulum substance of nerve cells) and leading to simple neural or pigmentary atrophy [8,11].

The neural hypertrophy is marked by the retraction of cell body with reduced functional capacity, probably due to changes in its cytoarchitecture as well as transmission capacity [8-9]. The pyramidal neurons of the hippocampus are altered presenting basal dilations in the dendrites, with reduction or loss of dendritic spines, which are neurotransmitter synthesizers, and a decrease in synapses contact surfaces impairing learning and memory functions [9,11,13].

The decline of glucose metabolism, after 10-15 year-old is considered a precursor of irreversible morphological changes caused by the loss of dendrites and synapses as it is followed by a reduction of about 20% of medium sized nerve cells of the cerebral cortex [9]. The content of brain myelin begins to decrease from 20 years of age and, together with the above changes and others in the neurotransmitters, degeneration of myelin can lead to a decrease in afferent and efferent nerve conduction velocity, and difficulty of processing regions of the cerebral cortex where the speed is very important. The nerve conduction velocity is 10 to 15% slower in elderly [8,10,15].

3.1.2.2. Granulovacuolar Degeneration

The granulovacuolar degeneration is characterized by the presence of single or multiple vacuoles, probably formed after the process of autophagy (partial degradation of tau protein in the lysosomes). These vacuoles are found in the pericardium of the pyramidal cells of the hippocampus, subiculum and more rarely in the parahippocampal cortex, amygdala and substantia innominata. Granulovacuolar degeneration is scarcely found before the age of 65 and occurs in greater quantity in Alzheimer's disease. [16]

3.1.2.3. Lipofuscin Granules

The residual bodies or lipofuscin pigment granules appear as intracytoplasmic granules and accumulate in neurons, glial cells and capillary endothelium [9,11,13]. These pigments are formed by phagocytosis of cellular constituents (autophagocytosis) to be polymerized and peroxidized in secondary lysosomes. Some of these lysosomes will generate such pigments lipofuscin and accumulation occurs differently depending on region and age.

In the inferior and olivary nucleus and dentate nucleus, from the earliest years of life, lipofuscin is to be found. While in the motor nuclei of cranial nerves and large neurons in the pre-central gyrus, the nucleus of the thalamus, globus pallidus and red nucleus, only are to be found at about middle age [11]. The effect of this accumulation is not yet known and not necessarily associated with cell death, but could contribute to cell degeneration [9,11].

3.1.2.4. Neurotransmitter System

During aging there is a decrease in the production, release and metabolism of neurotransmitters. Furthermore, significant reductions occur in the concentrations of the secondary messengers and enzymes involved in signal transductions. Calcium regulates functions such as synthesis and release of neurotransmitters, neuronal excitability and protein phosphorylation. Among the enzymes which concentration is reduced with aging are those responsible for abducting the free radicals and regulate calcium homeostasis. In the aging brain there is a significant increase in intracellular calcium that can lead to cell death [8].

The neurotransmitter dopamine is the most affected with aging [8] and disturbances in dopaminergic activity are responsible for motor abnormalities present in advanced ages [9,14,17]. Other neurotransmitters such as gamma-aminobutyric acid (GABA), serotonin, catecholamines, acetylcholine and cholinergic receptors are also found in low concentration in the elderly. The central cholinergic function is reduced in aging due to the reduction of the acetylcholine's production, of the plasticity of muscarinic cholinergic receptors and the function of these receptors [9,17]. The synthesis of acetylcholine is affected by reducing the turnover of glucose, as the key substrate in its synthesis is acetylcoenzyme A which is synthesized in the brain solely by anaerobic glycolysis. Studies show that the cholinergic system in the hippocampus is located where major changes occur and that the reduction of acetylcholinesterase and choline acetyl-transferase is present in the cortex and caudate nucleus compared with cells of the putamen [10].

It is also observed an increase in the concentration of noradrenaline, 3-methoxy-4-hydroxyphenylglycol (MHPG, a metabolite of noradrenaline) and 5-hydroxyindole (5-HIAA, a metabolite of serotonin) in the cerebrospinal fluid of elderly [9].

With respect to noradrenaline there is a controversy because some studies say that there is a 40% reduction in the levels of noradrenaline in the brain of the 70 year-old [8] and that the loss of noradrenergic neurons may be responsible in part for deficits in locomotor and cognitive functions in the elderly [17].

Another controversial issue in the literature is regarding the serotonergic system, and although many authors reported serotonin levels to decrease with aging, other studies show no significant change of this substance [10].

There are still conflicting data regarding the levels of glutamate in the cerebral cortex of elderly subjects, because this substance has been found at high, reduced or unchanged levels [14].

Regarding GABA, it was found in a decreased or normal level. Studies show that the levels of GABA are decreased in the caudate and inferior olivary nucleus in the aging process [9].

3.1.3. Free Radicals and Aging

Free radicals are species that have unpaired electrons and therefore react easily with other molecules. They are considered the main causes of the aging processes and general decline in bodily functions. They are responsible for both physical and mental aging, and in the brain they act more intensively and leading precociously to problems from minor memory loss to as much as neurodegenerative diseases.

How much of the free radicals that are formed as a natural consequence of breathing or working as a byproduct of cellular energy production, the body naturally takes care of by producing antioxidants to fight them. However when there are imbalances in the action of these natural antioxidants, they are not enough to keep systems free of their biochemical oxidative action.

The brain is more affected by the action of free radicals because it is an organ rich in polyunsaturated acids and thus consumes a high amount of oxygen inhaled, thus generating free radicals formed not only by reactive oxygen species, but also by metal ions such as iron and copper that accumulate in this process.

With respect to energy production at the cellular level, free radicals are formed because the mitochondria, which are the specialized parts in producing energy, need oxygen for this process using about 20% of the oxygen consumed. Free radicals, as by-products, are also generated, as particular reactive oxygen species (ROS).

These ROS are generated not only as a consequence of cellular metabolism, but also by cytokines and oxidases (specific plasma membrane enzymes) [18].

The main targets of ROS include DNA, lipids, proteins and sugars, the preference order of attack depending on many factors, including the location where the reactive species is generated, the relative ability of a biomolecule to be oxidized and the availability of metal ions associated with this biomolecule [19]. The metal ions often potentialize the action of free radicals. The brain is highlighted as a target because free radicals bind preferentially to fat cells and as the brain and nerve cells are composed primarily of fats (especially cell membranes, which are the protection of cells) the brain region is strongly affected by the action of free radicals.

The action of free radicals can inhibit the production of neurotransmitters such as acetylcholine (involved in memory and learning), dopamine (balance and physical movement) and serotonin (regulates mood and appetite), among others. Disturbances in the concentration of neurotransmitters compromise all functions of the brain.

The excess of free radicals leads to the so-called "oxidative stress" that causes a defensive response of the immune system, which ends up by producing more free radicals which results in an inflammatory process which is another harm that free radicals can impose in a living body.

In recent years, the inflammation produced by free radicals, has been linked to virtually all chronic diseases of the brain, including Parkinson's disease, Alzheimer's disease, multiple sclerosis and dementia.

Antioxidants are substances that protect the body against the formation and attack by free radicals, some may be obtained from food or medication, but many of them are produced by the body. Among the antioxidants that the body produces stand out glutathione (important for brain and nervous system), the enzymes catalase and peroxidase, carnitine and coenzyme Q10 which is liposoluble and can penetrate the fat cells of the cell membrane providing protection against the onslaught of free radicals [7].

The blood antioxidant system is classified into enzymatic and non-enzymatic. The enzymatic is represented mainly by superoxide dismutase (SOD) which catalyzes the dismutation of radical superoxide anion (O_2*) to hydrogen peroxide (H_2O_2) and O_2 by catalase (CAT) which acts on the decomposition of H_2O_2 to O_2 and H_2O and by glutathione peroxidase (GPx), which acts on peroxides in general, using glutathione as a cofactor.

The non-enzymatic antioxidant system is composed of many substances, especially glutathione (GSH), the main intracellular antioxidant compound, tocopherol, ascorbate, uric acid and β-carotene, and transport proteins of transition metals, such as transferrin (transport of iron) and ceruloplasmin (copper transport and oxidation of iron to be captured by transferrin) [20].

With aging, the brain becomes more vulnerable to degeneration mainly because the production of antioxidants decreases and therefore paving the way to a common fact observed in all neurodegenerative diseases which is the increased production of free radicals and oxidative stress.

Studies show that the combination of folate, vitamin E and acetylcarnitine provides a synergistic protection against oxidative stress [21].

3.2. Imbalance in the Concentration of Metal Ions and Neurodegeneration

Several biological processes need the presence of metal ions [22-23]. In the brain, in normal concentrations, metal ions are essential to the functioning of synapses (transmission of signals between neurons). Transition metals such as Mn, Fe, Co, Ni, Cu and Zn are present in the cell cytoplasm of brain cells, each with specific functions and within the appropriate concentrations.

Zinc, for example, is a component of several metalloenzymes and it is not uniformly distributed in the brain [24] and it is assumed that it acts along with glutamate as a neurotransmitter. Over 90% of zinc present in the body is bound to metalloproteins and the largest group is the proteases (MMPs - metalloproteinases), for example, collagenases [25], that are necessary for changes in the matrix where the cells live and multiply.

Iron is present in key enzymes for the synthesis of neurotransmitters such as serotonin, dopamine and nitric oxide. Enzymes that require hydroxylases, for example, hydroxy-tyrosinase (for dopamine) and hydroxytriptophanase (for serotonin) require a non-heme iron, and oxygen. [26]

Copper has an important role in the brain because it is essential for some enzymes that play a central role in brain metabolism as CuZn superoxide dismutase, ceruloplasmin, cytochrome c oxidase, tyrosinase and dopamina–hydroxylase [27]. This metal is also needed in the oxidative processes controlling certain steps in the synthesis of cognitive tissue and in the production of certain messengers such as norepinephrine from the dopamine and peptides with amide groups (more than 50% of all peptides messengers are derived from amide groups, including endorphins).

However, the imbalance in the concentration of any one of these metal ions causes toxicity because it enhances the formation of free radicals known as extremely harmful and further leading to precipitation of toxic aggregates of protein fragments that contribute to the excessive formation of free radicals. Once more free radicals accumulate, the consequence is a reduction in the antioxidant defense capacity [28].

The body has biological redox agents such as cysteine, ascorbic acid, glutathione, methionine, homocysteine that assist in neuromodulation, provide substrates for the synthesis of anti-oxidants and protect against intracellular oxidative stress, mainly because they capture the metal ions. In normal concentrations they can maintain the redox balance, preventing the excessive deposition of free radicals of metal ions. However, with aging, due to a reduction or loss of cellular homeostasis there is a dramatic decrease in the levels of these biological redox agents that in combination with the imbalance in the concentration of metal ions such as iron and copper [26], which also occur with age, resulting in an enhancement in the formation of neurotoxic free radicals [29,30] which end up increasing oxidative stress and leading to cell death.

During the course of these damaging processes, metal ions are involved in depletion of cellular antioxidants, activation of phagocytes, mitochondrial dysfunction, activation of microsomal enzymes and tissue damage related to fibrogenesis and atherogenesis.

Free radicals and metal ions (Zn(II), Fe(II), Cu(II) and Al(III)) participate in the processes of Alzheimer's disease. These metal ions accumulate in the neuropil of the AD brain and are enriched at certain levels in the amyloid deposits. Aß, amyloid beta, the major component of these deposits, avidly binds these metal ions. The global oxidative stress observed in AD may well be focused at the amyloid deposits due to accumulation of Aß and the presence of redox-active metal ions. Oxidative events convert Aß deposits to amyloid plaques. If so, then antioxidants and metal chelators may be of therapeutic value. Complexing agents presenting redox capabilities are being studied as potential antioxidants. [30 and references cited therein].

Moreover, these metal ions in the form of free radicals can catalyze the oxidation of proteins, which is an exothermic process, further contributing to increased oxidative stress and loss of neural functions. The ease of oxidation in the brain channel proteins depends on the structure of the amino acid, being sulfur amino acids, like methionine and cysteine, easy and reversibly oxidizable in mild conditions to disulfides and sulfoxides, respectively [31].

For these disturbances not to happen there must be a balance among these redox systems. The redox balance in biological fluids, organelles, cells or tissues is determined by the presence of redox couples which are responsible for the flow of electrons and if any change in this redox balance occurs, can cause oxidative stress [32]. The intensity and pathogenicity of these imbalances will depend on the local concentrations of pro- and antioxidant species, on the rate constants for the reaction with target molecules and the cellular compartmentalization of processes in which solubility and diffusivity factors are determinants [33].

When production of reactive oxygen species or other free radical is generated, the body has an efficient antioxidant system that can manage and restore the balance. Oxidative stress results from an imbalance between pro- and antioxidant system [32,34] mostly as a result of an imbalance in metal ion concentration.

Studies [30] show that not only aging, but any neurodegenerative disorder substantially increases the concentration of biological redox agents and also extracellular concentrations of transition metals such as Fe and Cu, thus causing imbalance in the redox system. With an increase of free radicals, the antioxidant capacity of the brain system becomes less and less, leaving the brain more vulnerable to degeneration.

Besides presenting themselves as free radicals, metal ions have been associated with protein deposits characteristic of all neurodegenerative diseases, enhancing the formation of amyloid plaques that further contribute to an increase in the oxidative stress. In Alzheimer's disease the generation of reactive oxygen species has been identified as a result of the coordination of the ß -amyloid (Aß) peptide with metal ions [35]. NMR studies show that copper binds anomalously to ß-amyloid plaque of AD [36]. So, metal ions presenting redox activities such as copper, iron, zinc and manganese are considered as possible pathogens [36-37] involved not only in Alzheimer's, but in several other neural disorders.

Markers of oxidative stress, such as carbonyl protein, nitrotyrosine, lipid peroxidation and oxidized DNA bases are found in increased concentrations in tissues of patients and model animals with Alzheimer's disease, Parkinson's disease and Amyotrophic Lateral Sclerosis. [38,39]

3.3. Malformation of Proteins: Origin of Degeneration

The brain is rich in amino acids and metal ions, forming enzymatic and protein chains that constitute the complex neural systems. Imbalance in the concentration of metal ions, the presence of free radicals leading to oxidative stress and other factors may contribute to changes in the genes responsible for encoding certain proteins. Problems in protein synthesis lead to production of isoforms, different structural conformations from the original protein (Fig. 1).

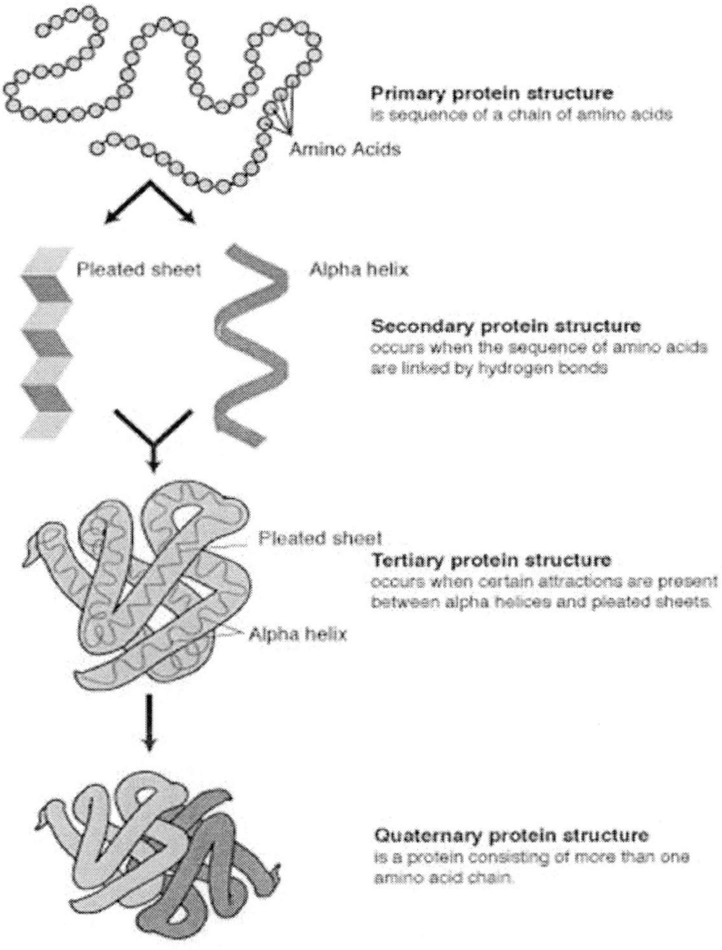

Figure 1 Primary to quaternary structure of a protein.
(http://upload.wikimedia.org/wikipedia/commons/a/a6/Protein-structure.png)

These modified proteins have been implicated as the origin of neurodegenerations because they are responsible for the formation of neurofibrillary tangles and insoluble proteic plaques that have been the main physiological changes found in the brain of patients with different neurodegenerative diseases.

The changes in proteins can occur in any of these levels (Fig. 1) from structural changes in the amino acid sequence to the three-dimensional structure or folding (bending) of the protein while the amino acid sequence remains unchanged. The metal ions may be responsible for mutations because protein can form internal cross-links by changing its structural conformation.

3.3.1. Protein Misfolding

All protein molecules are chains of amino acids, heterogeneous and unbranched. The folding is the process by which a protein assumes its functional configuration [40]. When bending and wrapping, the protein acquires a specific three-dimensional shape that makes it able to perform their biological functions [41]. The three-dimensional conformation of a protein is what gives it its specific function and is determined by the sequence of the amino acids in the protein [42].

Therefore, the structure of a protein and its functional properties are closely related and the opposite process is called denaturation of proteins, where an original protein is forced to lose its functional configuration, making it an amorphous and non-functional chain of amino acids. Denaturing occurs with the three-dimensional protein deformation without breaking the peptide bonds. In some cases, denaturation is reversible, and proteins can re-fold. However, the distortion is, in most cases, an irreversible process that can lead to inactivity, failure or conformational instability of the protein [43-45]. Denatured proteins can lose their solubility and precipitate. Thus, the denaturation of proteins is one of the causes of the formation of insoluble plaques.

Although denatured structures contribute to the formation of insoluble plaques, the protein isoforms associated with neurodegenerations structures are not amorphous structures, but rather proteins that somehow had undergone modifications. These changes compromise the folding of the proteins leading to incorrectly folded structures and therefore to different three-dimensional normal isoform, although with a well-defined structure, not at all amorphous. These proteins are called misfolded proteins and they are often used to describe the proteins associated with neurodegenerative disorders.

Sometimes the misfolded proteins are formed because changes in the amino acid sequence of proteins alter their folding pattern. Sometimes, even the amino acid sequence remaining intact, some changes may occur directly in the spatial structure of the protein also affecting its folding.

One way or another, when taking any other structure other than its normal isoform, the protein becomes inactive and can no longer perform its intended purpose, even forming insoluble plaques that may deposit and become pathogenic. When these plaques are formed in the brain they accumulate in the form of neurofibrillary tangles and insoluble plaques that are the main microscopic physiological alterations found in the brain of patients with different neurodegenerative diseases. And with that, the pathologies associated with different neurodegenerative diseases originate from the accumulation of misfolded proteins [46-47].

3.3.2. Neurofibrillary Tangles and Amyloid Plaques: Physiological Changes Associated with Neurodegenerative Diseases

It was seen in the above sections that some of the microscopic physiological changes that occur in the central nervous system are regarded as a natural consequence of aging and not always accompanied by neurodegenerative diseases. However, these physiological changes make the brain more vulnerable to the action of free radicals and other neurotoxic factors that can trigger the neural inactivation and neural death.

The brain of patients with neurodegenerative dementia presents some microscopic physiological alterations distinct and specifically associated with dementia processes which are the neurofibrillary tangles and deposits of protein plaques, the amyloid protein plaques. These neurofibrillary tangles and amyloid plaques are composed of misfolded proteins and have been described as the main responsible for pathological processes in all the neurodegenerative diseases as in depositing, they prevent the synapses and contribute to increased oxidative stress and neural death.

In small amounts they may occur diffusely in some brain regions (neocortex, caudate nucleus and putamen) of healthy elderly as normal consequence of aging, not producing damage in these concentrations. They occur more often after 60 years of age and mental capacity is preserved in 25 to 40% of individuals [11]. In Alzheimer's disease (AD) the density is higher in limbic areas and associative zones of neocortex [15].

3.3.2.1. Neurofibrillary Tangles

Neurofibrillary tangles are virtually absent in non-demented subjects [9,11,14]. Also called Alzheimer's neurofibrillary change and neuritic plaques, are characteristic of neurodegenerations. The main component of these tangles is an abnormally, phosphorylated form of an microtubule-associated protein - TAU protein [48-49]. Ubiquitin and ß-amyloid (also associated with microtubules) can be seen in lower concentrations in neurofibrillary tangles.

Neurofibrillary tangles are strongly associated with the neuropathology of Alzheimer's disease, where the phosphorylated forms of TAU protein are widely found. This process of phosphorylation of TAU will be discussed below.

3.3.2.2. Amyloid Plaques

Amyloids are molecules that have the ability to naturally form large insoluble aggregates of protein. They are formed by proteins that for some reason, had their structures modified. The plaques, also referred to as neuritic, dendritic or senile plaques, form extracellular deposits of protein plaques called amyloid fibrils. They also incorporate neuritic material (which are axons, dendrites and degenerated synapses) and appear as spherical structures located on the exterior of neuron [8,9], but are heterogeneous in the formation and composition.

Found in all neurodegenerative diseases, amyloids have been identified, more than the neurofibrils, as the cause of cell toxicity and origin of pathological symptoms, because they can cause the inactivation and death of neural cells when deposited among these last ones.

Unlike neurofibrillary tangles that are composed almost exclusively by phosphorylated Tau protein, amyloid deposits are formed by different proteins. For each neurodegenerative disease there is a specific protein involved in the formation of the amyloid substance.

The amyloid in Alzheimer's disease [50-55] are composed of a protein called amyloid (Aß42) that derives from the enzymatic cleavage of a larger protein called amyloid precursor protein (APP). The main physiological change found in the brain of patients is the presence of these amyloid plaques.

The Bovine Spongiform Encephalopathy (BSE), commonly called mad cow disease and its variants such as Creutzfeldt-Jakob disease, the human form, is associated with amyloid plaque deposits formed by the modified prion protein PrPSc (scrapie prion protein) that underwent changes in their structural configuration. [56-57]

The common aspect shown by the prion protein disorder (PrD) is the accumulation in the central nervous system of an abnormally folded, proteinase K resistant, isoform of the cellular prion protein (PrP^{Sc}), which is the post-translationally modified form of the normal cellular protein (PrP^{C}). The former is responsible for the formation of the insoluble amyloidogenic aggregates while the latter (PrP^{C}) is a soluble monomer. [58]

Modified proteins are therefore the causes of physiological changes that trigger all the pathological symptoms of neurodegenerative diseases.

3.4. Modified Proteins: Pathological Markers of Neurodegenerative Diseases

All neurodegenerative diseases are associated with deposits in the form of amyloids and neurofibrils besides the imbalance in the concentration of metal ions and oxidative stress generated by excess free radicals. The proteins that constitute the specific amyloid plaques in the main neurodegenerative diseases can be seen below and are pathological markers of each disease.

3.4.1. Tau and ß-Amyloid Proteins: Alzheimer's Disease (AD)

Alzheimer's disease (AD) is the most common form of dementia, being the fourth most frequent cause of death in developed countries [59]. The disease results in loss of neuronal function and synaptic damage with subsequent memory impairment, motor coordination and reasoning, besides loss of cognitive ability and dementia.

Among the most evident causes of the disease are the occurrence and extracellular deposition of amyloid plaques and the erratic formation of intracellular neurofibrils containing an abnormal, phosphorylated form of TAU protein, associated with microtubules [49,60].

The amyloid deposits in AD are characterized by extracellular deposition of protein in the fibrillar form [61], comprising a sequence of 42 amino acids [50] (Figure 2), the ß-amyloid protein of 4500 dalton [51,62], also called Aß1-42, ß/A4 [63] or still, ßAP.

H_2N-Asp-Ala-Glu-Phe-Arg-His-Asp-Ser-Gly-Tyr-Glu-Val-His-His-Gly-Lys-
Leu-Val-Phe-Phe-Ala-Glu-Asp-Val-Gly-Ser-Asn-Lys-Gly-Ala-Ile-Ile-Gly-
Leu-Met-Val-Gly-Gly-Val-Val-Ile-Ala-COOH

Figure 2. Protein aminoacids of Aß1-42, prevalent in Alzheimer's disease.

The fragment Aß1-40, generated by cleavage of ß- and γ-secretase comprises 672-711 residues of APP 770, while the Aß1-42 fragment comprises 672-713 fragments.

Reference [64] shows important figures for the description below. One of them shows the APP 770 isoform being fragmented. Another [64] shows the protein precursor APP 770 proteolysis, giving rise to fragment Aß1-42. Other, [64] shows an scheme of the membrane which seizures the cleavage sites of APP770, showing how the ß-amyloid precursor (APP) is divided into 3 regions, where the first consists of a long extracellular region (N-terminus), the second is the region that crosses the cell membrane and the third region is intracellular (the C-terminal). The fragment Aß1-40 (soluble) is generated by cleavage of ß- and γ-secretase and comprises 672-711 residues of APP 770, while the Aß1-42 (insoluble) fragment is generated by cleavage of ß- and γ-secretase, comprising 672-713 residues. The first 28 residues of ß-amyloid match the extracellular region of APP, while the last 15 (700-714) correspond to the transmembrane domain part, comprising the 700-723 amino acids. It is also found, to a lesser extent, the formation of intracellular neurofibril containing an abnormal phosphorylated form, a protein associated with microtubules – the TAU protein [48-49].

The neurofibrillary tangles are the result of intracellular changes that occur in the neuron cytoplasm due to hyperfosforilation of TAU protein (microtubule stabilizer) and correspond to the loss of the normal cytoskeleton of microtubules and neurofilaments [65]. TAU protein [66] is synthesized from a single gene on chromosome 17. It is known six human isoforms of this protein and all of them are also present in the intraneuronal neurofibrillary tangles and presenting a state of abnormal phosphorylation [67]. This abnormal phosphorylation changes the normally performed functions of the TAU protein [68] like participation in the assembly and stabilization of microtubules [69]. The destabilization of microtubules could lead to the complete dissolution of the cytoskeleton in the affected neurons by neurofibrillary tangles [66], contributing to neuronal loss that are observed in the later stages of the disease [70-71].

Postmortem studies (51,72-76) show that in all cases of Alzheimer's disease there is the coincidence of the presence of the toxic protein-amyloid, and neurofibrillary tangles and of metal ions with redox properties and imbalances in trace elements such as aluminum, lead, mercury, cadmium, copper, iron and zinc in the brain of patients.

In addition to pathological proteins and metal ion concentration imbalance, it is important to note that Alzheimer's patients show a general reduction in the concentration of all neurotransmitter substances and nerve tissue and brain constituents. It is also worth noting the changes in metallothionein III (MT-III) and acetylcholine levels. [77-79]

The MT-III is one metallothioneins found in the brain, having neuronal inhibitory activity and is involved in the recovery of neuronal injury. Studies show that the concentration of MT-III is decreased in AD [80].

Acetylcholine is a substance responsible for transmitting impulses between nerve and muscle cells. It is believed that the clinical features of AD are also related to a decrease in the levels of acetylcholine by the fact that patients have low levels of neurochemical response and reduced transmission of stimuli and messages. Acetylcholinesterase inhibitors are being tested in Alzheimer's disease and some results shown improvements in cognitive functions and dementia [81-82].

3.4.2. Prion Protein Prp Sc: Transmissible Spongiform Encephalopathies (TSEs)

The prion diseases are a related group of fatal neurodegenerative disorders called Transmissible Spongiform Encephalopathies (TSEs) or prion diseases, which affect humans and other mammals. As already mentioned above, in cattle is called Bovine Spongiform

Encephalopathy (BSE), commonly known as mad cow disease; in sheep and goats is called scrapie, and humans came to be known as Creutzfeld-Jacob disease.

The Creutzfeldt-Jakob disease presents loss of brain function, clinical and spongiform neuropathology, which is similar to Alzheimer's disease. However the progress of TSEs is much faster, causing dementia and total death in a period of up to six months.

In references [83–84] figures of the spongiform pathology of the brain in Alzheimer's disease and in another with Creutzfeld-Jakob disease can be seen.

All these variations of transmissible spongiform encephalopathies (TSE) are also called prion diseases or prion. The word "prion" is the acronym for proteinaceous and infectious [85]: it is an infectious protein that does not encodes the information by a nucleic acid [86] and is a protein constitutively expressed in various cell types, particularly in neurons where it can accumulate in pre- and postsynaptic membranes [86-87]. Prions are proteins capable of modifying other proteins, making copies of proteins that are parts of themselves. Transmissible spongiform encephalopathies are neurodegenerative disorders associated with transmissive malformation of these prion proteins.

These diseases are caused by an autoreplicative protein, the prion protein PrPsc (prion scrapie). Protein PrPSc is the pathogenic form of PrPc, cellular prion protein [87] which had its structure changed. PrPSc is a misfolded prion protein. The cellular prion protein PrPc is the product of a single gene, called Prnp, that has 252 (in some species) [88] amino acid residues and is present in all body cells, from nervous tissue to muscles.

There is another variant of the human form also called Creutzfeld-Jakob disease which is hereditary and progressive in which the prion protein presents a change in the amino acid sequence of methionine/valine, in fact replaced with methionine/methionine (Figure 3).

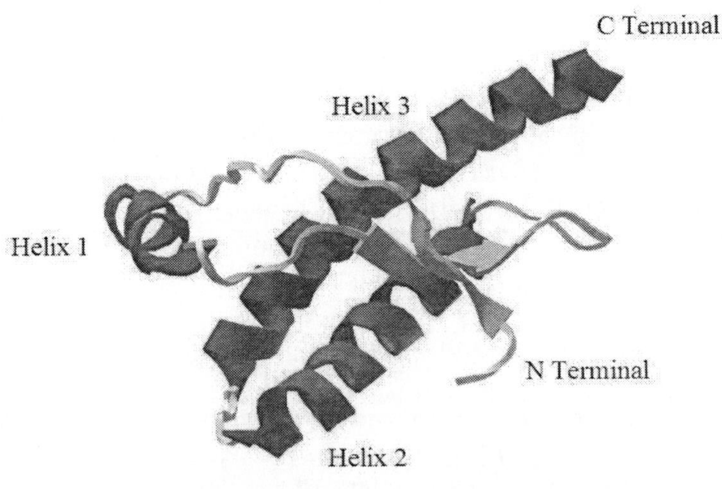

Figure 3. Structure of normal prion protein (PrP) in the hereditary Creutzfeld-Jakob disease. (http://en.wikipedia.org/wiki/File:PrPwildtypestructure.JPG) In reference [84] a figure of a pathogenic prion protein can be seen.

In all cases, metal ions which have redox activity play an active role causing abnormal oxidative processes in prions [88].

3.4.3. Biochemical Conversion Process of PrPc to PrPSc

Prions have autoreplicative nature, which easily explains how they can agglomerate forming plaques. The difference between the normal isoform PrPc and the pathogenic PrPSc, is only one malformation in secondary three-dimensional structure. The primary sequence of the two forms is the same, indicating that the conversion of the normal (which does not cause disease) to the infectious form of the cell is not caused by a covalent modification in the protein. In fact what is observed is the conversion of a structure rich in alpha-helices, in PrPC, into another structure that has higher content of beta sheets, PrPSc. Figure 4 shows the difference between these two isoforms. [89-90]

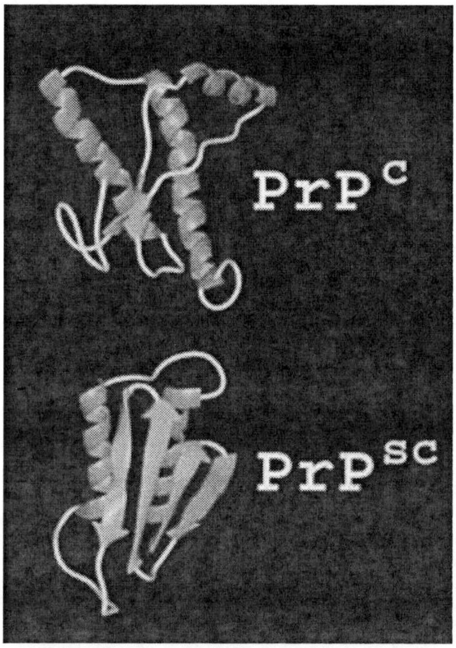

Figure 4. Normal (PrPc) and modified isoform (PrPSc) of prion protein. (http://4.bp.blogspot.com/_dt_tYyrEN_U/SGZzCE0jVMI/AAAAAAAAACM/BFVs2z6dLnE/s1600-h/prion.jpg)

The secondary structures of the two isoforms differ markedly, being the PrPc composed predominantly of α-helix (42%) and presenting itself in a cell cluster form, while the concentration of PrPSc is higher in regions containing β-pleated sheets (43%) which is the main characteristic of infectious molecules and being present in the amyloid form [91].

With the aid of synthetic peptides, it was possible to assess that the α-helices could be converted into β-pleated structures [92] and that they could induce conformational changes in peptides having helicoidal structure [93]. In addition, it was shown the conversion of PrPc into PrPSc *in vitro* [94], being the resulting protein from such conversion, one rich in 7 β-pleated structures, which would explain the deposition of toxic insoluble aggregates in the brain tissue [95].

The conversion of PrPC in its abnormal isoform PrPSc, is through a process in which a portion of α-helix structure is enovelated again in sheets. The conversion of PrPC into PrPSc undergoes through a self-catalytic mechanism. There are many grey areas on the kinetics of prion replication as well as the mechanism of toxicity associated with prions, but it is already known that self-catalysis is the most important mechanism in the prion kinetics, even if the spontaneous conversion of PrPC into PrPSc is not yet fully explained. Through a still not clear biological mechanism, once the abnormal isoform PrPSc is formed, it can promote the conversion of other normal PrPC to PrPSc form. The defective isoform, PrPSc prion, acts as a sort of template for the formation of new abnormal prion proteins by when in touch with the normal isoform PrPC in the cell, causes it to transform into the PrPSc form, which is pathogenic.

For a better understanding of the mechanism of propagation of prions, various mathematical models have been proposed and all were found to be the self-catalytic term (Michaelis-Menten term), the most important in the kinetic equations of prion. With respect to toxicity from the DLA (diffusion-limited aggregation) model it has been reinforced the hypothesis that for BSE (bovine spongiform encephalopathy) and vCJD (variant Creutzfeldt-Jakob disease) and naturally occuring prion diseases, toxicity is caused mainly by the formation of the amyloid plaque.

Only the protein PrPsc is amyloid type. That is to say, it has the ability to naturally form large insoluble aggregates of protein. In contrast, a normal cellular PrPC is globular and highly soluble. Protein PrPC is characterized as sensitive to proteinase K digestion while PrPSC is not.

3.4.3. Protein Alpha-Synuclein and Parkin: Parkinson's Disease (PD)

The Parkinsonism is characterized by a progressive disorder of movement due to dysfunction or death of neurons mainly those responsible for secreting dopamine, an essential substance for performance and control of movement. It attacks the brain region which is responsible for controlling the movements, causing muscle stiffness, tremors, slowness of movement and posture instability [96].

The proteins alpha-synuclein and parkin are directly related to Parkinson's disease. Defects in the formation of these proteins lead to the accumulation of the so-called Lewy bodies (named after the pathologist who first observed them), visible under a microscope.

The genetic form of Parkinson's disease is more severe and occurs by some defect in α-nuclein or the parkin genes. There are two distinct familiar forms of the disease that can occur together. The sporadic form of the disease (not genetic) is also related to subtle defects in the enzymes involved in the degradation of α-nuclein and / or parkin.

The gene for alpha-synuclein (PARK1) was first identified in Parkinson's disease [97]. It encodes a protein with only 144 amino acids which acts in the signaling between neurons. The pathological forms of this protein present tiny changes in the protein amino acid sequence. Several of these mutations are still not known, and the two most common results of these mutations result from a single amino acid in the sequence.

A second gene can mutate into another form of familiar Parkinson's disease, which encodes the protein parkin (PARK2) [98]. Parkin mutations appear to be more common than the alpha-synuclein. The parkin contains several amino acid sequences (or domains) that are common to many other proteins. One example is the so-called RING domains, which are involved in the system of protein degradation. New findings suggest that neuronal death in this form of Parkinson's disease stems in part from the failure of the component of the ubiquitinating system for protein degradation: parkin attaches to ubiquitin in the misfolded proteins, and with no ubiquitin, tagging and degradation are not possible. This mutation appears more frequently in individuals diagnosed before age 40. The earlier the disease appears more likely it is to be caused by a mutation in parkin and not in alpha-synuclein. [98]

The occurrence of Parkinson's disease is not only associated to pathological proteins but also related to the presence of metal ions and free radicals produced as a result of normal metabolic reactions. When in excess, both metal ions and free radicals can cause brain damage through oxidative processes and / or contribute to the protein disorders.

3.4.4. Huntingtin Protein: Huntington's Disease (HD)

The Huntington's disease is hereditary and rare. It affects people with about 40 years of age causing seizures and mental strain that kills after ten years. The cause can be the mutation of a gene localized in chromosome four region. [99]

In patients with Huntington's disease, in the chromosome four region, where the defective gene has been observed, there are more than 36 repetitions of the nucleotide sequence CAG (cytosine, adenine and guanine). In normal individuals in the same place, CAG sequence is usually less than 20. The nucleotides are like building blocks of DNA, the CAG is responsible for encoding glutamine.

The protein encoded by this defective gene is called huntingtin and has a sequence with abnormally high repetition of the amino acid glutamine. It can be found in every cell of the body [100].

3.5. Biochemistry of Protein Aggregates

One way to try to slow or stop the neurodegenerations is the knowledge of all biochemical processes that lead to protein deposits that impair synapses and increase oxidative stress. Thus, as important as knowing the biochemical processes that lead to formation of misfolded proteic isoforms is to understand how they proliferate generating the large deposits that become pathological. Aggregation is easily explained when speaking of prion proteins, due to the autoreplicative nature of prions, as previously described. Through a self-catalytic mechanism the defective prion isoform PrPSc acts as a kind of template for the formation of new abnormal prion protein. Once it is formed PrPSc, when in contact with the normal isoform PrPC in the cell causes it to get the PRPSc form, which is pathogenic.

For all other proteins that do not behave as prions, a useful model for understanding how the aggregations occur was found in serpins, which are serine protease inhibitors.

3.5.1. Serpins: A Useful Model for Understanding Aggregation of Proteins

The terms protease, proteinase and peptidase refer to a group of enzymes that break peptide bonds between amino acids, constituents of proteins. These enzymes act catalyzing the hydrolysis of peptide bonds. Excess protease leads to tissue destruction.

In serine proteases (the break of a peptide bond by one serine), the mechanism is based on a nucleophilic attack of peptide bonds by a serine (Ser), cysteine (Cys), threonine (Thr), water molecules associated with aspartate (Asp) or metals that can also play this role. In many cases the nucleophilic property of the group is enhanced by the presence of a histidine (His), kept in a "proton acceptor state" by an aspartate. The side chains of residues such as serine, histidine and aspartate form a catalytic triad commonly found in serine proteases. Among the serine proteases investigated further include the digestive enzymes chymotrypsin and trypsin, both synthesized and secreted by the exocrine pancreas in the inactive form (zymogen or zymogens). These inactive forms are chymotrypsinogen and trypsinogen and are activated in the duodenum. The serpins are proteins that act as inhibitors of serine proteases and belong to the serpin superfamily and can be found in several thermophilic organisms such as bacteria and mammals [101]. The serpins (**ser**in **p**rotease **in**hibitors) are described as the best model to explain the aggregation of non prion proteins.

The basic element of the structure of a serpin is composed of three β-sheets (A-C) and an exposed mobile reactive centre loop which has a peptide sequence that serves as a pseudo substrate for the protease target [102-103]. The critical amino acids within this loop act as a "hook" for the enzyme [104]. This happens because the reactive center of serpins is exposed and recognized by the target proteinase as a normal substrate to be cleaved. This is exactly the molecular mechanism of inhibition of serine proteases by serpins. Once the enzyme is arrested a serpin peptide bond P1-P1 is broken [105] and the protease is inactivated for a trap action-type, which makes it rotate from the upper pole to the bottom of the protein in association with insertion of the reactive loop as an extra line in the β-sheet A [106-110]. What follows is that the reactive center of the target protease is deformed resulting in irreversible inhibition [111-113].

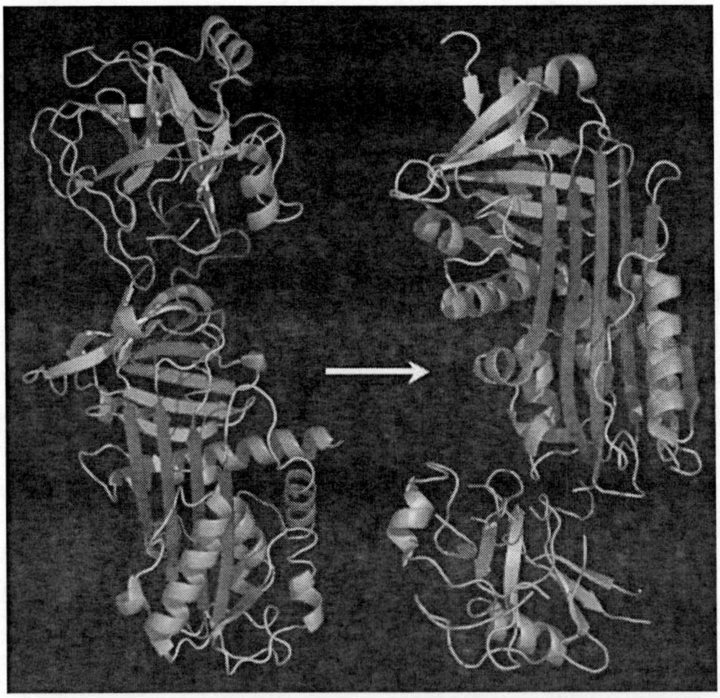

A) B)

Figure 5. Serine protease inhibitors alpha 1-antitrypsin (AAT) inhibiting elastase: the critical amino acids within the loop of AAT are the residues P1-P1 methionine-serine. They act as a "hook" for neutrophil elastase (31). AAT acts as a trap [110, 114-115]. Shortly after this arresting, elastase (Figure 5 A) of neutrophils (gray) is inactivated by a motion from the upper pole to the bottom of the protein (Figure 5 B). This is associated with the insertion of the reactive loop (red) as an extra line in the β-sheet A (green). (http://upload.wikimedia.org/wikipedia/commons/3/31/Mich_ezx.jpg)

Figure 5 shows an example of the serine protease alpha 1-antitrypsin (AAT) inhibiting elastase. One can see the AAT serine inhibitor in their normal protein form before inhibiting the enzyme (Figure 5 A) and the profound conformational transition it goes through after doing that (Figure 5 B).

The astonishing action of serpin trap is central to the effective inhibitor role of serine proteases. On the other hand, it is also its critical point since mutations in these mobile domains make the molecule vulnerable to conformational transitions as explained in Fig. 1.

Assuming another conformational form, the protein loses its specific function as well as becomes inactive forming insoluble deposits that accumulate. The AAT (alpha 1-antitrypsin) has undergone a type Z mutation, which occurs at residue P17 of serpin (17 residues near the active center) on the main line of the 5 β-sheet A and the base of mobile reactive centre loop. Undergoing mutation, the β-sheet A of the enzyme opens, thus favoring the insertion of the reactive loop of a second serpin molecule (Figure 5) leading to the formation of a dimer [114, 116-118]. It begins to repeat itself and soon, polymers that are formed, are trapped in the endoplasmic reticulum of hepatocytes, forming the inclusion bodies. Thus, the protein that should only have the inhibitory action of serine protease, while taking the new form Z, loses its specific function and polymer chains start to being formed as insoluble protein plaques, thus accumulating and becoming pathogenic.

The route of polymerization of AAT was determined by biochemical, biophysical and crystallographic analysis and seen in Figure 6 [117-119].

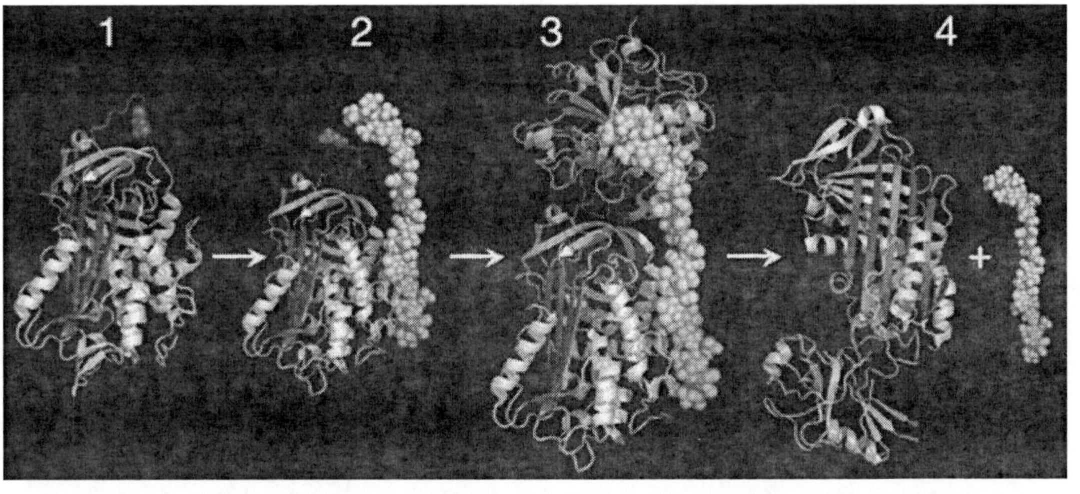

Figure 6: The structure of AAT focuses on β-sheet A and the loop that is the mobile active center (red). The formation of the polymer is produced from the Z variant AAT (Glu342Lis in P17-arrow) or mutations in "shutter" domain (blue circle) Siiyama (serine 53 (TCC)), Mmalton, S or I that open up the β-sheet partially favoring a partial insertion of the loop (step 1) and the formation of an unstable intermediate M*. This structure can then later accept the loop of another molecule (step 2) to form a D-dimer that soon extends to form a polymer P26, 52, or 54. It can also accept his own loop (step 3) to form a latent conformation L [125-126]. The individual molecules of AAT within the polymer are colored in red, yellow and blue. (http://upload.wikimedia.org/wikipedia/en/2/20/At3sequence2.jpg)

Step 1 represents the conformational change of AAT to a monomeric polymerogenic form (M*). Step 2 represents the formation of polymers (P) and step 3 represents a side way leading to the formation of a stable monomeric latent conformation (L). The Z mutation is the one that causes the most unstable part of the protein to form polymers. The presence of unstable intermediate polymerizing M* was predicted based upon a biophysical analysis of polymer formation [117], the demonstration of an unfolded intermediate [120-122] and solving the crystal structure of a mutant alpha 1-antichymotrypsin [119]. Recent studies suggest that the Z mutation forces the AAT into a conformation that approximates the unstable M* favoring the formation of polymers [123].

When incubated under physiological conditions the same formation of polymer chains is observed [116]. The rate of formation of the polymer is accelerated by increasing the temperature to 41° C and that was only blocked by peptides that competed for joining the β-sheet A [116-124].

However, this process can be repeated over and over again [127,128] and when this occurs the results are the formation of polymeric structures that are maintained in the cell leading to cell death and tissue damage [129], which can lead to diseases.

The aggregation of neuroserpins results in familial encephalopathy with neuroserpin inclusion bodies (familial encephalopathy with neuroserpin inclusion bodies-FENIB).

Familial encephalopathy with neuroserpin inclusion bodies is a mutation in the gene that encodes neuroserpin. It is a progressive dementia where a histological level of accumulation of cytoplasmic inclusions in the deeper cortical layers, substantia nigra and in the subcortical nuclei is observed. In FENIB, there is a direct relationship between the magnitude of the intracellular accumulation of neuroserpin and disease severity [130]. Although the aggregation of intracellular proteins alone is sufficient to cause neurodegeneration [130] the role of post-translational degradation of neuroserpin polymers in the development of neurotoxicity has yet to be explained [129].

4. Trends in Neurological Diseases Theories and Future Treatments and Design of Drugs

4.1. Old Hypothesis, New Theories

Dating back to 1987 [131], the literature has shown results supporting the hypothesis that glutamate disfunction was involved in the pathophysiology of Alzheimer's disease and was accounted for many of the neurochemical and behavioral deficits observed in this disorder. At that time, it was already known that this kind of dementia presented brain changes like senile plaques, neurofibrillary tangles, and granulovacuolar degeneration in cerebral cortex, amygdala, olfactory tubercle and hippocampus. Back in 1987, the hypothesis of abnormal levels of aluminum in the blood was already linked to the ethiology of this disorder.

It was also posed in this article [131] that although pathology of the ascending cholinergic system was undoubtedly present in Alzheimer's disease other neurochemical systems seem to be directly involved in the pathophysiology of this disease. Glutamate binding is also reduced in the subiculum of Alzheimer's brains. It is likely that glutamatergic neurotransmission in cortex and hippocampal formation is severely disrupted. Evidences at that time suggested that glutamate neurotoxicity can produce biochemical changes like those seen in this kind of disease.

N-methyl-D-aspartic acid (NMDA) applied to the cortex produced retrograde degeneration of cholinergic neurons in the nucleous basalis of rats suggesting that cortical glutamatergic neurotoxicity might be responsible for the subcortical atrophy previously observed in Alzheimer's disease. All data presented in this work [131], according to the authors, supported a possible connection between glutamate and the pathology of Alzheimer's disorder.

4.2. Metal Ions – Especially Copper and Zinc - and the Role in Neurodegenerative Diseases

Metal ions also play a role in the enzymatic biochemistry of the brain, like enzymes involved in the antioxidant processes (superoxidase dismutase – SOD), cellular respiration (cytochrome c oxidase), catecholamine synthesis (dopamine-ß-hydroxylase) among many others involved in growth , development and maintenance of the nervous system [132].

Zinc and copper bind to an aminoacid in order to alter protein function. Like zinc, copper accumulates in the brain. These metal ions are bound to proteins being some of them left as an extracellular metal concentration. In conditions during neurodegenerative diseases these external metal concentrations are present in 1000 to 2000 fold the normal concentration [132].

Synaptic vesicles, glutaminergic or adrenergic neurons, mainly in the hippocampus, olfactory bulb and locus coeruleous are the regions where complexed copper ions are mainly found. Like copper, zinc is essential for the correct development and functioning of the brain.

Current data [132 and references therein] indicate that ATP7A is responsible for the overall supply of copper to the brain due to its presence in the choroid plexus. Also ATP7B mediates the movement of copper to vesicles followed by expulsion of excess copper from cells, being these two ATPases the copper transporters in the central nervous system (CNS).

In the adult central nervous system (CNS), SOD1 (a cupro-enzyme) is notably expressed in the neurons of a) hippocampal (pyramidal ones); b) substantia nigra; c) dentate gyrus; d) pyramidal cortical cells and e) the motor neurons of spiral cord, all those neurons that are particularly vulnerable to oxidative damage [132]. In order to perform its antioxidant functions, SOD needs a certain level of zinc in the brain which quantity, when not enough, makes SOD function abnormally, acquiring peroxidase activity and thus generating peroxy nitrite and leading to neuronal death [132-135].

Continuing following the studies reported in the literature [132-134] it is also found that there are high concentrations of copper and zinc within the core and peripheral regions of amyloid plaques. Also Aß-metal interactions have an intimate biochemical relationship, revealing the importance of brain metal homeostasis as a potential factor in the evolution of Alzheimer disease pathology. According to the role of metal ions, (zinc, manganese, iron and copper for example), it is not only that of being involved in reactive oxygen species (ROS) [135-137] but also when neurodegeneration is at stake, it comprises the direct binding, of these metal ions, to the concerned proteins. Copper seems to be the most strongly connected metal to the biology of prions. This metal has also a primary role in the toxicity of Aß peptide action, and zinc seems to play a key role in aggregation of amyloid deposits [132].

A recent review [135] describes very well the delicate balance between the uptake and efflux of copper ions in and out the cell. Although only SOD1 is a cupro-enzyme, the amyloid precursor protein and its Aß fragments, the prion protein and α- synuclein, all have potentially binding sites to house Cu(II) ions. This review also describes several experimental evidences connecting copper to Alzheimer's disease [135]. SOD mutations can alter the protein folding and lead to a release of copper and zinc which then manifest toxic effects [58]. In the prion protein disease (PrD) several findings indicate that copper enhances the refolding of PrP^{Sc} into a protease-resistant, infectious form [58].

Another study in the literature [138] provided evidence that trace amounts of copper can induce neurotoxicity by interaction with elevated cholesterol in diet. Another [139] showed abnormal interaction of ß-amyloid 42 (Aß 42) with metal ions (copper, zinc and iron) inducing peptide aggregation and oxidation in Alzheimer's disease. Also in the absence of neurological diseases, Aß is degradated among others by extracellular metalloproteinases and insulin degrading enzyme (IDE). This study also reports investigation of the relationship between levels of Aß and biological metals in cerebrospinal fluid [139].

New trends in neurodegenerative diseases research have pursued answers to the question whether metabolites of the plaque (oligomers and peptides of the amyloid plaque) play a role in Alzheimer's disease. Since 1987, when the study of Maragos *et al*, [131] was released, some researchers have published some results indicating that "yes" could be a true answer.

It is accepted that metabolites of neural cells are known to have a significant effect on the normal physiology of neurons in brain. In 2009 [140] researchers have found experimental evidences enough to say that "metabolites of cerebellar neurons induced expression of Aß degrading enzymes and promoted Aß clearance in hippocampal neurons". Meaning that those metabolites of neural cells play a role in the pathogenesis of neurodegenerative diseases according to the authors [140].

In fact, in the first place, metabolites of cerebellar neurons significantly reduced brain Aß levels and reversed cognitive impairment in different stages of Alzheimer's disease. In the second place, metabolites of hippocampal neurons reduced the expression of Aß-degrading enzymes in the cerebellum thus causing cerebellar neurodegeneration according to results *in vivo*. The role of neurons metabolites is thus either required for maintaining the normal neuronal physiology and also promoting some pathogenical processes of Alzheimer's disease contributing for the regional specificity of Aß deposition [140].

Reports in the literature [141,142] have shown some successful treatments of Alzheimer's disease with drugs challenging the main accepted Alzheimer's theory. The results of some reported work showed an increase in the levels of ß-amyloid and at the same time improved, in phase II clinical trials [142], cognition and memory in Alzheimer's patients. Those results can lead to an aternative hypothesis that other forms of amyloid (like some metabolites, amyloid oligomers and peptides) may actually act as neurotoxins and be more toxic than insoluble plaques [141-142]. These amyloid oligomers have found recent experimental grounds to be thought of being neurotoxic and play a role in the development of Alzheimer's disease. In a recent work [143] not only the extracellular amyloid plaque is considered as a central pathological hallmark of Alzheimer's disease but also the formation of neurotoxic, soluble Aß oligomers which in turn interacts with copper further promoting the aggregation of Aß, the plaques. Copper dysregulation is also implicated in the hyperphosphorilation and aggregation of *Tau,* the main component of neurofibrillary tangles, also another hallmark of Alzheimer's disease.

Some researchers presented studies on the impact between the physiological signaling effects of soluble low molcular weight amyloid – ß species and the more global cellular damage that could derive from highly concentrated aggregated amyloid [144] and the study of

Aß peptides and oligomers in the mechanisms of cellular disfunction, aggregation and maladaptative responses [145]. In this last study [145] $Aß_{25\text{-}35}$ was used to provide data in cellular dysfunction studies. The results showed that the peptides are responsible for multiple disturbances, like participating in oxidative stress and causing deleterious effects in multiple enzymes.

In that direction studies are needed to be developed considering metal dyshomeostasis, Aß, *Tau* and metals to restore copper and other metal homeostasis in the treatment of neurological disorders. Researchers employing drugs that either reestablish the communication among neurons are targeting these news aspects concerning the development of neurodegenerative disorders. Either by chance [142] or by utilizing a combination of therapeutic agents, like insulin and another one [146], the reported results achieved an amelioration of some of Alzheimer's symptoms. Insulin was chosen in one of these works [146] because it was already found out that this molecule, apart from contributing for the obtention of energy in order to make the brain work, insulin also played a role *in vitro* in the memory neurological processes.

Concluding Remarks

To conclude on the results published so far concerning some ways of treating and understanding neurodegenerative disorders, although much more is yet to be done, nanotechnology may bring some extra help. This recent science has a promised capability of deliver materials or devices to interact with biological systems at a molecular level, while minimizing side-effects. A review recently published [147] reports current and future applications of nano-enabled drugs delivered systems. Those nano drugs would be able to cross the blood-brain barrier (BBB) employing different tools of the nanotechnology: molecular and the polymeric carrier approaches. Upon the molecular approach drugs can be targeted to brain cells and become activated once inside the targeted cell by specific enzymes. In the polymeric carrier approach, polymeric nanoparticles can be administered intravenously, intrathecally or as implantable cerebral device [147]. The nanostructures employed for the development of nano-enabled drug delivery systems in the treatment of neurodegenerative disorders are: a) polymeric nanoparticles; b) polymeric nanospheres and nanosuspensions; c) polymeric nanogels; d) carbon nanotubes and nanofibres; e) polymeric nano-micelles; f) polymeric nano-liposomes, all having different forms of delivering the drug and different approaches to transposing the blood-brain barrier [147].

Nanotechnology is a science that can aid neuroscience in helping the treatment of neurological degenerative diseases by affording targeting drugs to be delivered and to enhance the bioavailability and efficacy of various drugs: existing or to be discovered bioactive agents [147].

Other proeminent method to help neurodegenerative disease patients is chelation therapy. Long used to chelate toxic agents in the body, this technique can help in the complexation possible therapy of biometals related to neurodegenerative diseases. Regarding the gathered evidence that some metal ions play double role in the brain biochemistry, the chelation therapy can be an alternative strategy for the treatment of these diseases. Chelating agents can scavenge free metals which are present in excess in the brain or that are experiencing severe dishomeostasis [148]. The main problems encountered in this promising therapeutic option are first the requirement of this chelating agent to cross the blood brain barrier (BBB) excluding a list of common metal ligands that are essentially hydrophilic. Second the preferred nature required for the design of these complexing agents could rather be specific and moderate than reacting indiscriminately and as a massive chelators once a generalized metal depletion could cause a highly toxic effect [148]. This recent review [148] lists a series of possible chelators and their known side effects so far for the metal ions iron, copper, aluminum, zinc and calcium.

It is important to bear in mind that there are still some controversies in the processes of neurodegenerative diseases and the role metal ions play on these processes [149]. In them the results presented in the literature so far either show metals as adjuvants in the degenerating process or as contributing for an amelioration of the symptons of those diseases, so it is clear that any therapy should be conducted with great care and full surveillance.

References

[1] Cummings J.L.; Jest D.V. Alzheimer´s disease and its management in the year 2010. *Psychiatr Serv* 50(1999) 1173-1177.

[2] Eveling, L. D. *Private Communication*, 2009.

[3] Pamplona R.; Barja G. Mitochondrial oxidative stress, aging and caloric restriction: The protein and methionine connection. *Biochim Biophys Acta* 1757 (5-6) (2006) 496-508.

[4] Chowdhury A.; Santra A.; Bhattacharjee K.; Ghatak S.; Saha D.R.; Dhali G.K. Mitochondrial oxidative stress and permeability transition in Isoniazid and Rifamicin induced liver injury in mice. *J Hepatol* 45 (2006) 117-26.

[5] Seshari S.; Beiser A.; Selhub J.; Jacques P.F.; Rosenberg I.H.; D'Agostino R.B.; Wilson P.W.F.; Wolf P.A. Plasma homocysteine as a risk factor for dementia and Alzheimer's disease. *N Engl J Med* 346(2002) 476-483.

[6] Dhitavat S.; Ortiz D.; Rogers E.; Rivera E.; Shea T.B. Folate, vitamin E, and acetyl-L-carnitine provide synergistic protection against oxidative stress resulting from exposure of human neuroblastoma cells to amyloid-beta. *Brain Res.* 1061 (2005) 114-117.

[7] Perelmutter D.; Colman C. *The better brain book.* Riverhead Books, Penguin Group, USA, 2004.

[8] Garbellini D. Fisioterapia aplicada aos idosos portadores de disfunções neurológicas. In: Rebelatto J.R.; Morelli J.G.S. *Fisioterapia geriátrica: a prática da assistência ao idoso.* 2nd ed., Manole, SP, Brazil, 2007, pp. 237-242.

[9] [Cançado F.A.X.; Horta M.L. *Envelhecimento cerebral.* In: Freitas E. V. ed. *Tratado de geriatria e gerontologia.* Guanabara Koogan, Rio de Janeiro, Brazil, 2002, pp.112-127.

[10] Netto M.P.; Brito F.C. *Urgências em Geriatria.* Atheneu, São Paulo, Brazil, 2001, pp. 1-21.

[11] Pittela J.E.H. *Envelhecimento Cerebral normal: Morfologia.* In: Cançado F.A.X. ed. *Noções práticas de geriatria.* Coopmed, Belo Horizonte, Brazil, 1994, pp. 69-81

[12] Andrade L. *Alguns aspectos neurobiológicos do envelhecimento.* Apresentado em: Reunião Anual da Sociedade Brasileira para o Progresso da Ciência, 39, Brasília, Brazil, 1987.

[13] Carvalho Filho E.T.; Netto P.M. *Geriatria: fundamentos, clínica e terapêutica*. Atheneu, SP, Brazil, 1994.

[14] Ribeiro A.M. *Envelhecimento Cerebral normal: Bioquímica*. In: Cançado F.A.X. ed. *Noções práticas de geriatria*. Coopmed, Belo Horizonte, Brazil, 1994, pp. 61-67.

[15] Mendonça L.I.Z. *Envelhecimento neurológico*. In: Carvalho Filho E.T.; Netto M.P. *Geriatria: fundamentos, clínica e terapêutica*. 2nd ed, Atheneu, SP, Brazil, 2006, pp. 95-102.

[16] Okamoto, K.; Hirai, S.; Iizuka, T.; Yanagisawa T.; Watanabe. M. Reexamination of granulovacuolar degeneration. *Acta Neurophatol* 82 (1991) 340-345.

[17] Coutinho E.; Brasileiro O. *Reabilitação Fisioterapêutica do Paciente Idoso*. In: Pereira C.U.; Andrade Filho A.D. *Neurogeriatria*. Revinter, Rio de Janeiro, Brazil, 2001, pp. 372-375.

[18] Barzilai A.; Yamamoto K-I. DNA damage responses to oxidative stress. *DNA Repair* 3 (2004) 1109-1115.

[19] Evans M.D.; Dizdaroglu M.; Cooke M.S. Oxidative DNA damage and disease: induction, repair and significance. *Mutat Res/ Rev in Mutat Res* 567 (2004) 1-61.

[20] Halliwell B.; Gutteridge J.M.C. *Free Radical in Biology and Medicine*, 3rd ed., Oxford University Press, Oxford, (2002).

[21] Dhitavat S.; Ortiz D.; Rogers E.; Rivera E.; Shea T.B. Folate, vitamin E, and acetyl-L-carnitine provide synergistic protection against oxidative stress resulting from exposure of human neuroblastoma cells to amyloid-beta. *Brain Res* 1061 (2005) 114-117.

[22] Fraústo da Silva J.J.R.; Williams R.J.P. *The Biological Chemistry of the Elements*. Clarendon Press, Oxford, UK, 1991.

[23] Lippard S.J.; Berg J.M. *Principles of Bioinorganic Chemistry*. University Science Books, Mill Valley, California, USA, 1994.

[24] Dreosti I.E.; Mills C.F. (eds.). *Zinc in human biology*. Chap 15, Springer-Verlag, Berlim, Germany, 1989.

[25] Williams R.J.P.; Fraústo da Silva J.J.R. *The natural selection of the chemical elements*. Oxford University Press, Oxford, UK, 1993.

[26] Williams R.J.P. *The biological chemistry of the brain and its possible evolution*. *Inorg Chim Acta* 356 (2003) 27-40.

[27] Gybina A.A.; Prohaska J.R. Fructose-2,6-Bisphosphate Is Lower in Copper Deficient Rat Cerebellum Despite Higher Content of Phosphorylated AMP-Activated Protein Kinase. Fructose-2,6-Bisphosphate Is Lower in Copper Deficient Rat Cerebellum Despite Higher Content of Phosphorylated AMP-Activated Protein Kinase. *Experim Biol Medicine* 223 (2008) 1262-1270.

[28] Garrod A.E. The Croonian Lecture on Inborn errors of metabolism. *Lancet* 172 (1908) 73-79, 142-148, 214-220.

[29] Cuanjunco M.P; Fagét K. Y.; Huang X.; Tanzi R.E.; Bush A.I. Metal Chelation as a Potential Therapy for Alzheimer's Disease. *Annals New York Academy of Sciences* 920 (2000) 292-304.

[30] Bush A.I. The metallobiology of Alzheimer's disease. *Trends Neurosci* 26 (2003) 207-214.

[31] Levine R.L.; Mosoni L.; Berlett B.S.; Stadtman E.R. Methionine residues as endogenous antioxidants in proteins. *Proc Natl Acad Sci USA* 93(26) (1996)15036-15040.

[32] Schafer F.Q.; Buettner G.R. Redox state of the cell as viewed though the glutathione disulfide/glutathione couple. *Free Radic Biol Med.* 30 (2001) 1191-1212.

[33] Grow A. J.; Ischiropoulos H. Nitric oxide chemistry and cellular signaling. *J Cell Physiol* 187(2001) 277-282.

[34] Finkel T.; Holbrook, N. J. Oxidants, oxidativc stress and the biology of ageing. *Nature* 408 (2000) 239-247.

[35] Seshari S.; Beiser A.; Selhub J.; Jacques P.F.; Rosenberg I.H.; D'Agostino R.B.; Wilson P.W.F.; Wolf P.A. Plasma homocysteine as a risk factor for dementia and Alzheimer's disease. *N Engl J Med* 346(7) (2002) 476-483.

[36] Hou L.; Zagorski M.G. NMR reveals anomalous copper (II) binding to the amyloid Aβ peptide of Alzheimer's disease. *J Am Chem Soc* 128 (2006) 9620-9621.

[37] Hozumi I.; Asanuma M.; Yamada M.; Uchida, Y. Metallothioneins and Neurodegenerative Diseases. *J Health Sci* 50 (4) (2004) 323-331.

[38] Halliwell B. Role of Free Radicals in the Neurodegenerative Diseases: Therapeutic Implications for Antioxidant Treatment. *Drugs Aging* 18 (2001) 685-716.

[39] Ischiropoulos H.; Beckman J.S. Oxidative stress and nitration in neurodegeneration: cause, effect or association? [Perspective for Neurodegeneration Series]. *J Clin Invest* 111 (2003) 163-169.

[40] *[Alberts B.; Alexander J.; Julian L.; Martin R.; Keith R.; Peter W.* The Shape and Structure of Proteins. Molecular Biology of the Cell. *4th Ed., Garland Science, New York and London, 2002.*

[41] *Berg J.M.; Tymoczko J.L.; Stryer L. Protein Structure and Function. In:* Biochemistry. *W.H. Freeman, New York, USA, 2007.*

[42] *Anfinsen C. The formation and stabilization of protein structure.* Biochem. J. *128 (1972) 737–749.*

[43] Kopito R.R.; Ron D. Conformational disease. *Nat Cell Biol* 2 (2000) E207 - E209.

[44] Carrell R.W.; Lomas D.A. Alpha 1-Antitrypsin Deficiency — A Model for Conformational Diseases. *N Engl J Med* 346 (2002) 45-53.

[45] Lomas D.A.; Carrell R.W. Serpinopathies and the conformational dementias. *Nat Rev Genetics* 3 (2002) 759-768.

[46] Glenner G. Amyloid deposits and amyloidosis. *N Engl J Med 302 (*1980) 1283-1292.

[47] Blake C.C.F.; Serpell L.C. Synchrotron X-ray studies suggest that the core of the transthyretin amyloid fibril is a continuous beta-sheet helix. *Structure 4 (*1996) 989-998.

[48] Francis P.T.; Palmer A.M.; Snape M.; Wilcock G.K. The cholinergic hypothesis of Alzheimer's disease: a review of progress. *J Neurol Neurosurg Psychiatry* 66(1999) 137-147.

[49] Gooch M.D.; Stennett D.J. The Molecular Basis of Alzheimer's Disease. *Am J Health Syst Pharm* 53 (1996) 1545-1557.

[50] Citron M.; Westaway D.; Xia W.; Carlson G.; Diehl T.; Levesque G.; Wood K.J.; Lee M.; Suebert P.; Davis A.; Kholodenko D.; Motter R.; Sherrington R.; Perry B.; Yao H.; Strome R.; Lieberburg I.; Rommens J.; Kim S.; Shenk D.; Fraser P.; St. George-Hyslop P.H.; Selkoe D.J. Mutant presenilins of Alzhemer's disease increase production of 42-residue β-protein in both transfected cells and transgenic mice. *Nat Med* 3(1997) 67-72.

[51] Glenner G.G.; Wong C. W. Alzheimer's disease: initial report of the purification and characterization of a novel cerebrovascular amyloid protein. *Biochem Biophys Res Commun* 120 (1984) 885-890.

[52] Masters C.L.; Simms G.; Weinman N.A.; Multhaup G.; McDonald B.L.; Beyreuther K. Amyloid plaque core protein in Alzheimer's disease and Down syndrome. *Proc Natl Acad Sci USA* 82 (1985) 4245-4249.

[53] St. George-Hyslop P.H.; Tanzi R.E.; Polinsky R.J.; Harnes J.L.; Nee L.; Watkins P.C.; Myers R.H.; Feldman R.G.; Pollen D.; Drachman D.; Growdon J.; Brune A.; Foncin J-F.; Salmos D.; Frommelt P.; Amaducci L.; Sorbi S.; Piancentini S.; Stewart G.D.; Hobbs W.J.; Conneally F.M.; Gusella J.F. The genetic defect causing familial Alzheimer's disease maps on chromosome 21. *Science* 235(1987) 885-890.

[54] Shoji M.; Golde T.E.; Ghiiso J.; Cheug T.T.; Estus S.; Shaffer L.M.; Cai X.D.; McKay D.M.; Tintner R.; Frangione B.; Younkin S.G. Production of the Alzheimer amyloid beta protein by normal proteolytic processing. *Science* 258 (1992) 126-129.

[55] Vigo-Pelfrey C.; Lee D.; Keim P.; Leiberburg I.; Schenk D.B. Characterization of ß-amyloid peptide from human cerebrospinal fluid. *J Neurochem* 61 (1993) 1965-1968.

[56] Prusiner S.B. Príons. *Proc Natl Acad Science USA* 95(23) (1998) 13363-13383.

[57] Soto C. *Prions: the new biology of proteins*. Taylor & Francis, Boca Raton, Fl, USA, 2006.

[58] Du H.; Yan S.S. Mitochondrial medicine for neurodegenerative diseases. Int J *Biochem Cell Biol* 42 (2010) 560–572.

[59] Cação J.C.; Souza D.R.S.; Tognola W.A.; Godoy M.R.P.;.Pinhel M.A.S. Polimorfismo da Apolipoproteína E nos familiares em primeiro grau de pacientes com doença de Alzheimer familial ou esporádica. *Arq Neuropsiquiatr* 65 (2-A) (2007) 295-298.

[60] Shevtsov P.N.; Shevtsova E.F.; Burbaeva G. Sh.; Bachurin S.O. Disturbed Assembly of Human Cerebral Microtubules in Alzheimer's Disease. *Bull Experim Biol Medicine, Morph Pathomorph 141 (2006) 265-268.*

[61] Berg L.; Morris J.C. *Clinical presentation: diagnosis* In: Terry R.D.; Katsman R.; Bush A.I; Bick K.L. (eds), Alzheimer disease, Rave, Nova York, (1994) pp. 9-25.

[62] Van Vickle G.D.; Esh C.L.; Kokjohn T.A.; Patton R.L.; Kalback W.M.; Luehrs D.C.; Beach T.G.; Newel A.J.; Lopera F.; Ghetti B.; Vidal R.; Castaño E.M.; Roher A.E. Presenilin-1 280Glu→Ala Mutation Alters C-Terminal APP Processing Yielding Longer Aβ Peptides: Implications for Alzheimer's Disease. *Mol Med* 14 (3-4) (2008) 184–194.

[63] Masters C.L.; Simms G.; Weinman N.A.; Multhaup G.; McDonald B.L.; Beyreuther K. Amyloid plaque core protein in Alzheimer's disease and Down syndrome. *Proc Natl Acad Sci USA* 82 (1985) 4245-4249.

[64] Frank M. LaFerla, Kim N. Green & Salvatore Oddo. Intracellular amyloid-β in Alzheimer's disease. *Nature Rev Neurosci* 8 (2007) 499-509. (http://www.nature.com/nrn/journal/v8/n7/fig_tab/nrn2168_F1.html)

[65] Götz J.; Chen F.; van Dorpe J.; Nitsch R.M. Formation of Neurofibrillary Tangles in P301L Tau Transgenic Mice Induced by Aß42 Fibrils. *Science* 293 (2001) 1491- 1495.

[66] Kosik K.S. Alzheimer's Disease: a cell biological perspective. *Science* 256 (1992) 780-782.

[67] Brion J-P.; Hanger D.P.; Bruce M.T.; Lorek A-M.; Flamente-Durand J.; Anderson B.H. Tau in Alzheimer neurofibrillary tangles. *Biochem J* 273 (1991) 127-133.

[68] Mandelkow E-M.; Mandelkow, E. Tau as a marker for Alzheimer's disease. *Trends Biochem.Sci* 18 (1993) 480-483.
[69] Baas P.W.; Pienkowiski T.P.; Kosik K.S. Processes induced by tau expression in Sf9 cells have an axon-like microtubule organization. *J.Cell Biol* 115(1991) 1333-1344.
[70] Wild K.V.; Kay J.A. The rate of progression of Alzheimer's disease in the later stages: Evidence from the Severe Impairment Battery. *J Int Neuropsychol Soc* 4 (1998) 512-516.
[71] Holtzmann D.M.; Mobley W.C. Molecular studies in Alzheimer's disease. *Trends Biochem Sci* 16 (1991) 140-144.
[72] Ball M.J.; Murdoch G.H. Consensus recommendations for the postmortem diagnosis of Alzheimer's disease. *Neurobiol Aging* 18(4) (1997) S1-S2.
[73] Tiraboschi P.; Hansen L.A.; Thal L.J. The importance of neuritic plaques and tangles to the development and evolution of AD. *Neurology* 62(11) (2004) 1984-1989.
[74] Hardy J.; Allsop D. Amyloid deposition as the central event in the aetiology of Alzheimer's disease. *Trends Pharmacol. Sci* 12(10) (1991) 383–388.
[75] Polvikoski T.; Sulkava R.; Haltia M.; Kainulainen K.; Vuorio A.; Verkkoniemi A.; Niinisto L.; Halonen P.; Kontula K. Apolipoprotein E, dementia, and cortical deposition of beta-amyloid protein. *N Engl J Med* 333(19) (1995) 1242–1247.
[76] Ikonomovic M.D.; Klunk W.E.; Abrahamson E.E.; Klunk W.E.; Abrahamson E.E.; Mathis C.A.; Price J.C.; Tsopelas N.D.; Lopresti B.J.; Ziolko S.; Bi W.; Paljug W.R.; Debnath M.L.; Hope C.E.; Isanski B.A.; Hamilton R.L.; DeKosky S.T. Post-mortem correlates of in vivo PiB-PET amyloid imaging in a typical case of Alzheimer's disease . *Brain* 131(6) (2008) 1630–1645.
[77] Shcherbatykh I.; Carpenter D.O. The role of metals in the etiology of Alzheimer's disease. *J Alzheimer Dis* 11 (2) (2007) 191–205.
[78] Geula C.; Mesulam M.M. Cholinesterases and the pathology of Alzheimer disease. *Alzheimer Dis Assoc Disord* 9 (2) (1995) 23–28.
[79] Yu W. H.; Lukiw W. J.; Bergeron C.; Niznik H.B.; Fraser P.E. Metallothionein III is reduced in Alzheimer's disease. Brain Res 894(1-9) (2001) 37-45.
[80] Meloni G.; Knipp M.; Vasak M. Detection of neuronal growth factor (metallothionein-3) in polyacrilamide gels and by Wetern blot analysis. *J Biochem Biophys Methods* 64(1) (2005) 76-81.
[81] O'Brien J.T.; Ballard C.G. Drugs for Alzheimer's disease. *BMJ* 323 (2001) 123-24.
[82] Wilcock G.K.; Llianfeld S.; Gaens E. Efficacy and safety of galantamine in patients with mild to moderate Alzheimer's disease: multicentre randomised controlled trial. *BMJ* 321 (2000) 1-7.
[83] Cooper J.R.; Blom F.E.; Roth R.H. Amino-acid. The biochemical basis of neuropharmacology. Oxford University Press, Oxford. 4th ed, New York, 1982, p. 250.
[84] Kopp N. Maladie de Creutzfeldt-Jakob. Maladie de la vache folle. Editions Ellipses, Paris, 1999 (collection Vivre et Comprendre).
[85] Prusiner SB. *Prions.Sci AM*, 1984, vol.251, n^0 4, p. 48-57.
[86] Prusiner SB. *Prions.Proc Natl Acad Sci USA*, 1998, vol.95, p. 13363-13383.
[87] Soto C. *Prions: the new biology of proteins*. Boca Raton: Editora Taylor & Francis, 2006.
[88] Petersen R.B.; Siedlak S.L.; Lee H.; Kim Y.; Nunomura A.; Tagliavini F.; Ghetti B.; Cras P.; Moreira P.I.; Castellani R.J.; Guentchev M.; Budka H.; Ironside J.W.;

Gambetti P.; Smith M.A.; Perry G. Redox metals and oxidative abnormalities in human prion diseases. *Acta Neuropathol* 110 (2005) 232-238.

[89] Prusiner. ;Prions. *P. Natl Acad. Sci. USA*, 1998, 95, 23, 13363-13383.

[90] Riek R.; Hornemann S.; Wider G.; Billeter M.; Glockshuber R.; Wuthrich K. NMR structure of the mouse prion protein domain PrP (121-231). *Nature* 382 (1996) 180-182.

[91] Pan K.M.; Stahl N.; Prusiner S.B. Purification and properties of cellular prion protein from Syrian hamster brain. *Protein Sci,* 1(1992) 1343-1352.

[92] Gasset M.; Baldwin M.A.; Lloyd D.H.; Gabriel L.M.; Oltzman F.C.; Cohen F.; Fletterick R.; Prusiner S.B. Predicted alfa-helical regions of the prion protein when synthesized as peptides form amyloid. *Proc Natl Acad Sci, USA* 89 (1992) 10940-10944.

[93] Nguyen J.; Baldwin M.A.; Cohen F.E.; Prusiner S.B. Prion protein peptides induce alpha-helix to beta-sheet conformation transitions. *Biochemistry* 34 (1995) 186-192.

[94] Kocisko D.A.; Come J.H.; Priola S.A.; Chesebro B.; Raymond G.J.; Lansbury P.T.; Caughey B. Cell-free formation of protease-resistant prion protein. *Nature* 370 (1994) 471-474.

[95] [Kakizuka A. Protein precipitation: a common etiology in neurodegenerative disorders? *Trends Genet* 14 (1998) 396-402.

[96] Savitt J.M.; Dawson V.L.; Dawson T.M. Diagnosis and treatment of Parkinson disease: molecules to medicine. *J Clin Invest* 116(7) (2006) 1744-1754.

[97] Kitada T. Mutations in the PARKIN gene cause autosomal recessive juvenil parkinsonism. *Nature* 392 (1998) 605-608.

[98] Polymeropoulos M.H. Mutation in the alpha-synucle-in gene identified in families with Parkinson's disease. *Science* 276 (1997) 2045-2047.

[99] Harper B. Huntington disease. *J Royal Soc Med* 98 (2005) 550-550.

[100] Goswami A.; Dikshit P.; Mishra A.; Mulherkar S.; Nukina N.; Jana N.R. Oxidative stress promotes mutant huntingtin aggregation and mutant huntingtin-dependent cell death by mimicking proteasomal malfunction. *Biochem Biophys Res Commun* 342 (2006) 184-190.

[101] Quinsey N.S.; Greedy A.L.; Bottomley S.P.; Whisstock J.C.; Pike R.N. Antithrombin: in control of coagulation. *Int. J. Biochem. Cell Biol* 36 (2004)386-389.

[102] Stein P.E.; Carrel R.W. What do dysfunctional serpins tell us about molecular mobility and disease? *Nat Struct Biol* 2 (1995) 96-113.

[103] Kim S-J.; Woo J-R.; Seo E.J.; Yu M-H.; Ryu S-E. A 2.1A° resolution structure of an uncleaved α1-antitrypsin shows variability of the reactive centre and other loops. *J Mol Biol* 306 (2001) 109–119.

[104] Johnson D.; Travis J. Structural evidence for methionine at the reactive site of human α-1-proteinase inhibitor. *J Biol Chem* 253 (1978) 7142-7144.

[105] Wilczynska M.; Fa M.; Ohlsson P-I.; Ny T. The inhibition mechanism of serpins. Evidence that the mobile reactive centre loop is cleaved in the native protease inhibitor complex. *J Biol Chem* 270 (1995) 29652–29655.

[106] Wilczynska M.; Fa M.; Karolin J.; Ohlsson P-I.; Johansson L.B-A.; Ny T. Structural insights into serpin-protease complexes reveal the inhibitory mechanism of serpins. *Nat Struc Biol* 4 (1997) 354–357.

[107] Stratikos E.; Gettins P.G.W. Major proteinase movement upon stable serpinproteinase complex formation. *Proc Natl Acad Sci USA* 4 (1997) 453–458.

[108] Stratikos E.; Gettins P.G.W. Mapping the serpin-proteinase complex using single cysteine variants of α1-antitrypsin inhibitor Pittsburgh. *J Biol Chem* 273 (1998) 15582–15589.

[109] Stratikos E.; Gettins P.G.W. Formation of the covalent serpin-proteinase complex involves translocation of the proteinase by more than 70A° and full insertion of the reactive centre loop into b-sheet A. *Proc Natl Acad Sci USA* 96 (1999) 4808-4813.

[110] Huntington J.A.; Read R.J.; Carrell R.W. Structure of a serpin-protease complex shows inhibition by deformation. *Nature* 407 (2000) 923–926.

[111] STRATIKOS, E.; GETTINS, P. G. W. Formation of the covalent serpin-proteinase complex involves translocation of the proteinase by more than 70 Å and full insertion of the reactive center loop into β-sheet A. Proc. Natl. Acad. Sci.(USA), Washigton, v. 96, p. 4808-4813, 1999.

[112] HUNTINGTON, J. A.; READ, R. J.; CARREL, R. W. Structure of a serpin-protease complex shows inhibition by deformation. Nature (London), London, v. 407, p. 923-926, 2000.

[113] Lawrence D.A.; Olson S.T.; Muhammad S. Partitioning of serpin-proteinase reactions between stable inhibition and substrate cleavage is regulated by the rate of serpin reactive center loop insertion into β-sheet A. *J. Biol. Chem* 275 (2000) 5839-5844.

[114] Elliott P.R.; Lomas D.A.; Carrell R.W.; Abrahams J.P. Inhibitory conformation of the reactive loop of a1-antitrypsin. *Nat Struct Biol* 3 (1996) 676–681.

[115] Mahadeva R.; Lomas D.A. Alpha1-antitrypsin deficiency, cirrhosis and emphysema. *Thorax* 53 (1998) 501–505.

[116] Lomas D.A.; Evans D.L.; Finch J.T.; Carrell R.W. The mechanism of Z α1-antitrypsin accumulation in the liver. *Nature* 357 (1992) 605–607.

[117] Dafforn T.R.; Mahadeva R.; Elliott P.R.; Sivasothy P.; Lomas, D.A. A kinetic description of the polymerisation of α1-antitrypsin. *J Biol Chem* 274 (1999) 9548-9555.

[118] Sivasothy P.; Dafforn T.R.; Gettins P.G.W.; Lomas D.A. Pathogenic α1-antitrypsin polymers are formed by reactive loop-β-sheet A linkage. *J Biol Chem* 275 (2000) 33663–33668.

[119] Gooptu B.; Hazes B.; Chang W-S. W.; Dafforn T.R.; Carrell R.W.; Read R.J.; Lomas D.A. Inactive conformation of the serpin α1-antichymotrypsin indicates two stage insertion of the reactive loop; implications for inhibitory function and conformational disease. *Proc Natl Acad Sci USA* 97 (2000) 67–72.

[120] Yu M-H.; Lee K.N.; Kim J. The Z type variation of human a1-antitrypsin causes a protein folding defect. *Nat Struc Biol* 2 (1995) 363–367.

[121] Kim J.; Lee K.N.; Yi G-S.; Yu M-H. A thermostable mutation located at the hydrophobic core of α1-antitrypsin suppresses the folding defect of the Z-type variant. *J Biol Chem* 270 (1995) 8597–8601.

[122] Kang H.A.; Lee K.N.; Yu M-H. Folding and stability of the Z and Siiyama genetic variants of human α1-antitrypsin. *J Biol Chem* 272 (1997) 510–516.

[123] Mahadeva R.; Dafforn T.R.; Carrell R.W.; Lomas D.A. 6-mer peptide selectively anneals to a pathogenic serpin conformation and blocks polymerisation: implications

for the prevention of Z α1-antitrypsin related cirrhosis. *J Biol Chem* 277(2002) 6771-6774.

[124] Lomas D.A.; Evans D.L.; Stone S.R.; Chang W.S.W.; Carrell R.W. Effect of the Z mutation on the physical and inhibitory properties of α1-antitrypsin. *Biochemistry* 32 (1993) 500–508.

[125] Lomas D.A.; Elliott P.R.; Chang W-S.W.; Wardell M.R.; Carrell R.W. Preparation and characterisation of latent α1-antitrypsin. J Biol Chem 270 (1995) 5282–5288.

[126] Im H.; Woo M-S.; Hwang K.Y.; Yu M-H. Interactions causing the kinetic trap in serpin protein folding. *J Biol Chem* 277 (2002) 46347–46354.

[127] Stein P.E.; Carrell R.W. What do dysfunctional serpins tell us about molecular mobility and disease? *Nat. Struct. Biol* 2 (1995) 96-113.

[128] Davis R.L.; Shrimpton A.E.; Carrell R.W.; Lomas D.A.; Gerhard L.; Baumann B.; Lawrence D.A.; Yepes M.; Kim T.S.; Ghetti B.; Piccardo P.; Takao M.; Lacbawan F.; Muenke M.; Sifers R.N.; Bradshaw C.B.; Kent P.F.; Collins G.H.; Larocca D.; Holohan P.D. Association between conformational mutations in neuroserpin and onset and severity of dementia. *Lancet* 359 (2002) 2242-2247.

[129] Lomas D.A.; Carrell R.W. Serpinopathies and the conformational dementias. *Nat Rev Genet* 3 (2002) 759-768.

[130] [130] Abrahamson M.; Grubb A. Increased body temperature accelerates aggregation of the Leu-68-->Gln mutant cystatin C, the amyloid-forming protein in hereditary cystatin C amyloid angiopathy. *Proc Natl Acad Sci USA* 91 (1994) 1416-1420.

[131] Maragos W.F.; Greenamyre J.T.; Penney, Jr J.B.; Young A.B. Glutamate dysfunction in Alzheimer's disease: an hypothesis. *Trends Neurosci* 10 (2) (1987) 65-68.

[132] Mathie A.; Sutton G.L.; Clarke C.E.; Veale E.L. Zinc and copper: Pharmacological probes and endogenous modulators of neuronal excitability. *Pharmacol Ther* 111 (2006) 567–583.

[133] Strozyk D.; Launer L.J.; Adlard P.A.; Cherny R.A.; Tsatsanis A.; Volitakis I.; Blennow K.; Petrovitch H.; White L.R.; Bush A.I. Zinc and copper modulate Alzheimer A_ levels in human cerebrospinal fluid. *Neurobiol Aging* 30 (2009) 1069–1077.

[134] Barnham K.J.; Bush A.I. Metals in Alzheimer's and Parkinson's Diseases. *Curr Opinion Chem Biol* 12 (2008) 222–228.

[135] Kozlowski H.; Janicka-Klos A.; Brasun J.; Gaggelli E.; Valensin D.; Valensin G. Copper, iron, and zinc ions homeostasis and their role in neurodegenerative disorders (metal uptake, transport, distribution and regulation). *Coord Chem Rev* 253 (2009) 2665–2685.

[136] Leiva J.; Palestini M.; Infante C.; Goldschmidt A.; Motles E. Copper suppresses hippocampus LTP in the rat, but does not alter learning or memory in the morris water maze. *Brain Res* 1256 (2009) 69-75.

[137] Zatta P.; Drago D.; Bolognin S.; Sensi S.L. Alzheimer's disease, metal ions and metal homeostatic therapy. *Trends Pharmacol Sci* 30 (7) (2009) 346-355.

[138] Lu J.; Wu D-M.; Zheng Y-L.; Sun D-X.; Hu B.; Shan Q.; Zhang Z-F.; Fan S-H. Trace amounts of copper exacerbate beta amyloid-induced neurotoxicity in the cholesterol-fed mice through TNF-mediated inflammatory pathway. *Brain Behav Immun* 23 (2009) 193–203.

[139] Strozyk D.; Launer L.J.; Adlard P.A.; Cherny R.A.; Tsatsanis A.; Volitakis I.; Blennow K.; Petrovitch H.; White L.R.; Bush A.I. Zinc and copper modulate Alzheimer Aß levels in human cerebrospinal fluid. *Neurobiol Aging* 30 (2009) 1069–1077.

[140] Du J.; Sun B.; Chen K.; Zhang L.; Liu S.; Gu Q.; Fan Li.; Zhao N.; Wang Z. Metabolites of Cerebellar Neurons and Hippocampal Neurons Play Opposite Roles in Pathogenesis of Alzheimer's Disease. *PLoS ONE* 4 (5) (2009) e5530.

[141] Penner J.; Rupsingh R.; Smith M.; Wells J.L.; Borrie M.J.; Bartha R. Increased glutamate in the hippocampus after galantamine treatment for Alzheimer disease. *Progr Neuro-Psychopharmacol Biol Psych* 34 (2010) 104–110

[142] Doody R.S.; Gavrilova S.I.; Sano M.; Thomas R.G.; Aisen P.S.; Bachurin S.O.; Seely L.; Hung D. Effect of dimebon on cognition, activities of daily living, behaviour, and global function in patients with mild-to-moderate Alzheimer's disease: a randomised, double-blind, placebo-controlled study. *Lancet* 372 (2008) 207–215.

[143] Hung Y.H.; Bush A.I.; Cherny R.A. Copper in the brain and Alzheimer's disease. *J Biol Inorg Chem* 15 (2010) 61–76.

[144] Tabaton M.; Zhu X.; Perry G.; Smith M.A.; Giliberto L. Signaling effect of amyloid-β42 on the processing of AβPP. *Experim Neurol* 221 (2010) 18–25.

[145] Kaminsky Y.G.; Marlatt M.W.; Smith M.A.; Kosenko E.A . Subcellular and metabolic examination of amyloid-β peptides in Alzheimer disease pathogenesis: Evidence for Aβ25–35. *Experim Neurol* 221 (2010) 26–37.

[146] De Felice F.G.; Vieira M.N.N.; Bomfim T.R.; Decker H.; Velasco P.T.; Lambert M.P.; Viola K.L.; Zhao W-Q.; Ferreira S.T.; Klein W.L. Protection of synapses against Alzheimer's-linked toxins: Insulin signaling prevents the pathogenic binding of Aß oligomers. *PNAS* 106 (2009) 1971-1976.

[147] Modi G.; Pillay V.; Choonara Y.E.; Ndesendo V.M.K.; du Toit L.C.; Naidoo D. Nanotechnological applications for the treatment of neurodegenerative disorders. *Progr Neurobiol* 88 (2009) 272–285.

[148] Bolognin S.; Drago D.; Messori L.; Zatta P. ChelationTherapy for Neurodegenerative Diseases. *Medicinal Res Rev* 29(4) (2009) 547-570.

[149] Emily House, Matthew Mold, Joanna Collingwood, Alex Baldwin, Steven Goodwin, Christopher Exley. Copper Abolishes the β-Sheet Secondary Structure of Preformed Amyloid Fibrils of Amyloid-β42 . *J Alzheimer disease* 18 (2009) 811-817.

In: Molecular and Supramolecular Bioinorganic Chemistry ISBN: 978-1-61728-785-5
Ed: A. L. R. Mercê, J. Felcman and M. A. L. Recio

Chapter 7

RADICALS AND TRANSITION METAL COMPLEXES IN AQUEOUS SOLUTIONS

Alexandra Masarwa[1] and Dan Meyerstein[1,2]
[1]Department of Chemistry, Ben-Gurion University of the Negev, Beer-Sheva, Israel
[2]Department of Biological Chemistry, Ariel University Center of Samaria, Ariel, Israel

INTRODUCTION

Electron transfer reactions are of central importance to many branches of chemistry and biochemistry. Redox processes mediated by transition metal complexes also play an increasing role in organic chemistry. As nature prefers single-electron transfer reactions, many redox reactions with main group compounds involve the formation and/or reaction of radicals. These processes include the formation of organic radicals via the reaction of an organic substrate with a transition metal complex, reactions of radicals with transition metal complexes and reactions of radicals with organic substrates bound to a central metal cation. [1-10]

Radicals play an important role in combustion, atmospheric chemistry, catalysis, polymerization, plasma chemistry, biochemistry, and many other chemical processes, including human physiology. Evidence implies that radicals play a key role in both normal biological functions and in the pathogenesis of certain diseases. [3,9]

Radicals are implicated in certain vital biological processes, e.g. several essential enzymatic processes such as photosynthesis and the biosynthesis of DNA (ribonucleotide reductases), in B_{12} catalyzed processes, cell-signaling (neuromodulators), xenobiotic metabolism (fate of foreign compounds in biological systems), mediators in inflammatory processes or killing of bacteria by neutrophil granulocytes (antimicrobial defense). [3,9]

Alternatively radical reactions may result in cell damage. It is commonly accepted that radical reactions play a major role in aging and in a variety of catalytic and biological processes. These processes include among others, reactions involving ROS (Reactive Oxygen Species; including $^{\cdot}OH$, superoxide and alkyl peroxyl radicals, and peroxides, e.g. H_2O_2, alkyl peroxides, the latter acting as a source of radicals) and RNS (Reactive Nitrogen Species, such as $NO^{\cdot}$ (nitric oxide) and its by-products $NO_2^{\cdot}$ (nitrogen dioxide), nitrate (NO_3^-), nitrite (NO_2^-), peroxynitrite ($ONOO^-$), and 3-nitrotyrosine)). [3,9]

In biology 'Fenton like' reactions (i.e. reactions via iron(II) or other low valent transition metal complexes) are believed to be the main source of ROS in the body causing a variety of diseases, e.g. cancer, atherosclerosis, essential hypertension, Alzheimer's disease, Parkinson's disease, amyloidosis, osteoarthritis, etc.. Superoxide radicals are also produced as a byproduct of the dioxygen reduction in the mitochondria. (It should be mentioned that O_2 is a biradical triplet. The discussion of its properties are beyond the scope of this review).

The role of nitric oxide (NO) in mammalian biology include cytotoxic immune response to pathogen invasion and cellular signaling in the cardiovascular and nervous systems. Metal centers are primary targets for NO binding and reactions of NO with dioxygen and other reactive oxygen species produce NO_x intermediates, which have important physiological roles. Nitric oxide has also been reported to inhibit metalloenzymes such as catalase and cytochrome c oxidase, and its vasodilator properties have been implicated in blood pressure regulation by virtue of its action on the vascular smooth muscle. The interaction of NO with hemoglobin leads to the formation of S-nitrosothiols (SNO). SNO is formed when oxidized NO interacts with the highly reactive thiol groups on the two cystein residues in the hemoglobin molecule. $RS^{\cdot}$ radicals and bonded RS react with NO through $RS^{\cdot} + NO \rightarrow RSNO$ and $RSM^{n+1} \leftrightarrows RS^{\cdot}M^n \rightarrow RSNO + M^n$. [11,12] Numerous disease states have been shown to involve the over- or under-production of NO. [9]

Transition metal complexes are key participants in biological and catalytic radical processes due to two major reasons:

a) radicals are usually generated by redox reactions involving transition metal complexes;
b) the majority of secondary radicals formed in biological and/or catalytic systems react much faster with transition metal complexes than with organic substrates. Therefore the participation of the transition metal complexes determines the nature of the final products.

Examples for catalysis are e.g oxidations by Co^{III}_{aq}, [5,8,13] Ullman reactions, [14,15] reactions involving Cu [16,17]. Recently investigation of radical reactions with surfaces, especially with nano-particles, immersed in aqueous solutions, has emerged as a new research field. Radicals are formed at the surface of metals, semi-conductors and insulating oxides at a variety of processes (see below). Therefore, the mechanism of reaction of radicals with surfaces is of importance and interest.

FORMATION OF RADICALS

Radicals can be formed by several different mechanisms: Absorption of ionizing radiation in the medium, homolytic bond breakage of specific molecules induced by thermolysis, ultrasound cavitation, by photochemistry via excited molecules, by electrochemistry and by redox reactions involving transition metals.

Pulse-Radiolysis is one of the most powerful techniques for the study of the kinetics of reaction of a variety of radicals in aqueous solutions (but also in organic solvents) and enables the observation of short-lived intermediates.

a) Absorption of ionizing radiation in the medium, e.g. in dilute aqueous solutions initially forming:[18-20]

$$H_2O \xrightarrow{\gamma\,;\,e^-} e^-_{aq}\ (2.65);\ {}^{\cdot}OH\ (2.65);\ H^{\cdot}\ (0.60);\ H_2\ (0.45);\ H_2O_2\ (0.75);\ H_3O^+\ (2.65) \quad (1)$$

where the values in parentheses are the number of molecules of a given product formed by the absorption of 100 eV in the medium. The radicals thus formed are powerful redox reagents, *i.e.* e^-_{aq} (E° = -2.87 V) is a strong reducing agent; $H^{\cdot}$ (E° = ±2.31 V) is an equally strong reducing and oxidizing agent and $^{\cdot}OH$ (E° = +2.73 V) is a powerful oxidizing agent.[21] Thus a mixture of strong single electron oxidizing- and reducing- agents is formed.

The primary radicals formed can be transformed into a variety of secondary inorganic radicals determined by the solutiõn composition (*e.g.* $O_2^{\cdot -}$; $HO_2^{\cdot}$; $O_3^{\cdot -}$; $CO_2^{\cdot -}$; $CO_3^{\cdot -}$; $CN^{\cdot}$; $N_3^{\cdot}$; $^{\cdot}NH_2$; $^{\cdot}NO$; $^{\cdot}NO_2$; $^{\cdot}NO_3$; $NCO^{\cdot}$; $PO_3^{\cdot 2-}$; $PO_4^{\cdot 2-}$; $HS^{\cdot}$; $RSSR^{\cdot}$; $SO_2^{\cdot -}$; $SO_3^{\cdot -}$; $SO_4^{\cdot -}$; $SO_5^{\cdot -}$; $(SCN)_2^{\cdot -}$; $SeO_3^{\cdot -}$; $HSeO_4^{\cdot 2-}$; $(SeCN)_2^{\cdot -}$; $Cl_2^{\cdot -}$; $Br_2^{\cdot -}$; $I_2^{\cdot -}$; $ClO_2^{\cdot}$; $BrO_2^{\cdot}$; $IO_2^{\cdot}$) [22] and mainly via hydrogen abstraction from suitable organic solutes into organic radicals (*e.g.* $^{\cdot}CH_3$; $^{\cdot}CH_2CH_3$; $^{\cdot}CH_2CH_2CH_3$; $^{\cdot}CH(CH_3)_2$; $^{\cdot}C(CH_3)_3$; c-$^{\cdot}C_5H_9$; $^{\cdot}CH_2Cl$; $^{\cdot}CH_2Br$; $^{\cdot}CF_3$; $^{\cdot}CCl_3$; $^{\cdot}CBr_3$; $^{\cdot}CH_2OH$; $^{\cdot}CH(CH_3)OH$; $^{\cdot}C(CH_3)_2OH$; $^{\cdot}CH_2CH_2OH$; $^{\cdot}CH_2C(CH_3)_2OH$; $^{\cdot}CH_2OCH_3$; $^{\cdot}CH(CH_3)OC_2H_5$; $^{\cdot}CH(OH)CH_2OH$; $^{\cdot}CH_2CHO$; $^{\cdot}CH_2CO_2H$; $^{\cdot}CH(CH_3)CO_2H$; $^{\cdot}CH(OH)CO_2H$; $^{\cdot}C(OH)(CH_3)CO_2H$; $^{\cdot}CH(CH_2NH_3^+)CO_2^-$; $^{\cdot}CH(CH_3)NH_2$; $^{\cdot}CH_2C(CH_3)_2NH_3^+$; $^{\cdot}CH_2CN$; $^{\cdot}CH_2C_6H_5$; $^{\cdot}SC_2H_5$; $CH_3OO^{\cdot}$: $CCl_3OO^{\cdot}$; $NCCH_2OO^{\cdot}$; $HO_2CCH_2OO^{\cdot}$; *etc.*)[23,24.21-26] Furthermore radicals can react with aromatics or unsaturated compounds by addition to double bonds.

b) Cosmic radiation as the source of radicals in the atmosphere;
c) Homolytic bond breakage of specific molecules induced by thermolysis or ultrasound cavitation, the latter is in principle due to the high temperatures induced by cavitation;
d) Radical production by photochemistry;
e) Ultrasonic cavitation in aqueous solutions yields hydroxyl radicals and hydrogen atoms. The major products escaping the cavity are $^{\cdot}OH$ radicals. The yield of radicals depends strongly on the nature of the gases in equilibrium with the solution and the temperature,[27] i.e. cleavage of bonds

of all compounds in the gas phase of the micro-bubble. Ultrasonic cavitation produces localized heat in excess of 5000°C;

f) Electrochemistry and photo-electrochemistry;
g) Pyrolysis (combustion), though usually no metal ions are involved;
e.g. $2H_2 + O_2 \rightarrow 2H_2O$ (2)
h) Reactions of O_3 and peroxides;
i) Alternatively, radicals can be generated by redox reactions involving transition metals (e.g. Fenton like reactions, vide infra). A variety of redox reactions of transition metal complexes with organic molecules producing radicals exist:
Aliphatic radicals can be formed in oxidation reactions of alkanes by high valent transition metal complexes of e.g. Cu(III), Co(III), Fe(IV) or 'Fe(V)':[2,7]

$$M^{n+1}L_m + RH + H_2O \rightarrow M^nL_m + H_3O^+ + {}^{\cdot}R \qquad (3)$$

or reduction of organic halides (RX), e.g. $CHCl_3$, by low valent transition metal complexes:[2,7]

$$M^nL_m + RX \rightarrow L_{m-1}M^{n+1}X + R^{\cdot} + L \qquad (4)$$

Organohalides are severe environmental pollutants. Their toxicity is due to reactions (4), which produce radicals which initiate deleterious biological processes. Also their dehalogenation via reaction (4) is important in pollution control.[28]

Furthermore reaction of alkenes with transition metal complexes produces radicals and presenting as endproducts, usually, the dimerization products (vide infra).[29]

$$M^nL \rightarrow M^{n-1}L^{\cdot} \text{ (with L = alkene)} \qquad (5)$$

$$2\ M^{n-1}L^{\cdot} \rightarrow \text{dimer} \qquad (6)$$

$$M^nL \rightarrow M^{n+1}L^{\cdot} \qquad (7)$$

$$2\ M^{n+1}L^{\cdot} \rightarrow \text{dimer} \qquad (8)$$

A further mechanism, which yields $R^{\cdot}$ radicals is:

$$M^{n+1} + RCO_2^- \leftrightarrows M^{n+1}\text{-O-C(O)R} \rightarrow M^n + CO_2 + R^{\cdot} \qquad (9)$$

i.e. a mechanism analogous to the Kolbe process.

Reaction (10) was studied in efforts to mimic the Cytochrome P-450 chemistry.[30,31]

$$M^{n+2}\text{=O} + RH \rightarrow M^{n+1}\text{-OH} + R^{\cdot} \qquad (10)$$

Though radicals are often implicated in these systems they are often very short lived as they react in the cage formed *via*:[30,31]

$$M^{n+1}\text{-OH} + R^{\cdot} \rightarrow M^{n} + ROH \qquad (11)$$

Production of radicals via transition metal complexes is also realized in a variety of enzymatic processes in biological systems, e.g. radicals are a byproduct of cellular respiration (stepwise transfer of electrons from NADH (and $FADH_2$) to oxygen molecules to form water molecules in a 4 electron reduction). Some electrons always leak in the early stages of the respiratory chain and reduce dioxygen molecules to the superoxide anion. $NO^{\cdot}$ is synthesized in cells by Nitric Oxide Synthases. Radicals like $NO^{\cdot}$ and $O_2^{\cdot-}$ are synthesized by dedicated enzymes in phagocytic cells like neutrophils and macrophages.

The Role of Transition Metal Complexes in Radical Chemistry

The role of transition metal complexes pertaining to radical reactions and biologically deleterious processes is diverse. On one hand, these compounds are capable of interacting with radicals present in solution in a variety of possible reactions; on the other hand, transition metal complexes can be responsible for the generation of radicals. The reaction with the inorganic radicals $^{\cdot}OH$, $X_2^{\cdot-}$, $NO_2^{\cdot}$, $CO_3^{\cdot-}$, etc. usually results in the formation of high valent complexes, e.g. Ni(III) and Cu(III); whereas the reaction with $H^{\cdot}$, $O_2^{\cdot-}$ usually results in the reduction of the transition metal complex yielding a low valent complex, e.g. Cu(I) and Fe(II). However in some systems $O_2^{\cdot-}$ and especially $HO_2^{\cdot}$ act as an oxidizing agent. [2,7]

Reactions relevant to biological systems can be grouped into two major classes:

a) reactions of transition metal complexes with radicals (mainly aliphatic radicals or aliphatic peroxyl radicals formed in the presence of oxygen), and
b) reactions producing radicals like reactions (2) to (5) and in particular 'Fenton like' reactions producing $^{\cdot}OH$ radicals or endogenous oxidizing agents. The 'Fenton like' reactions play an important role in the production of $^{\cdot}OH$ radicals or analogous oxidizing agents in the cell, and they as well as the first two reactions direct radical, initiated processes to the biological sites where transition metal complexes are located (site specific effect).

Lately metal complexes incorporating ligand radicals[32-37] gain increased interest, due to their use as model systems for biological systems (Cytochrome P450 and Galactose Oxidase being prominent examples), including their role as catalysts, and based on their magnetic properties.

Fenton and Fenton Like Reactions

In 1894 Fenton reported that alcohols are oxidized in the presence of H_2O_2 and $Fe(H_2O)_6^{2+}$.[38] 'Fenton Reagents' refer thus to a mixture of hydrogen peroxide and ferrous salts and 'Fenton like reactions' are analogous reactions of peroxides with metal complexes, M^nL_m, in their low valent oxidation states (e.g. Fe(II), Cu(I), Ti(III), Cr(II), Co(II), V(IV)).[39-42] 'Fenton like' reactions play an important role in a variety of environmental,[43] catalytic [44,45] and biological [3,42,46-49] processes. In biology these reactions are believed to be the main source of ROS (Reactive Oxygen Species) in the body. The detailed mechanism of these kinds of reactions and their active intermediate are still ambiguous and repeatedly discussed in many publications and reviews.[40,50-67] In 1934 Haber and Weiss suggested that the key step in this process is: [68]

$$H_2O_2 + Fe(H_2O)_6^{2+} \rightarrow {}^{\cdot}OH + OH^- + Fe(H_2O)_6^{3+} \quad k = (42\text{-}79)\ M^{-1}s^{-1}\ [66] \qquad (12)$$

i.e. involves the formation of $^{\cdot}OH$ radicals via reaction (12), and that the $^{\cdot}OH$ radicals induce the oxidation of the substrates. Reaction (12) and its general form

$$M^nL_m + H_2O_2 \rightarrow M^{n+1}L_m + {}^{\cdot}OH + OH^- \qquad (13)$$

are often referred to as the Fenton reaction.

The rates of Fenton like reactions depend on the nature of M, L, ROOR', and on the medium. They might be orders of magnitude faster than reaction (12) e.g. for Fe^{2+} with suitable ligands: [66,67,69]

$$ROOH + M^nL_m \rightarrow {}^{\cdot}OR + OH^- + M^{n+1}L_m \qquad (14)$$

Thus the $^{\cdot}OH/RO^{\cdot}$ radical was suggested to be formed as the active intermediate via an outer sphere mechanism and radical pathways were generally assumed throughout the remaining mechanism.

However, detailed kinetic studies and thermodynamic considerations indicate that not in all these reactions hydroxyl radicals or $RO^{\cdot}$ radicals are formed. [40,66]

Already Barb et al [70] and earlier Bray and Gorin proposed [71] that the mechanism of the reaction involves reaction (15)

$$Fe(H_2O)_6^{2+} + H_2O_2 \rightarrow (H_2O)_5Fe^{IV}{=}O^{2+} + 2H_2O \qquad (15)$$

Recently, PMD calculations confirmed the viability of an Fe(IV) intermediate. [72-75]

The detailed mechanism seems to involve the formation of a transient low-valent transition-metal-peroxide complex via a ligand exchange mechanism:

$$H_2O_2 + M^nL_m \leftrightarrows L_mM^{n}{\cdot}H_2O_2 \text{ or } L_{m-1}M^{n}{\cdot}H_2O_2 + L \qquad (16)$$

or

$$H_2O_2 + Fe(H_2O)_6^{2+} \leftrightarrows Fe(H_2O)_5(H_2O_2)^{2+} + H_2O \qquad (16')$$

which means that the redox process follows the inner sphere mechanism. Reaction (16) is followed by one of the ensuing routes (where RH is a substrate): [40,66]

$$\xrightarrow{+L} M^{n+1}L_m + {}^{\cdot}OH + OH^- \qquad (17a)$$

$$L_{m-1}M^{n\cdot}H_2O_2 \longrightarrow L_{m-1}M^{n+2}{=}O + H_2O \qquad (17b)$$

$$\xrightarrow{+RH} M^{n+1}L_m + R^{\cdot} + H_2O + OH^- \qquad (17c)$$

$${}^{\cdot}OH + RH \rightarrow R^{\cdot} + H_2O \qquad (18)$$

$$L_{m-1}M^{n+2}{=}O + RH \xrightarrow{+L} M^{n+1}L_m + R^{\cdot} + OH^- \qquad (19)$$

(H_2O_2 is used in these equations for simplicity purposes, but other peroxides might be involved).

Accordingly either the hydroxyl radical (RO$^{\cdot}$ radical) or a high valent transition-metal species, e.g. a ferryl species, can be formed, or the organic substrate can be directly oxidized by the $L_{m-1}M^{n\cdot}H_2O_2$ transient. Obviously the rates of reaction of those species might well differ from those of the $^{\cdot}OH$ radicals with the substrate. In the presence of an organic substrate, RH, often the same radicals $R^{\cdot}$, are ultimately formed.

Furthermore, it was suggested that the choice between these pathways depends on: the nature of the low valent metal ion, M^{n+}, if it differs from Fe^{2+}; the nature of its ligands, if they differ from water; the nature of the substrate, RH, and its concentration; the pH, etc. [40,51,76-78]

Recent results suggest that for $M^nL_m = Fe(H_2O)_6^{2+}$ in strongly acidic solutions reaction (17b) is the dominant pathway. [58,64] However it was recently reported that $[(H_2O)_5Fe^{IV}{=}O]^{2+}$, which can be independently generated by the reaction of $[Fe(H_2O)_6]^{2+}$ with ozone (O_3) and characterized by Moessbauer spectroscopy, X-ray absorption and transfer of the =O atom to DMSO, differs from the Fenton intermediate in acidic and neutral solutions. This information was derived by comparing kinetic patterns and products for different substrates in the Fenton reaction and in the reaction with $[(H_2O)_5Fe^{IV}{=}O]^{2+}$.[65,79] However Kremer detailed new evidence and theoretical grounds in support of an $[(H_2O)_5Fe^{IV}{=}O]^{2+}$ intermediate. [80] For Fe(edta) the Fenton product was shown to be undistinguishable from the $^{\cdot}OH$ radical under the majority of conditions, [77] whereas for Fe(nta) oxidations by ferryl species are more favored under most conditions. [78] Analogously, it was shown that also in the aqueous Cu(I) and Cu(I)phen /H_2O_2 system the main product is not the hydroxyl radical. [76,81,82]

Recently, while the kinetics of the Fenton reaction were studied in more detail, it was shown that the reaction between Fe(II) and the intermediate $Fe^{III}(^-O_2H)_{aq}$ complex might play an important role in the Fenton reaction. In the presence of an excess of H_2O_2, a second reaction step observed is attributed to the formation of the rather stable complex $Fe^{III}(^-O_2H)_{aq}$, as verified by studying this reaction directly.[69]

$$Fe^{III}_{aq} + H_2O_2 \leftrightarrows Fe^{III}(^-O_2H)_{aq} + H^+ \qquad (20)$$

Therefore, the reaction

$$Fe(H_2O)_6^{2+} + Fe^{III}(^-O_2H)_{aq} \longrightarrow 3\ Fe^{III}_{aq} \quad k = (7.7 \pm 1.5) \times 10^5\ M^{-1}\,s^{-1}\ [83] \qquad (21)$$

may contribute to the overall Fenton reaction.

Formation of complexes of the type $(L)Fe^{III}(^{-}O_2H)_{aq}$ has been reported,[84] e.g. for L = edta[85] and for other high valent transition metal ions.[44,48,86] The $Fe^{III}(^{-}O_2H)_{aq}$ has been reported as an intermediate in the catalytic decomposition of H_2O_2 with Fe(III) and is relevant to Catalase activity.[87,88]

The formation of a relatively stable $Fe^{III}(^{-}O_2H)_{aq}$ complex in the Fenton reaction has major consequences as it was shown that reaction (21) is four orders of magnitude faster than $k_{17}K_{16}$.[83] Clearly it is reasonable to assume that also reaction (21) does not proceed via an outer sphere mechanism, i.e. that a transient of the type $Fe^{III}(^{-}O_2H)Fe^{II}(H_2O)_5$ is formed in analogy to reaction (16). [69]

Furthermore, this could account for some of the debates in the literature concerning the detailed mechanism of the Fenton reaction. It is plausible that the transient formed in reaction (16) differs from that formed in reaction (21). In other words, $^{\cdot}OH$ radicals may be the major product under given conditions, whereas $Fe^{IV}{=}O_{aq}{}^{2+}$ may be the main product under different conditions.

If this is correct also for $LFe^{III}(^{-}O_2H)_{aq}$ then it might be of significant biological importance.

In aerated aqueous solutions $R^{\cdot}$ radicals produced in Fenton reactions, naturally react diffusion controlled with dioxygen to form the respective peroxyl radicals $ROO^{\cdot}$, and subsequently to form the transient complexes $L_{m-1}M^{n+1}$-OOR (vide supra). The decomposition reactions of these transient peroxo metal complexes determine the products of reaction.

An important source of H_2O_2 in biological and catalytic systems is dioxygen, which can react with low-valent transition metal complexes via the following sequence:

$$M^{n}L_{m} + O_2 \rightarrow M^{n+1}L_{m} + O_2^{\cdot -} \tag{22}$$

$$2O_2^{\cdot -} + 2H^{+} \rightarrow H_2O_2 + O_2 \tag{23}$$

$$M^{n}L_{m} + O_2^{\cdot -} \xrightarrow{H+} M^{n+1}L_{m} + H_2O_2 \tag{24}$$

or analogous reactions.

Low-valent transition metal complexes can be formed in systems containing a reducing agent (e.g. ascorbate) and a higher valent transition metal complex. These kinds of systems are often referred to as Udenfriend reagents and serve as models for oxygenases.[89,90] Reactions (22-24) are also the main source of $O_2^{\cdot -}$ and H_2O_2 in biological systems and thus the source of biological deleterious processes.

REACTIONS WITH ALIPHATIC AND ALIPHATIC PEROXYL RADICALS

The kinetics and mechanisms of reactions of many carbon-centered radicals with a large variety of transition metal complexes were extensively studied. [2,7,24]

$$R^{\cdot} + M^{n}L_{m} \rightarrow \text{products} \tag{25}$$

The specific rates of these types of reactions are usually considerably faster than those of the same radicals with organic constituents present in the biological systems. Therefore carbon-centered radicals are expected to react in biological systems selectively with the transition metal complexes present in it.

Alternatively, in the presence of dioxygen, aliphatic radicals react diffusion controlled with dioxygen to yield the corresponding aliphatic peroxyl radicals $RO_2^{\cdot}$, which then further react with the transition metal complexes in solution (vide infra).

$$RO_2^{\cdot} + M^nL_m \rightarrow \text{products} \quad (26)$$

or releases $O_2^{\cdot}$.These types of reactions direct the radical-initiated processes to biological sites where the transition metal complexes are located. They provide an alternative to the explanation offered by the site-specific effect.[91-93] The products of radical reactions with transition metal complexes, especially those resulting from the decomposition of the transient complexes (L_mM^{n+1}-R and L_mM^{n+1}-OOR), are probably the cause of many biologically deleterious effects initiated by radicals.

Reactions with Aliphatic Radicals

In principle the reaction of transition metal complexes with aliphatic radicals might proceed via one of the following mechanisms:[2,7]

$$M^nL_m + R^{\cdot} \rightarrow M^{n\pm1}L_m + R^{-/+} \quad \text{outer sphere redox process} \quad (27)$$

$$M^nL_m + R^{\cdot} \rightarrow L_mM^{n+1}\text{-R or } L_{m-1}M^{n+1}\text{-R} + L \quad \text{formation of M-C } \sigma\text{-bond} \quad (28)$$

$$M^nL_m + R^{\cdot} \rightarrow M^{n-1}L_{m-1} + \text{L-R or } L^{\pm} + R^{-/+} \quad \text{ligand abstraction via homolytic bond breakage} \quad (29)$$

$$M^nL_m + R^{\cdot} \rightarrow L_{m-1}M^n\text{-LR}^{\cdot} \rightarrow M^{n\pm1}L_{m-1} + \text{L-R or} + L^{\pm} + R^{-/+} \quad \text{ligand abstraction via transient formation} \quad (30)$$

$$M^nL_m + R^{\cdot} \rightarrow L_{m-1}M^n(L^{\pm}) + R^{-/+} \rightarrow M^{n\pm1}L_m + R^{-/+} \quad \text{metal through ligand redox reaction} \quad (31)$$

These equations describe possible reactions of the aliphatic radicals with metal complexes in outer- (equation 27) or inner-sphere redox processes including those incorporating the ligands (equations 28-31) including the formation of metal-carbon σ-bonds between the aliphatic radical and the central metal cation (equation 28) in which the metal is formally oxidized. The choice between inner sphere and outer sphere mechanism is mainly determined by the nature of the reactants (charge, steric hindrance, solvation). In outer sphere reactions an electron is transferred from reductant to oxidant while no chemical bond between the reaction couple is formed in the transition state. In the inner sphere mechanism, a bridged complex between oxidant and reductant is formed before electron transfer occurs. The latter is, vide infra, the prevailing mechanism due to the low rates of the self exchange reactions of the R· radicals and as for most of these reactions their exothermicity is due to the formation of RH or ROH while the formation of R^+ and R^- are endothermic. [2,7]

Reaction (29) is realized for several alkyl complexes [2,7]

$$L_mM^{n+1}\text{-}R + R'^{\cdot} \rightarrow M^nL_m + R\text{-}R' \text{ or } RH + R_{-H} \quad (32)$$

Reaction (30) is mainly observed for halide or analogous complexes as the oxidizing species, e.g. [7]

$$M(NH_3)_5X^{2+} + R^{\cdot} \rightarrow RX + M^{2+}{}_{aq} + 5NH_4^+ \quad (33)$$

and also plays a role in the oxidation of benzoatopentaamminecobalt(III). [94]

Processes following the mechanism outlined in reaction (31) are for example reductions by $R^1R^2CO^-$ [2,7]

TRANSITION METAL COMPLEXES WITH METAL-CARBON □-BONDS.

Processes following the mechanism described in equation (28) are the common mechanism of reaction for "low valent" complexes with ligand exchange rates, and/or steric structures, which enăble a bond formation between the central metal cation and the attacking radical. The products of these reactions are transient or stable complexes with metal-carbon σ-bonds.[2]

These reactions are well known for a series of metal complexes in various oxidation states (e.g. Cr(II),[95-105] Mn(II),[106,107] Mn(III),[107] Fe(II),[106,108-114] Fe(III),[112,114,115] Co(II),[101,106,116-129] Ni(I), [108,130-132] Ni(II),[108,130,133-135] Cu(I),[136-145] Cu(II) [138,145-153]. It should be noted that the rates of these reactions in general follow the rate of ligand exchange for the specific metal complex in accordance with the above mechanism.

Numerous studies concerning the formation, decomposition mechanism and characteristics of these types of complexes have been performed and are described in detail in a recent review.[2] The majority of complexes containing metal carbon σ-bonds are unstable under physiological conditions and the products of their decomposition might cause deleterious biological processes. It is of interest to note that alkyl radicals react also with metal powders immersed in aqueous solutions forming analogous transients.

MECHANISMS OF DECOMPOSITION OF THE TRANSIENT COMPLEXES LMMN+1-R

The mechanisms and kinetics of decomposition of the transient complexes L_mM^{n+1}-R in aqueous solutions depend on the nature of the central cation, of the ligands, L, of the substituents on the aliphatic residue, R, on the pH and on the presence and nature of various substrates, S, in the medium, *e.g.* O_2.[2,7] The major mechanisms are:

1) Heterolysis of the metal carbon σ-bond to form RH or ROH/R_{-H} products.

$$M^{n+1}L_m + RH + OH^- \qquad (34)$$

$$L_mM^{n+1}\text{-}R + H_2O \rightarrow$$

$$M^{n-1}L_m + ROH/R_{-H} + H_3O^+ \qquad (35)$$

Reaction (34) is observed mainly for complexes which are not expected to be reduced (e.g. M^n = Cr(II), Mn(II), Fe(II), Co(II), Ni(I), Cu(I)). Reaction (35) is mainly observed for complexes which are difficult to oxidize but can be reduced (e.g. M^n = Fe(III), Cu(II)).[2,7]

2) Homolysis of the metal carbon σ-bond

In many systems studied the results point out that equation (28) is an equilibrium process and that the mechanism of decomposition of the transient complex L_mM^{n+1}-R involves radical processes initiated by the homolytic bond breakage.

$$L_{m-1}M^{n+1}\text{-}R + L \leftrightarrows M^nL_m + R^{\cdot} \qquad (36)$$

This mechanism of decomposition has been observed for a large variety of complexes with M^{n+1} = Cr(III), Mn(III), Fe(III), Fe(IV), Co(III), Ni(III) and Cu(II). [2,7]

The radicals $R^{\cdot}$ obtained by the homolytic bond breakage can further react in a variety of possible pathways: [2,7]

Radical-radical reactions leading to combination or disproportionation reactions:

$$R^{\cdot} + R^{\cdot} \rightarrow R_2 \text{ or } RH + R_{-H} \qquad (37)$$

Reaction of the radicals with the transient complex with the metal-carbon bond:

$$L_{m-1}M^{n+1}\text{-}R + R^{\cdot} \xrightarrow{L} M^nL_m + R_2 \text{ or } RH + R_{-H} \qquad (38)$$

i.e. leading to analogous organic products as radical - radical reactions.

Reaction of radicals with suitable substrates or scavengers:

$$R^{\cdot} + S \rightarrow \text{products} \qquad (39)$$

Reactions with dioxygen:

In the presence of dioxygen, which acts as a scavenger for the radicals $^{\cdot}R$, homolytic decomposition of an M-R complex is followed by:[122,123,134,135]

$$R^{\cdot} + O_2 \rightarrow RO_2^{\cdot} \qquad (40)$$

Which is usually followed by:[122,123,134,135]

$$RO_2^{\cdot} + M^{n}L_m \rightarrow L_{m-1}M^{n+1}\text{-OOR} + L \quad (41)$$

The transient complexes $L_{m-1}M^{n+1}$-OOR thus formed are an important class of intermediates by themselves.[7,8,154,155] Moreover in several systems, where M-R transient complexes decompose via homolysis, oxygen was used as a scavenger for the methyl radical [122,134] and thus aided in the elucidation of the mechanism and the determination of the associated rate constants [122,134].

It should be noted here that, in principle, also polymerization reactions or living polymerizations involving metal complexes as initiators and/or radical carriers could arise under any conditions where radicals are present in solutions. However most of these processes are not carried out in aqueous solutions and are therefore outside the scope of this report.

3) β-elimination reactions

Complexes with metal-carbon bonds, containing a good leaving group X in the β position to the bonded carbon are likely to undergo this type of decomposition reaction.

$$L_mM^{n+1}\text{-}CR^1R^2CR^3R^4X \rightarrow M^{n+1}L_m + R^1R^2C{=}CR^3R^4 + X^- \quad (42)$$

Good leaving groups are –OR, NR_2, NHC(O)R, and halides. [2,7]

4) Other decomposition reactions

Transient complexes with metal-carbon σ-bonds can decompose in a variety of additional possible decomposition reactions. Decomposition reactions include reactions with oxidants or reductants, β-hydride shift reactions, CO insertion, methyl migration, rearrangement of the carbon skeleton of R, bimolecular decomposition, methyl transfer reactions, etc.[2,7]

Participation of Complexes with Metal-Carbon σ-Bonds in the Reduction of Alkenes

A range of transition metal complexes act as catalysts or electrocatalysts for various reduction processes, due to the ability of these cations to exist in a variety of oxidation states, stabilised by suitable ligands.[156-162] $Ni^{II}L^{i}$ complexes, with macrocyclic ligands, are reduced reversibly to relatively stable nickel(I) species in aqueous solutions.[163,164] Nickel(I) macrocyclic complexes have attracted considerable attention because they enable the electrocatalytic reduction of alkyl halides.[28,158,165-172]

Halo-alkanes are common pollutants. Therefore there is an ongoing study on plausible approaches to the dehalogenation of halo-alkanes. One of these approaches involves their electrocatalytic reduction. $Ni^{II}L^{2+}$ (L = a tetraaza macrocyclic ligand) complexes were proposed as plausible electrocatalysts.[173] A pulse radiolytic study on the mechanism and kinetics of the reaction:

$$Ni^{I}L^{+} + RX \longrightarrow Ni^{II}L^{2+} + {}^{\cdot}R + X^{-} \qquad (43)$$

indicated that it is often followed by

$$Ni^{II}L^{2+} + {}^{\cdot}R \longrightarrow LNi^{III}\text{-}R^{2+} \qquad (44)$$

$$Ni^{I}L^{+} + LNi^{III}\text{-}R^{2+} \longrightarrow LNi^{II}\text{-}R^{+} + Ni^{II}L^{2+} \qquad (45)$$

$$LNi^{II}\text{-}R^{+} + H^{+} \longrightarrow Ni^{II}L^{2+} + HR \qquad (46)$$

i.e. the kinetics of the latter reactions affect the nature of the final products of the dehalogenation process.[173] This mechanism is analogous to the Cu(I)L catalyzed Ullman process.[14]

Furthermore Nickel complexes have recently increasingly been utilised as catalysts for ethylene polymerisations and oligomerisations.[174] Alkenes are known to form relatively stable d→π complexes with low valent transition metal ions (or atoms), thus stabilizing their low valent complexes.[175] For example ethylene was shown to stabilize several Ni(I) complexes. [16]

Recently while trying to stabilize $(Ni^{I}L')^{+}$ (L'=1,4,8,11-tetraazacyclotetradecane) in neutral aqueous solutions it was noted that allyl substituents on the macrocycle indeed thermodynamically stabilize $(Ni^{I}L')^{+}$ but kinetically destabilize it as the central cation reduces the allyl substituents.[176,177]

$(Ni^{I}L')^{+}$ reduces also alkenes, present homogeneously in the solution to form transient radicals. Two such reactions were studied up to date: [176,178-180]

Reaction of Ni(I)L with Ethylene

The quantitative reduction of ethylene yields mainly ethane with butane as a side product (traces of hexane are also obtained). The addition of ethylene accelerates the disappearance of the mono-valent nickel complex in a general acid catalyzed process. The following radical mechanism of reduction of ethylene was proposed. The first step probably involves the formation of a d→π* complex between the low valent NiL^{+} complex and the ethylene:[180]

$$NiL(H_2O)_2^{+} + C_2H_4 \overset{K_1}{\rightleftharpoons} ((H_2O)LNi\text{-}\|)^{+} \qquad (47)$$

followed by

$$((H_2O)LNi\text{-}\|)^{+} \xrightarrow{H^{+}} \begin{cases} \longrightarrow NiL^{2+} + {}^{\cdot}CH_2CH_3 \overset{K_{2b}}{\rightleftharpoons} (H_2O)LNi^{III}\text{-}CH_2CH_3^{2+} \\ \longrightarrow (H_2O)LNi^{III}\text{-}CH_2CH_3^{2+} \end{cases} \qquad (48)$$

Ethane is then most probably formed via

$$NiL(H_2O)_2^{+} + (H_2O)LNi^{III}\text{-}CH_2CH_3^{2+} \rightarrow NiL^{2+} + LNi^{II}\text{-}CH_2CH_3{}^{+}_{aq} \quad (49)$$

$$LNi^{II}\text{-}CH_2CH_3{}^{+}_{aq} \xrightarrow{H^+} NiL^{2+} + C_2H_6 \quad (50)$$

and butane via

$$2\,{}^{\cdot}C_2H_5 \rightarrow C_4H_{10} \quad (51)$$

or via

$$(H_2O)LNi^{III}\text{-}CH_2CH_3{}^{2+}_{aq} + {}^{\cdot}C_2H_5 \rightarrow NiL^{2+} + C_4H_{10} \quad (52)$$

or alternative reactions. As under the experimental conditions (low dose rate γ-irradiations) $NiL(H_2O)_2^{+}$ is formed continuously, reactions (49) and (50) compete well with reactions (51) and (52) and therefore the yield of C_2H_6 is considerably larger than that of C_4H_{10}.

REACTION OF NI(I)L WITH MALEATE

Studying the effect of electron withdrawing groups, e.g. CO_2^{-}, on the mechanism of reduction of alkenes by $Ni(I)L^{+}$ yielded the following results.[179] The mechanism of reduction of maleate and fumarate by $[Ni(I)L]^{+}$ differ significantly from each other despite of the similarity of the two substrates. The difference is probably mainly due to the fact that the radical anions formed in the reduction of both substrates have different characteristics due to the strength of the hydrogen bond in the radical anion of the maleic acid.[181-184]

Catalytically of interest is the reduction of maleic acid, which proceeds via a significantly different mechanism than the reduction by ethylene.[179] The first reaction observed is attributed to the equilibrium reaction between radiolytically produced NiL^{+} and maleate in the solution to produce the Ni-maleate d$\rightarrow\pi$ complex in analogy to the Ni(I)-ethylene system:

$$[Ni(I)L]^{+} + {}^{-}OOC\text{-}HC{=}CH\text{-}COO^{-} \leftrightarrows [Ni(I)L\text{-}({}^{-}OOC\text{-}HC{=}CH\text{-}COO^{-})]^{-} \quad (53)$$
$$K = (2.3\pm0.5)\times10^{4}\ M^{-1}$$

This reaction is followed by reduction of the maleic acid by the Ni(I) complex to form the maleate radical, which then forms a maleate intermediate with excess Ni(I):

$$[Ni(I)L\text{-}({}^{-}OOC\text{-}HC{=}CH\text{-}COO^{-})]^{-} \leftrightarrows I / II + (NiL)^{2+} \quad (54)$$

With I/II =

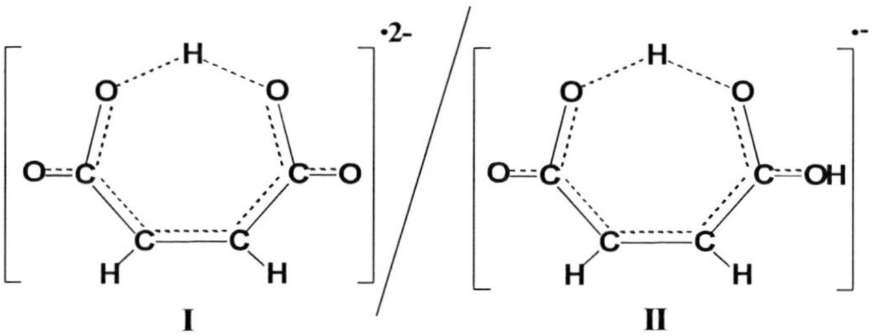

I II

Under the conditions of excess $[Ni(I)L]^+$ or its maleate complex, it is expected that the following reaction takes place to yield a succinate intermediate

$$I / II + [Ni(I)L]^+ \rightarrow [Ni(II)L\text{-}CH(CO_2^-)CH_2(CO_2^-)]^- \quad (55)$$

The final decomposition of the succinate intermediate is anticipated to proceed mainly according to the following reactions:

$$[Ni(II)L\text{-}CH(CO_2^-)CH_2(CO_2^-)]^- \leftrightarrows [Ni(I)L]^+ + \cdot CH(CO_2^-)CH_2(CO_2^-) \quad (56)$$

$$[Ni(II)L\text{-}CH(CO_2^-)CH_2(CO_2^-)]^- + \cdot CH(CO_2^-)CH_2(CO_2^-) \rightarrow \quad (57)$$

$$[Ni(I)L]^+ + (\text{-}CH(CO_2^-)CH_2CO_2^-)_2$$

with $(\text{-}CH(CO_2^-)CH_2CO_2^-)_2$ being the dimer.

The reaction mechanism of the reduction of maleic acid by $Ni(I)L^+$ was substantiated by analysis of the organic products by HPLC. The disappearance of maleic acid, and formation of fumaric, succinic acid and dimer were followed and their respective yields evaluated.

These results thus suggest that the detailed mechanism of reduction of ethylene and maleate by Ni^IL^+ yields as transients complexes with Ni-carbon bonds. The mechanism of decomposition of the latter depends on the nature of substituents on the alkyl radical.

$(NiL)^{2+}$ was also shown to be an electrocatalyst for the reduction of maleic acid in aqueous solutions.[179]

Reactions of Radicals with Surfaces, Especially with Nano-Particles, Immersed in Aqueous Solutions.

Radicals are formed at the surface of metals, semi-conductors and insulating oxides at a variety of processes, *e.g.:*

1. Heterogeneous radical induced catalytic processes: these include between others oxidation by dioxygen, probably many hydrogenation processes *etc.*[185[
2. Electrochemical processes: Radicals or radical ions, are formed as primary intermediates in a large variety of organic electrochemical processes including oxidations (*e.g* the Kolbe Reaction[186]) and reductions (*e.g.* reduction of alkyl halides[167] and diazonium salts[187]).

3. Photochemical processes in which the light is absorbed by the surface: As an example photocatalysis on TiO_2, electrodes and nano-particles, for environmental pollutant degradation[188-192] and solar energy conversion.[193,194]
4. Reduction of halo-organic compounds by metals, which is of environmental importance.[195]

Therefore the study of the mechanisms of reaction of radicals with these surfaces is of importance. Only few studies of this type were performed up-to-date though electrochemical studies point out that when the radical is formed at the electrode surface, *e.g.* via reduction of a diazonium salt at the open circuit potential, then the radical is expected to react with the electrode forming an intermediate, or a stable, product with a covalently bound organic molecule to the electrode[196]. Thus the reduction of aryl-diazonium salts, mainly in aprotic media, is used to prepare covalently bound organic layers to conductive or semi-conductive surfaces[187]. For metals forming stable organometallic compounds, *e.g.* Hg, Cd and Pb, results point out that "in a few electrochemical reactions, the initial electron transfer step does generate a σ radical at the electrode surface and organometallic compounds are formed."[197] However, it was suggested that usually "In the majority of electroorganic reactions, the working electrode is an inert material. Electron transfer generates a radical-ion species with sufficient lifetime to migrate away from the electrode surface. Further reactions then generate more reactive free radical species and these undergo terminal reactions before they are able to react with the electrode surface." [197] Recent results[195,198-200] suggest that the latter assumption has to be reconsidered as radicals formed at the surface of metals, even in water, react with them forming short lived intermediates with metal-carbon σ bonds. The mechanism of these reactions depends on the nature of the metal, the substituents on the radical, the composition of the solution *etc*. Thus for example methyl radicals react with iron powder to form methane via: [198]

58)

59)

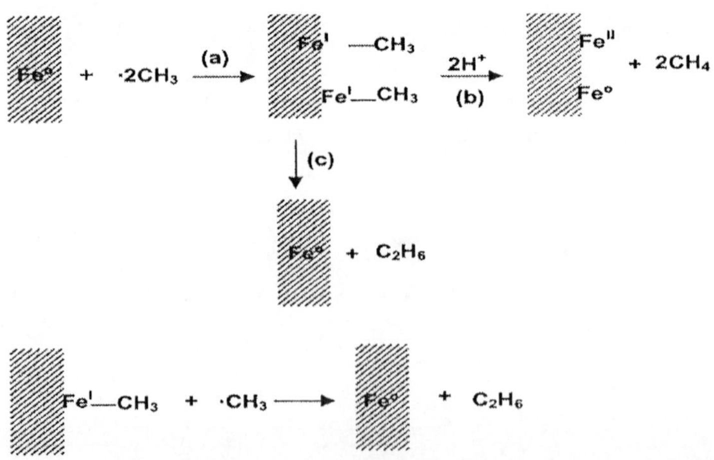

60) $\cdot CH_3 + \cdot CH_3 \longrightarrow C_2H_6$ $K_{60} = 1.2 \times 10^9\ M^{-1}s^{-1}$ [23] (t = 25°C, pH 5.5)

Scheme I

A similar mechanism is observed for Cr°, Mn°, Co°, Ni° and Zn°. However for Cu° the following mechanism was observed: [198]

61)

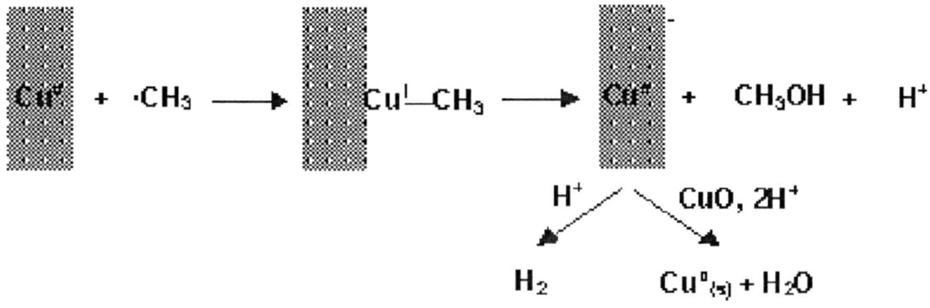

Scheme II

For Au° and Ag° colloids the reactions proceed via: [201]

62)

63)

64)

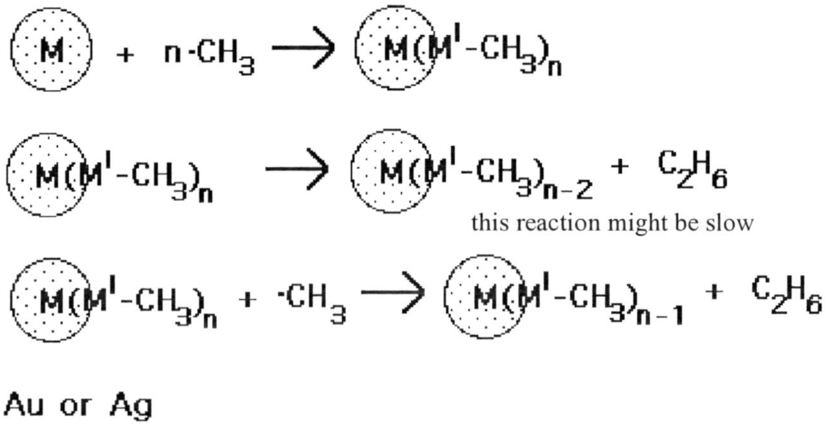

Scheme III

All the reactions of methyl radicals with the metal surfaces are very fast and thus all the radicals formed near the surface of the metals react with them forming intermediates with metal-carbon σ bonds. The nature of the metal affects the mechanism of decomposition of these intermediates. For the Au° nano-particles the results suggest that the intermediate Au-CH_3 is relatively long lived, $t_{½} > 15$ s.[202]

For the radical $^{\cdot}CH_2C(CH_3)_2OH$ the mechanism outlined in Scheme IV was observed.[200] The relative rates of reactions (66) and (67) depend on the metal, the pH and the temperature.[200] Similar reactions were observed for analogous radicals, *e.g.* $^{\cdot}CH_2CH_2OH$; $^{\cdot}CH_2CH(CH_3)OH$ and $^{\cdot}CH_2C(CH_3)_2NH_3^+$.[200] The relative rates of the reactions analogous to reactions (66) and (67) for these radicals differ from those observed for the $^{\cdot}CH_2C(CH_3)_2OH$ radical.[200]

65)

66)

67)

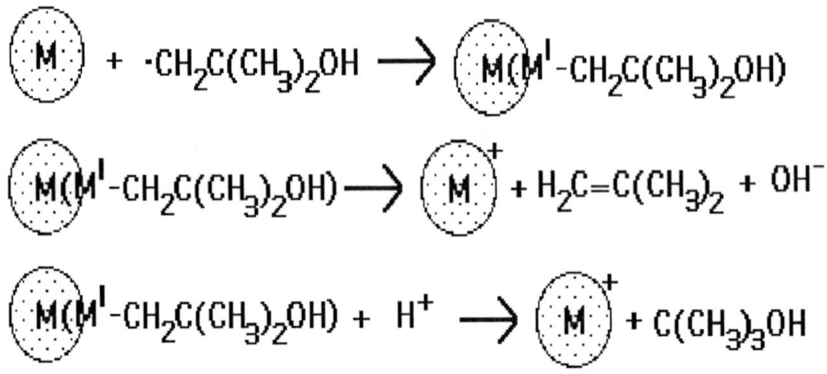

M = Fe; Co; Ni

Scheme IV

These results suggest that the reactions of most, if not all, alkyl radicals with metal surfaces have to be considered when radicals are formed near the surface of a metal. The mechanism of decomposition of the transients, with metal carbon σ bonds, thus formed depends on the nature of the metal, the substituents on the radical, the composition of the solution and the temperature.

The results suggested that probably also other types of radicals react with metal surfaces via the formation of short lived transient of the type. (M-)...... R, where R is the radical. For this purpose the methyl-peroxyl radical, $CH_3O_2^{\cdot}$ was reacted with Au° and Ag° nano-particles and with Fe° powders immersed in aqueous solutions. The preliminary results point out that also these reactions are fast and that their products are M^{2+} cations and methanol.[203] These results suggest that indeed also alkyl-peroxyl radicals react with metal surfaces via the formation of short lived transient with a metal-radical σ bond.

For reactions of radicals with semiconductors, *e.g.* TiO_2, only electron transfer processes were reported, see for example references [189,190,204]. However recent unpublished results in our laboratory show that methyl radicals react in fast reactions with TiO_2, anatase, immersed in aqueous solutions to form ethane. This result clearly proves that the methyl radicals react with the TiO_2 to form a short-lived transient in which the methyl is bound to the TiO_2. This conclusion is not so surprising as recent DFT calculations show that hydrogen atom, which is electronically similar to the methyl radical, adsorption to TiO_2 is so strong that it affects the preferred site of adsorption of gold nano-particles to this surface.[205]

Recently it was observed that the mechanism of reaction of the strong reducing radical $^{\cdot}C(CH_3)_2OH$ with Ag° nano-particles deposited on SiO_2 nano-particles differs considerably from that of the reaction of the same radical with "free" Ag° nano-particles.[206] This finding, though it is not fully understood, clearly demonstrates that the binding of the Ag° nano-particles to an "inert" substrate affects its chemical properties. This finding is of interest as it is known that metal nano-particles, especially Au° [207] but also other metals[208-216], deposited on different metal oxides,[212,216-218] including TiO_2 [207-216,218,219] are better catalysts than the pure metals. Thus it seems that the study of the mechanisms and kinetics of reaction of radicals with metal nano-particles deposited on different supports might shed light on the factors enhancing, or retarding, catalytic activity.

MECHANISMS OF REACTION OF RADICALS IN AQUEOUS SOLUTIONS.

Radicals and specifically alkyl-radicals ($R^{\cdot}$) are constrained to undergo one-electron reduction or oxidation processes due to their paramagnetic nature:

$$R^- \overset{e^-}{\leftrightarrows} R^{\cdot} \overset{e^-}{\leftrightarrows} R^+ \qquad (68)$$

The distinction of these types of processes into outer or inner sphere electron transfer reactions often determines the nature of the final products.

REDOX PROPERTIES OF RADICALS

Many radicals are highly reactive, as their HOMO has only one electron. Thus they are relatively strong oxidizing agents (i.e. electron acceptors) and reducing agents as they tend to donate the single electron occupying the high energy HOMO.

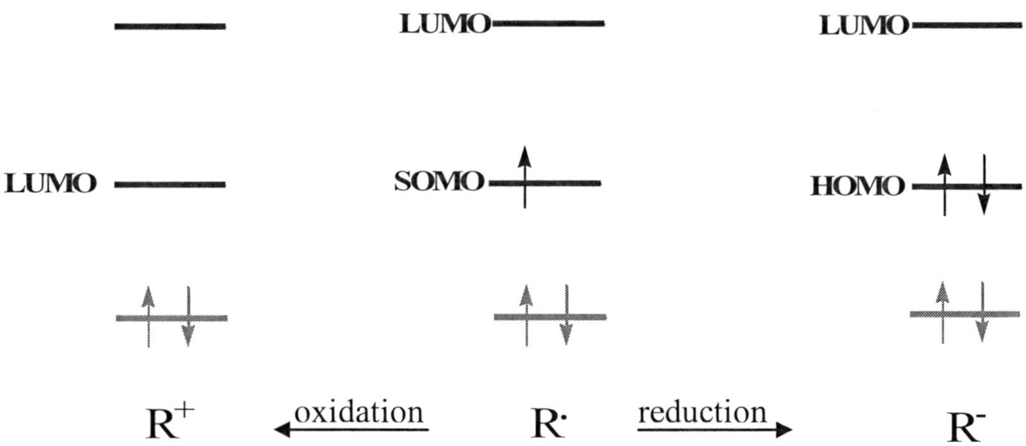

Thus the reactivity of radicals stems mainly from the fact that they are powerful one-electron redox reagents.

All the organic radicals listed above are relatively strong single electron oxidizing and reducing agents. The redox properties of aliphatic carbon-centered radicals depend on the substituents on the α-carbon. Thus for example radicals of the type $^{\cdot}CR^1R^2(OH)$ are relatively strong reducing agents; however they do oxidize low-valent transition metal complexes, e.g., $Cr(H_2O)_6^{2+}$ and $V(H_2O)_6^{2+}$.[2] On the other hand, $^{\cdot}CCl_3$ and $^{\cdot}CH_2CO_2H$ are relatively strong oxidizing agents. The known redox potentials of many radicals are summed up in two reviews[21,26].

From a compilation of redox-potentials for a wide variety of radicals in aqueous solutions, Table 1 presents an overview of redox-potentials for some selected important inorganic and organic radicals.[21,26]

A wealth of studies on reactions of radicals has been performed in aqueous solutions and rate constants have been compiled.[22,24] Rate constants for the reaction of e^-_{aq} or $^{\cdot}OH$ radicals are very high, often approaching the diffusion controlled limit, in accord with their high standard potentials (high driving force). $H^{\cdot}$ atoms and most organic radicals also react at high rates, though often considerably slower than $^{\cdot}OH$ radicals, with organic or inorganic compounds. Unfortunately the reduction potentials of many organic radicals are not known.

Table 1. Redox potentials of selected radicals in aqueous solutions

Compound/couple	E/ V
aq/e^-_{aq}	-2.87
$^{\cdot}H/H_2$	+2.31
$^{\cdot}H/H^+$	-2.31
$^{\cdot}OH/OH^-$	+1.9
$^{\cdot}OH, H^+/H_2O$	+2.73
$O_2^{\cdot-}, H^+/HO_2^-$	+1.0
$O_2^{\cdot-}, 2H^+/H_2O_2$	+1.76
$HO_2^{\cdot}, H^+/H_2O_2$	+1.48
$HO_2^{\cdot}/HO_2^-$	+0.79
$O_2/O_2^{\cdot-}$	-0.33
NO/NO^-	~+0.8
NO^+/NO	+0.33
$NO_2^{\cdot}/NO_2^-$	+1.0
$CO_2^{\cdot-}, H^+/HCO_2^-$	+1.07
$CO_2/CO_2^{\cdot-}$	-1.9
$Br_2^{\cdot-}/2Br^-$	+1.66
$Cl_2^{\cdot-}/2Cl^-$	+2.3
$I_2^{\cdot-}/2I^-$	+1.1
$CH_2O/^{\cdot}CH_2O^-$	-1.81
$CH_2O, H^+/^{\cdot}CH_2OH$	-1.18
$CH_3CH_2O, H^+/CH_3{^{\cdot}}CHOH$	-1.25
$(CH_3)_2CO, H^+/(CH_3)_2{^{\cdot}}COH$	-1.39
$RS^{\cdot}, H^+/RSH$	+1.3
$HS^{\cdot}/HS^-$	+1.15
$RO_2^{\cdot}, H^+/RO_2H$	
$CH_3O_2^{\cdot}, H^+/CH_3O_2H$	(+0.6)-(+0.7)

Inner-Sphere vs. Outer Sphere Electron-Transfer Mechanism

In theory electron transfer reactions may proceed via two distinct possible mechanisms, an inner sphere or outer sphere mechanism.

Outer sphere mechanism: the electron is transferred between the redox couples, while the inner coordination shells, if existing, of the respective reactants remain intact. No chemical bond between the reactants is formed, i.e. no chemical bonds are formed or broken during the process.

Inner sphere mechanism: A chemical bond is formed between the reactants prior to the electron transfer or basically there is a substantial interaction between the donor and acceptor species in the transition state (e.g. formation of an encounter complex with charge transfer character, i.e. strongly coupled donor and acceptor).

DEPENDENCE OF THE ELECTRON TRANSFER RATE CONSTANT ON THE DRIVING FORCE

Marcus introduced a powerful correlation concerning outer sphere electron transfer reactions, in its simplified form:[220,221]

$$A^+ + B^- \xrightarrow{k_{12}} A + B \quad (69)$$

$$A + {}^*A^+ \xrightarrow{k_{11}} A^+ + A^* \quad (70)$$

$$B + {}^*B^- \xrightarrow{k_{22}} B^- + {}^*B \quad (71)$$

$$k_{12} = \sqrt{k_{11}k_{22}K_{12}f} \quad (72)$$

with k_{11} and k_{22} being the self exchange rate constants for either reaction partner and k_{12}, K_{12} the rate constant and equilibrium constant for the cross reaction.

$$\ln f = (\ln K_{12})^2/(4\ln(k_{11}k_{22}/Z^2)) \quad (73)$$

and Z is the collision frequency, usually taken to be $10^{11}M^{-1}s^{-1}$. Thus the reaction rate is mainly controlled by the free energy change in the so-called normal region,[220,221] and the self exchange rate constants. For accurate calculations work term corrections have to be applied.[220,221]

In order to calculate electron transfer rates from tabulated redox potentials, it has to be ascertained that tabulated redox potentials are indeed due to the pure redox reaction and do not incorporate follow-up reactions. The last statement seems to be especially true for reactions involving radicals, which usually are accompanied by fast follow-up reactions with solution components. Thus for example the redox potentials reported for $^{\cdot}CH_3$ radicals are for the reactions:

$$^{\cdot}CH_3 + e^- + H^+ \rightarrow CH_4 \quad (74)$$

$$^{\cdot}CH_3 - e^- + OH^- \rightarrow CH_3OH \quad (75)$$

and not for the outer-sphere reactions:

$$^{\cdot}CH_3 + e^- \rightarrow CH_3^- \quad (76)$$

$$^{\cdot}CH_3 - e^- \rightarrow CH_3^+ \quad (77)$$

respectively. Thus the redox potentials, which have to be taken into account for evaluating K_{12}, are not those reported and they have to be corrected by the free energy gain in the reactions

$$CH_3^- + H_2O \rightarrow CH_4 + OH^- \quad (78)$$

$$CH_3^+ + H_2O \rightarrow CH_3OH + H^+ \quad (79)$$

respectively. Similarly, even for the reductions by $^{\cdot}CR^1R^2OH$ radicals the potentials are given for the couples $(CR^1R^2O + H^+)/^{\cdot}CR^1R^2OH$ and not for the outer-sphere couples $CR^1R^2OH^+)/^{\cdot}CR^1R^2OH$, which should be used for evaluating K_{12}. The difference between the two values is 0.059(pH-pK_a) V, where pK_a is the pK_a of the acid $CR^1R^2OH^+$, which is not known.

Kochi et al (studying electron transfer reactions in excited donor/acceptor (arene/quinone) redox pairs) stated, that inner sphere mechanisms should be considered in all cases where charge transfer complexes, exiplexes, or contact ion-radical pairs are actually observed as reaction intermediates.[222] Thus only interactions in the transition state of less than 1 kcal/mol (350 cm^{-1}) are arbitrarily defined to result in outer sphere electron transfer.[222] It should be pointed out that linear free energy relations were reported also for hydrogen⁻ or halide abstraction reactions, which result in radical formation.[167,223]

k_{12} rate constants higher than those predicted by Marcus theory were reported for inner sphere mechanisms and their deviation correlated with the strength of the corresponding encounter complexes. Fast rate constants in inner sphere reactions are explained by the adiabatic pathway. Strong electronic coupling between donor and acceptor leads to an avoided crossing of the potential surfaces resulting in a substantial energy gap.[222]

Eberson and Shaik analyzed electron transfer reactions of radical anions employing the valence bond curve-crossing model. They reached the conclusion that the reaction between an organic radical anion or cation and an even-electron organic species should proceed by an inner sphere mechanism, unless forced by structural features into the outer-sphere extreme. Generally outer sphere reactions are considered rare for this class of reactions. Thus the inner sphere mechanism appears to be an optimum choice between the tendency of transition states to minimize intermolecular repulsion and maximize bonding. Due to avoided crossing a factor B(inner sphere) of 2.3 kcal/mol ≤ B(inner sphere) ≤ 7 kcal/mole (related to about 3 Å distance of the respective atoms in the transition state) is gained in those reactions vs. comparative outer sphere reactions. It is shown, that inner sphere processes can adhere perfectly well to the Marcus equation, provided that B(inner sphere) is a constant.[224]

While it is commonly perceived that most electron transfer processes in biological systems proceed via outer sphere electron transfer mechanisms, these kinds of reactions are in fact rare for radicals. It has to be noted here, that most electron transfer processes of radicals do not transpire via simple outer sphere electron transfer processes. Generally electron transfer processes proceed via an outer sphere mechanism, only when the reaction is not followed by a very exothermic step, see above, and does not require major bond rearrangements and therefore the self exchange rates of the couples are high.

Even most of the simple radicals described in table 1 do not adhere to the requirements for outer sphere electron transfer. Most uncharged radicals will be converted into charged entities due to the electron transfer and thus intrinsically, have in polar solvents, e.g. water, distinctly higher solvation energy than the originally uncharged species. Thus for example the $H^{\cdot}$ or $^{\cdot}OH$ radicals only have small hydration energies, while their one electron redox products H^+ or OH^- respectively possess high solvation energies. The reduced product of $H^{\cdot}$, H^-, which will require also considerable solvation, on the other hand is unstable and requires the follow up reaction with H^+ to form H_2 for the reaction to proceed as the latter reaction is very exothermic. Thus most, if not all, the reactions of these radicals proceed via the inner sphere mechanism.[1,2]

$CO_2^{\cdot -}$ for example is a powerful reducing agent but as it is bent and its oxidized product CO_2 is linear, most of its reactions proceed via the inner-sphere mechanism.[225] For the reactions of $X_2^{\cdot -}$ to form X_2^{2-} the latter are unstable and therefore the reactions do not proceed via outer sphere mechanisms. $N_3^{\cdot}$ on the other hand is a strong outer sphere oxidizing agent, as N_3^- does not have a high hydration energy.

All the organic radicals are relatively strong single electron oxidizing and reducing agents. The redox properties of aliphatic carbon-centered radicals depend on the substituents on the α-carbon. Thus for example radicals of the type $^{\cdot}CR^1R^2(OH)$ are relatively strong reducing agents but they reduce transition metal complexes, even $Co(NH_3)_6^{3+}$,[226] *via* the inner sphere mechanism; however they also oxidize low-valent transition metal complexes, e.g. $Cr(H_2O)_6^{2+}$,[227] and $V(H_2O)_6^{2+}$ [228] again *via* the inner sphere mechanism. On the other hand, $^{\cdot}CCl_3$ and $^{\cdot}CH_2CO_2H$ and alkyl-peroxyl radicals are relatively strong oxidizing agents. As the self exchange rates for the $R^{\cdot/-}$ and $R^{+/\cdot}$ couples are usually slow (the $R^{\cdot/-}$ couple involves radical steric changes from planar to tetrahedral), and as the products R^+ and R^- are highly unstable, outer sphere reactions are not abundant for these types of molecules. Indeed transients of the type M^{n+1}-R are observed in most reactions of the type

$$M^nL_m + R^{\cdot} \xrightarrow{-L} L_{m-1}M^{n+1}\text{-R} \tag{80}$$

$$M^nL_m + RO_2^{\cdot} \xrightarrow{-L} L_{m-1}M^{n+1}\text{-OOR} \tag{81}$$

and the activation parameters are analogous to those of ligand exchange reactions of the complexes M^nL_m. [2]

To conclude it is pointed out that most radical reactions, at least those involving transition metal complexes in protic media, proceed via the inner sphere mechanism.

ACKNOWLEDGMENTS

We are indebted to our co-workers whose work is cited in this manuscript. The financial help via grants from the Israel Science Foundation, The Planning and Granting Committee and the Israel Atomic Energy Committee, The German Israel Binational Science Foundation and The American Israel Binational Science Foundation and The Alexander von Humboldt Foundation enabled these studies.

REFERENCES

[1] Masarwa, A.; Meyerstein, D. In *Inorganic And Bio-Inorganic Chemistry 9. Inorganic Reaction Mechanisms*; Encyclopedia Of Life Support Systems (EOLSS), 2006; Vol. 2006.
[2] Masarwa, A.; Meyerstein, D. In *Advances In Inorganic Chemistry: Including Bioinorganic Studies, Vol 55*, 2004; Vol. 55, Pp 271-313.
[3] Halliwell, B.; Gutteridge, J. M. C. *Free Radicals In Biology And Medicine*; Oxford Univ. Press, 1999.
[4] Kochi, J. K. *Oxidation Reduction Reactions*; Wiley: New York, 1973; Vol. 1.
[5] Kochi, J. K. *Organometallic Mechanisms And Catalysis*; Academic Press: New York, 1978.
[6] Meyerstein, D. *Abstracts Of Papers Of The American Chemical Society* 1977, *174*, 163-163.
[7] Meyerstein, D. In *Metal Ions In Biological Systems, Vol 36*; Marcel Dekker: New York, 1999; Vol. 36, Pp 41-77.
[8] Sheldon, R. A.; Kochi, J. K. *Metal Catalyzed Oxidations Of Organic Compounds*; Academic Press: New York, 1981.
[9] Sigel, A.; Sigel, H. *Metal Ions In Biological Systems*; Marcel Dekker: New York, 1999.
[10] Stanbury, D. M. In *Advances In Inorganic Chemistry: Including Bioinorganic Studies, Vol 54*, 2003; Vol. 54, Pp 351-393.
[11] Marie, M.; Warren, T. H. In *The 230th ACS*, 2005.
[12] Kapoor, P.; Warren, T. H. In *The 230th ACS*, 2005.
[13] Sheldon, R. A.; Kochi, J. K. *J. Am. Chem. Soc.* 1968, 5196-5207.
[14] Rusonik, I.; Cohen, H.; Meyerstein, D. *Dalton Transactions* 2003, 2024-2028.
[15] Saphier, M.; Masarwa, A.; Cohen, H.; Meyerstein, D. *European Journal Of Inorganic Chemistry* 2002, 1226-1234.
[16] Navon, N.; Cohen, H.; Paoletti, P.; Valtancoli, B.; Bencini, A.; Meyerstein, D. *Industrial & Engineering Chemistry Research* 2000, *39*, 3536-3540.
[17] Navon, N.; Burg, A.; Cohen, H.; Van Eldik, R.; Meyerstein, D. *European Journal Of Inorganic Chemistry* 2002, 423-429.
[18] Tabata, Y. *Pulse Radiolysis*; CRC Press: Boca Raton, Florida, 1991.
[19] (Matheson, M. S.; Dorfman, L. *Pulse Radiolysis*; M.I.T. Press: Cambridge, MA, 1969.
[20] Spinks, J. W. T.; Woods, R. J. *An Introduction To Radiation Chemistry, 3rd Ed.*; John Wiley&Sons: New York, 1990.
[21] Wardman, P. *J. Phys. Chem. Ref. Data* 1989, *18*, 1637-1755.

[22] Neta, P.; Huie, R. E.; Ross, A. B. *J. Phys. Chem. Ref. Data* 1988, *17*, 1027-1284.
[23] Neta, P.; Huie, R. E.; Ross, A. B. *J. Phys. Chem. Ref. Data* 1990, *19*, 413-513.
[24] Neta, P.; Grodkowski, J.; Ross, A. B. *J. Phys. Chem. Ref. Data* 1996, *25*, 709-1050.
[25] Buxton, G. V.; Greenstock, C. V.; Helman, W. P.; Ross, A. B. *J. Phys. Chem. Ref. Data* 1988, *17*, 513-886.
[26] Stanbury, D. M. *Advances In Inorganic Chemistry* 1989, *33*, 69.
[27] Armour, E. P. A.; Corry, P. M. *Radiation Research* 1982, *89*, 369–380.
[28] Shandalov, E.; Zilbermann, I.; Maimon, E.; Nahmani, Y.; Cohen, H.; Meyerstein, D. *Tetrah. Lett.* 2004, *45*, 989-992.
[29] Cohen, H.; Gould, E. S.; Meyerstein, D.; Nutkovich, M.; Radlowski, C. A. *Inorganic Chemistry* 1983, *22*, 1374-1377.
[30] Watanabe, Y.; Fujii, H. In *Metal-Oxo And Metal-Peroxo Species In Catalytic Oxidations*, 2000; Vol. 97, Pp 61-89.
[31] Groves, J. T. *Journal Of Inorganic Biochemistry* 2006, *100*, 434-447.
[32] Chaudhuri, P.; Wieghardt, K.; Weyhermuller, T.; Paine, T. K.; Mukherjee, S.; Mukherjee, C. *Biological Chemistry* 2005, *386*, 1023-1033.
[33] Que, L.; Tolman, W. B. *Nature* 2008, *455*, 333-340.
[34] Thomas, F. *European Journal Of Inorganic Chemistry* 2007, 2379-2404.
[35] Chaudhuri, P.; Wieghardt, K. In *Progress In Inorganic Chemistry, Vol. 50*, 2001; Vol. 50, Pp 151-216.
[36] Herebian, D.; Wieghardt, K. E.; Neese, F. *Journal Of The American Chemical Society* 2003, *125*, 10997-11005.
[37] Marlin, D. S.; Bill, E.; Weyhermuller, T.; Bothe, E.; Wieghardt, K. *Journal Of The American Chemical Society* 2005, *127*, 6095-6108.
[38] Fenton, H. J. H. *J.Chem.Soc.* 1894, *65*, 899-910.
[39] Wardman, P.; Candeias, L. P. *Radiat.Res.* 1996, *145*, 523-531.
[40] Goldstein, S.; Meyerstein, D.; Czapski, G. *Free Radical Biology And Medicine* 1993, *15*, 435-445.
[41] Halliwell, B.; Gutteridge, J. M. C. *Meth. Enzymol.* 1990, *86*, 1-85.
[42] Liochev, S. I. In *Metal Ions In Biological Systems, Vol 36*; Marcel Dekker: New York, 1999; Vol. 36, Pp 1-40.
[43] Pignatello, J. J.; Oliveros, E.; Mackay, A. *Critical Reviews In Environmental Science And Technology* 2006, *36*, 1 - 84.
[44] Edwards, J. O.; Curci, R. *Catalytic Oxidations With Hydrogen Peroxide As Oxidant*; Kluwer Academic Publishers, 1992.
[45] Shi, F.; Tse, M. K.; Li, Z.; Beller, M. *Chem. Eur. J.* 2008, *14*, 8793-8797.
[46] Gutteridge, M. C.; Westermarck, T.; Halliwell, B. *Free Radicals, Aging And Degenerative Diseases*; Alan R. Liss, Inc.: New York, 1986; Vol. 8.
[47] Chiu, D.; Kuypers, F.; Lubin, B. *Seminars In Hematology* 1989, *26*, 257-276.
[48] Meunier, B. *Metal-Oxo And Metal-Peroxo Species In Catalytic Oxidations*; Springer Verlag: Berlin, 2000; Vol. 97.
[49] Prousek, J. *Pure And Applied Chemistry* 2007, *79*, 2325-2338.
[50] Dunford, H. B. *Coordination Chemistry Reviews* 2002, *233*, 311-318.
[51] Goldstein, S.; Meyerstein, D. *Accounts Of Chemical Research* 1999, *32*, 547-550.
[52] Gozzo, F. *Journal Of Molecular Catalysis A-Chemical* 2001, *171*, 1-22.

[53] Sawyer, D. T.; Sobkowiak, A.; Matsushita, T. *Accounts Of Chemical Research* 1996, *29*, 409-416.

[54] Sychev, A. Y. *Russian Chemical Reviews* 1995, *64*, 1105-1129.

[55] Walling, C. *Acc. Chem. Res.* 1975, *8*, 125-131.

[56] Walling, C. *Accounts Of Chemical Research* 1998, *31*, 155-157.

[57] Kremer, M. L. *Physical Chemistry Chemical Physics* 1999, *1*, 3595-3605.

[58] Kremer, M. L. *Journal Of Inorganic Biochemistry* 2000, *81*, 325-325.

[59] Kremer, M. L. *Journal Of Physical Chemistry A* 2003, *107*, 1734-1741.

[60] Kremer, M. L. *International Journal Of Chemical Kinetics* 2006, *38*, 725-736.

[61] Masarwa, M.; Cohen, H.; Meyerstein, D.; Hickman, D. L.; Bakac, A.; Espenson, J. H. *Journal Of The American Chemical Society* 1988, *110*, 4293-4297.

[62] Sutton, H. C.; Winterbourn, C. C. *Free Radical Biology And Medicine* 1989, *6*, 53-60.

[63] Wink, D. A.; Nims, R. W.; Saavedra, J. E.; Utermahlen, W. E.; Ford, P. C. *Proceedings Of The National Academy Of Sciences Of The United States Of America* 1994, *91*, 6604-6608.

[64] Wink, D. A.; Wink, C. B.; Nims, R. W.; Ford, P. C. *Environmental Health Perspectives* 1994, *102*, 11-15.

[65] Pestovsky, O.; Stoian, S.; Bominaar, E. L.; Shan, X. P.; Munck, E.; Que, L.; Bakac, A. *Angewandte Chemie-International Edition* 2005, *44*, 6871-6874.

[66] Masarwa, A.; Rachmilovich-Calis, S.; Meyerstein, N.; Meyerstein, D. *Coordination Chemistry Reviews* 2005, *249*, 1937-1943.

[67] Rachmilovich-Calis, S.; Masarwa, A.; Meyerstein, N.; Meyerstein, D. *European Journal Of Inorganic Chemistry* 2005, 2875-2880.

[68] Haber, F.; Weiss, J. J. *Proc. R. Soc. London* 1934, *A147*, 332-351.

[69] Rachmilovich-Calis, S.; Masarwa, A.; Meyerstein, N.; Meyerstein, D.; Van Eldik, R. *Chem. Eur. J.* 2009, Online.

[70] Barb, W. G.; Baxendale, J. H.; George, P.; Hargrave, K. R. *Transactions Of The Faraday Society* 1951, *47*, 462-500.

[71] Bray, W. C.; Gorin, M. H. *J. Am. Chem. Soc.* 1932, *54*, 2124.

[72] Buda, F.; Ensing, B.; Gribnau, M. C. M.; Baerends, E. J. *Chem. Eur. J.* 2003, *9*, 3436-3444.

[73] Buda, F.; Ensing, B.; Gribnau, M. C. M.; Baerends, E. J. *Chem. Eur. J.* 2001, *7*, 2775-2783.

[74] Ensing, B.; Buda, F.; Blochl, P. E.; Baerends, E. J. *Angew. Chem. Int. Ed.* 2001, *40*, 2893-2895.

[75] Ensing, B.; Buda, F.; Blochl, P. E.; Baerends, E. J. *Phys. Chem. Chem. Phys.* 2002, *4*, 3619-3627.

[76] Masarwa, M.; Cohen, H.; Meyerstein, D.; Hickman, D. L.; Bakac, A.; Espenson, J. H. *J. Am. Chem. Soc.* 1988, *110*, 4293-4297.

[77] Luzzatto, E.; Cohen, H.; Stockheim, C.; Wieghardt, K.; Meyerstein, D. *Free Radical Research* 1995, *23*, 453-463.

[78] Bamnolker, H.; Cohen, H.; Meyerstein, D. *Free Radical Research Communications* 1991, *15*, 231-241.

[79] Pestovsky, O.; Bakac, A. *Inorganic Chemistry* 2006, *45*, 814-820.

[80] Kremer, M. L. *Int. J. Chem. Kin* 2008, *40*, 541-553.

[81] Johnson, G. R. A.; Nazhat, N. B.; Saadalla-Nazhat, R. A. *J. Chem. Soc. Faraday Trans.I* 1988, *84*, 501-509.

[82] Johnson, G. R. A.; Nazhat, N. B. *J. Am. Chem. Soc.* 1987, *109*, 1990-1994.

[83] Mansano-Weiss, C.; Cohen, H.; Meyerstein, D. *Journal Of Inorganic Biochemistry* 2002, *91*, 199-204.

[84] Costas, M.; Mehn, M. P.; Jensen, M. P.; Que, L. *Chemical Reviews* 2004, *104*, 939-986.

[85] Sharma, V. K.; Szilagyi, P. A.; Homonnay, Z.; Kuzmann, E.; Vertes, A. *European Journal Of Inorganic Chemistry* 2005, 4393-4400.

[86] Clerici, M. G. In *Heterogeneous Catalysis And Fine Chemicals Iii*, 1993; Vol. 78, Pp 21-33.

[87] Kremer, M. L.; Stein, G. *Trans. Faraday Soc.* 1959, 959-973.

[88] Wada, G.; Nakamura, T.; Terauchi, K.; Nakai, T. *Bull. Chem. Soc. Jpn.* 1964, *37*, 447-455.

[89] Ulrich, V.; Staudinger, H. In *Microsoms And Drug Oxidation*; Academic Press: New York, 1969, Pp 199-223.

[90] Udenfriend, S.; Clark, C. T.; Axelrod, J.; Brodie, B. B. *J. Biol. Chem.* 1954, *208*, 731-740.

[91] Chevion, M. *Free Rad. Biol. Med.* 1988, *5*, 27.

[92] Czapski, G. *Isr. J. Chem.* 1984, *24*, 29.

[93] Lynch, R. C.; Fridovich, I. *Biochim. Biophys. Acta* 1979, *571*, 195.

[94] Cohen, H.; Meyerste.D *Journal Of The American Chemical Society* 1971, *93*, 4179-&.

[95] Cohen, H.; Meyerstein, D. *Inorg. Chem.* 1974, *13*, 2434-2443.

[96] Mulac, W. A.; Cohen, H.; Meyerstein, D. *Inorganic Chemistry* 1982, *21*, 4016-4020.

[97] Espenson, J. H. In *Inorganic And Bioinorganic Mechanisms*; Sykes, G., Ed.; Academic Press, 1982; Vol. 1, P 5.

[98] Espenson, J. H.; Connolly, P.; Meyerstein, D.; Cohen, H. *Inorganic Chemistry* 1983, *22*, 1009-1013.

[99] Cohen, H.; Meyerstein, D.; Shusterman, A. J.; Weiss, M. *Journal Of The Chemical Society, Chemical Communications* 1985, 424-425.

[100] Cohen, H.; Gaede, W.; Gerhard, A.; Meyerstein, D.; Vaneldik, R. *Inorganic Chemistry* 1992, *31*, 3805-3809.

[101] Bakac, A.; Espenson, J. H. *Inorganic Chemistry* 1989, *28*, 3901-3904.

[102] Cohen, H.; Feldman, A.; Ishshalom, R.; Meyerstein, D. *Journal Of The American Chemical Society* 1991, *113*, 5292-5299.

[103] Vaneldik, R.; Gaede, W.; Cohen, H.; Meyerstein, D. *Inorganic Chemistry* 1992, *31*, 3695-3696.

[104] Espenson, J. H. *Accounts Of Chemical Research* 1992, *25*, 222-227.

[105] Guldi, D. M.; Neta, P.; Hambright, P. *J. Chem. Soc., Faraday Trans. I* 1992, *88*, 2337.

[106] Cohen, H.; Meyerstein, D. *Inorganic Chemistry* 1988, *27*, 3429-3431.

[107] Morehouse, K. M.; Neta, P. *J. Phys. Chem.* 1984, *88*, 1575.

[108] Guldi, D. M.; Kumar, M.; Neta, P.; Hambright, P. *J. Phys. Chem.* 1992, *96*, 9576.

[109] Sorek, Y.; Cohen, H.; Meyerstein, D. *J. Chem. Soc., Faraday Trans. I* 1985, *81*, 233-239.

[110] Sorek, Y.; Cohen, H.; Meyerstein, D. *J. Chem. Soc., Faraday Trans. I* 1986, *82*, 3431-3438.

[111] Goldstein, S.; Czapski, G.; Cohen, H.; Meyerstein, D. *Journal Of The American Chemical Society* 1988, *110*, 3903-3907.
[112] Brault, D.; Neta, P. *J. Am. Chem. Soc.* 1981, *103*, 2705.
[113] Brault, D.; Neta, P. *J. Phys. Chem.* 1982, *86*, 3405.
[114] Brault, D.; Neta, P. *J. Phys. Chem.* 1987, *91*, 4156.
[115] Cabelli, D. E.; Rush, J. D.; Thomas, M. J.; Bielski, B. H. J. *J. Phys. Chem.* 1989, *93*, 3579.
[116] Meyerstein, D.; Schwarz, H. A. *J. Chem. Soc., Faraday Trans. I* 1988, *84*, 2933-2949.
[117] Shaham, N.; Masarwa, A.; Matana, Y.; Cohen, H.; Meyerstein, D. *Eur. J. Inorg. Chem.* 2002, 87-92.
[118] Roche, T. S.; Endicott, J. F. *Inorg. Chem.* 1974, *13*, 1575.
[119] Elroi, H.; Meyerstein, D. *Journal Of The American Chemical Society* 1978, *100*, 5540-.
[120] Meyerstein, D. *Pure And Applied Chemistry* 1989, *61*, 885-889.
[121] Sorek, Y.; Cohen, H.; Meyerstein, D. *J. Chem. Soc., Faraday Trans. I* 1989, *85*, 1169-1179.
[122] Sauer, A.; Cohen, H.; Meyerstein, D. *Inorganic Chemistry* 1989, *28*, 2511-2512.
[123] Vaneldik, R.; Cohen, H.; Meyerstein, D. *Angewandte Chemie-International Edition In English* 1991, *30*, 1158-1160.
[124] Blackburn, R.; Kyaw, M.; Phillips, G. O.; Swallow, A. J. *J. Chem. Soc., Faraday Trans. 1* 1975, *71*, 2277.
[125] Mchatton, R. C.; Espenson, J. H.; Bakac, A. *Journal Of The American Chemical Society* 1986, *108*, 5885-5890.
[126] Bakac, A.; Espenson, J. H. *Inorganic Chemistry* 1989, *28*, 4319-4322.
[127] Lee, S. Y.; Espenson, J. H.; Bakac, A. *Inorganic Chemistry* 1990, *29*, 3442-3447.
[128] Baral, S.; Neta, P. *J. Phys. Chem.* 1983, *87*, 1502.
[129] Kumar, M.; Natarajan, E.; Neta, P. *J. Phys. Chem.* 1994, *98*, 8024.
[130] Guldi, D. M.; Neta, P.; Hambright, P.; Rahimi, R. *Inorg. Chem.* 1992, *31*, 4849.
[131] Kelm, M.; Lilie, J.; Henglein, A.; Janata, E. *J. Phys. Chem.* 1974, *78*, 882.
[132] Ram, M. S.; Bakac, A.; Espenson, J. H. *Inorganic Chemistry* 1986, *25*, 3267-3272.
[133] Kelley, D. G.; Marchaj, A.; Bakac, A.; Espenson, J. H. *Journal Of The American Chemical Society* 1991, *113*, 7583-7587.
[134] Sauer, A.; Cohen, H.; Meyerstein, D. *Inorganic Chemistry* 1988, *27*, 4578-4581.
[135] Vaneldik, R.; Cohen, H.; Meshulam, A.; Meyerstein, D. *Inorganic Chemistry* 1990, *29*, 4156-4158.
[136] Freiberg, M.; Mulac, W. A.; Schmidt, K. H.; Meyerstein, D. *Journal Of The Chemical Society, Faraday Transactions I* 1980, *76*, 1838-1848.
[137] Cohen, H.; Meyerstein, D. *Inorg. Chem.* 1986, *25*, 1505-1506.
[138] Cohen, H.; Meyerstein, D. *Inorganic Chemistry* 1987, *26*, 2342-2344.
[139] Goldstein, S.; Czapski, G.; Cohen, H.; Meyerstein, D. *Inorganic Chemistry* 1988, *27*, 4130-4135.
[140] Masarwa, M.; Cohen, H.; Meyerstein, D. *Inorg. Chem.* 1991, *30*, 1849-1854.
[141] Goldstein, S.; Czapski, G.; Cohen, H.; Meyerstein, D. *Inorg. Chim. Acta* 1992, *192*, 87-93.
[142] Goldstein, S.; Czapski, G.; Cohen, H.; Meyerstein, D.; Cho, J. K.; Shaik, S. S. *Inorganic Chemistry* 1992, *31*, 798-803.

[143] Navon, N.; Golub, G.; Cohen, H.; Meyerstein, D. *Organometallics* 1995, *14*, 5670-5676.
[144] Szulc, A.; Meyerstein, D.; Cohen, H. *Inorg. Chim. Acta* 1998, *270*, 440-445.
[145] Buxton, G. V.; Green, J. C. *J. Chem. Soc., Faraday Trans. I* 1977, *74*, 697-714.
[146] Freiberg, M.; Meyerstein, D. *Journal Of The Chemical Society, Chemical Communications* 1977, 127-128.
[147] Mulac, W. A.; Meyerstein, D. *J. Chem. Soc., Chem. Comm.* 1979, 893-895.
[148] Freiberg, M.; Meyerstein, D. *J. Chem. Soc., Faraday Trans. I* 1980, *76*, 1825-1837.
[149] Kirschenbaum, L. J.; Meyerstein, D. *Inorganic Chemistry* 1980, *19*, 1373-1379.
[150] Masarwa, M.; Cohen, H.; Meyerstein, D. *Inorganic Chemistry* 1986, *25*, 4897-4900.
[151] Masarwa, M.; Cohen, H.; Glaser, R.; Meyerstein, D. *Inorg. Chem.* 1990, *29*, 5031-5035.
[152] Masarwa, M.; Cohen, H.; Saar, J.; Meyerstein, D. *Israel J. Chem.* 1990, *30*, 361-368.
[153] Goldstein, S.; Czapski, G.; Cohen, H.; Meyerstein, D. *Inorganic Chemistry* 1992, *31*, 2439-2444.
[154] Strukul, G., Ed. *Catalytic Oxidations With Hydrogen Peroxide As Oxidant*; Kluwer Academic Publishers: Netherlands, 1992.
[155] Sawyer, D. T. *Oxygen Chemistry*; Oxford University Press: New York, 1991.
[156] Fisher, B.; Eisenberg, R. *J. Am. Chem. Soc.* 1980, *102*, 7361-7363.
[157] Brown, G. M.; Brunschwig, B. S.; Creutz, C.; Endicott, J. F.; Sutin, N. *J. Am. Chem. Soc.* 1979, *101*, 1298-1300.
[158] Becker, J. Y.; Kerr, J. B.; Pletcher, D.; Rosas, R. *J. Electroanal. Chem.* 1981, *117*, 87-99.
[159] Pearce, D. J.; Pletcher, D. *J. Electroanal. Chem.* 1986, *197*, 317-330.
[160] Beley, M.; Collin, J. P.; Ruppert, R.; Sauvage, J. P. *J. Am. Chem. Soc.* 1986, *108*, 7461-7467.
[161] Taniguchi, I.; Nakashima, N.; Matsushita, K.; Yasukouchi, K. *J. Electroanal. Chem.* 1987, *224*, 199-209.
[162] Scibioh, M. A.; Rani, S.; Vijayaraghavan, V. R. *Trans. SAEST* 2001, *36*, 97-101.
[163] Jubran, N.; Ginzburg, G.; Cohen, H.; Koresh, Y.; Meyerstein, D. *Inorganic Chemistry* 1985, *24*, 251-258.
[164] Zilbermann, I.; Cohen, H.; Meyerstein, D. *Chem. Rev.* 2005, *105*, 2609-2625.
[165] Becker, J. Y.; Kerr, J. B.; Pletcher, D.; Rosas, R. *J. Electroanal. Chem. Eur. J.* 1981, *117*, 87-99.
[166] Gosden, C.; Kerr, J. B.; Pletcher, D.; Rosas, R. *J. Electroanal.Chem. Eur. J.* 1981, *117*, 101-107.
[167] Shandalov, E.; Zilbermann, I.; Maimon, E.; Nahmani, Y.; Cohen, H.; Adar, E.; Meyerstein, D. *European Journal Of Inorganic Chemistry* 2003, 4105-4109.
[168] Kashiwagi, Y.; Kikuchi, C.; Kurashima, F.; Ono, T. *Electrochim. Acta* 2007, *52*, 2045-2050.
[169] Esteves, A. P.; Freitas, A. M.; Medeiros, M. J.; Pletcher, D. *J. Electroanal. Chem.* 2001, *499*, 95-102.
[170] Semones, M. A.; Peters, D. G. *J. Electrochem. Soc.* 2000, *147*, 260-265.
[171] Olivero, S.; Rolland, J.-P.; Dunach, E. *Organometallics* 1998, *17*, 3747-3753.
[172] (172) Olivero, S. A.; Clinet, J. C.; Dunach, E. *Tetrah. Lett.* 1995, *36*, 4429-4432.

[173] (173) Shandalov, E.; Zilbermann, I.; Maimon, E.; Nahmani, Y.; Cohen, H.; Adar, E.; Meyerstein, D., Submitted For Publication.
[174] Wang, K.; Shen, M.; Sun, W.-H. *Dalton Trans.* 2009, 4085-4095.
[175] Cotton, F. A.; Wilkenson, G. In *Advanced Inorganic Chemistry*; 6th Ed.; Interscience Publishers: New York, 1999, P 680pp.
[176] Zilbermann, I.; Raznoshik, H.; Masarwa, A.; Maimon, E.; Cohen, H.; Meyerstein, D., Unpublished Results.
[177] Raznoshik, H.; Zilbermann, I.; Maimon, E.; Cohen, H.; Meyerstein, D. In *ICCC34*: Edinburgh, Scotland, 2000.
[178] Raznoshik, H.; Maimon, E.; Zilbermann, I.; Cohen, H.; Meyerstein, D. In *26th International Symposium On Macrocyclic Chemistry*: Fukuoka, Japan, 2001.
[179] Raznoshik, H.; Masarwa, A.; Cohen, H.; Zilbermann, I.; Maimon, E.; Meyerstein, D. *Dalton Trans.* 2009.
[180] Raznoshik, H.; Masarwa, A.; Cohen, H.; Meyerstein, D. *J. Phys. Chem. A.* 2008, *112*, 12769-12771.
[181] Neta, P.; Fessenden, R. W. *J. Phys. Chem.* 1972, *76*, 1957.
[182] Hayon, E.; Simic, M. *J. Am. Chem. Soc.* 1973, *95*, 2433-2439.
[183] (Anderson, N. H.; Dobbs, A. J.; Edge, D. J.; Norman, R. O. C.; West, P. R. *J. Chem. Soc. B* 1971, 1004.
[184] Schutz, O.; Meyerstein, D. *Tetrah. Lett.* 2006, *47*, 1093-1096.
[185] Rhodes, C. J. *Progress In Reaction Kinetics And Mechanism* 2005, *30*, 145-213.
[186] Shono, T. *Electroorganic Chemistry As A New Tool In Organic Chemistry*; Springer-Verlag: Berlin, 1984.
[187] (Pinson, J.; Podvorica, F. *Chemical Society Reviews* 2005, *34*, 429-439.
[188] Gerischer, H.; Heller, A. *Journal Of The Electrochemical Society* 1992, *139*, 113-118.
[189] Wang, C. Y.; Rabani, J.; Bahnemann, D. W.; Dohrmann, J. K. *Journal Of Photochemistry And Photobiology A-Chemistry* 2002, *148*, 169-176.
[190] Klauson, D.; Preis, S. *International Journal Of Photoenergy* 2005, *7*, 175-180.
[191] Portjanskaja, E.; Preis, S.; Kallas, J. *International Journal Of Photoenergy* 2006.
[192] Syoufian, A.; Nakashima, K. *Journal Of Colloid And Interface Science* 2007, *313*, 213-218.
[193] Rajeshwar, K. *Journal Of Applied Electrochemistry* 2007, *37*, 765-787.
[194] Adachi, M.; Jiu, J.; Isoda, S. *Current Nanoscience* 2007, *3*, 285-295.
[195] Rusonik, I.; Cohen, H.; Meyerstein, D. *European Journal Of Inorganic Chemistry* 2005, 1227-1229.
[196] Stewart, M. P.; Maya, F.; Kosynkin, D. V.; Dirk, S. M.; Stapleton, J. J.; Mcguiness, C. L.; Allara, D. L.; Tour, J. M. *Journal Of The American Chemical Society* 2004, *126*, 370-378.
[197] Grishaw, J. *Electrochemical Reactions And Mechanisms In Organic Chemistry*; Elsevier: Amsterdam, 2000.
[198] Rusonik, I.; Polat, H.; Cohen, H.; Meyerstein, D. *Eur.J.Inorg.Chem.* 2003, 4227-4233.
[199] Rusonik, I.; Zidky, T.; Cohen, H.; Meyerstein, D. *Glass Physics And Chemistry* 2005, *31*, 115-118.
[200] Rusonik, I.; Cohen, H.; Meyerstein, D. *Inorganic Chemistry* 2006, *45*, 7389-7396.
[201] Zidki, T.; Cohen, H.; Meyerstein, D. *Physical Chemistry Chemical Physics* 2006, *8*, 3552-3556.

[202] Zidki, T.; Ben Gurion University: Beer Sheva, 2009; Vol. 111, Pp 10461-10466.
[203] (Bar-Ziv, R.; Ben Gurion University Of The Negev, 2007.
[204] Gao, R. M.; Safrany, A.; Rabani, J. *Radiation Physics And Chemistry* 2002, *65*, 599-609.
[205] Fernandez, S.; Markovits, A.; Minot, C. *Journal Of Physical Chemistry C* 2008, *112*, 14010-14014.
[206] Zidki, T.; Cohen, H.; Meyerstein, D.; Meisel, D. *Journal Of Physical Chemistry C* 2007, *111*, 10461-10466.
[207] Chen, M. S.; Goodman, D. W. *Chemical Society Reviews* 2008, *37*, 1860-1870.
[208] Kaise, M.; Nagai, H.; Tokuhashi, K.; Kondo, S.; Nimura, S.; Kikuchi, O. *Langmuir* 1994, *10*, 1345-1347.
[209] Kasarevic-Popovic, Z.; Behar, D.; Rabani, J. *Journal Of Physical Chemistry B* 2004, *108*, 20291-20295.
[210] Behar, D.; Rabani, J. *Journal Of Physical Chemistry B* 2006, *110*, 8750-8755.
[211] Ui, K.; Fujita, T.; Koura, N.; Yamaguchi, F. *Journal Of The Electrochemical Society* 2006, *153*, C449-C454.
[212] Irie, H.; Miura, S.; Kamiya, K.; Hashimoto, K. *Chemical Physics Letters* 2008, *457*, 202-205.
[213] Li, Y. Z.; Zhang, H.; Guo, Z. M.; Han, J. J.; Zhao, X. J.; Zhao, Q. N.; Kim, S. J. *Langmuir* 2008, *24*, 8351-8357.
[214] Li, H. B.; Duan, X. C.; Liu, G. C.; Li, L. L. *Materials Research Bulletin* 2008, *43*, 1971-1981.
[215] Zhang, Z.; Wang, X.; Long, J.; Fu, X.; Ding, Z.; Li, Z.; Wu, L.; Fu, X. *Res. Lett. Phys. Chem.* 2008, 1.
[216] Shigeru, K.; Kazuhro, Y.; Toshiyasu, M.; Akihide, I.; Akihiko, A.; Hideto, M.; Ryoichi, N. *Phys. Chem. Chem. Phys.* 2008, *10*, 2986.
[217] Herzing, A. A.; Kiely, C. J.; Carley, A. F.; Landon, P.; Hutchings, G. J. *Science* 2008, *321*, 1331-1335.
[218] Chen, X.; Zhu, H. Y.; Zhao, J. C.; Zheng, Z. T.; Gao, X. P. *Angewandte Chemie-International Edition* 2008, *47*, 5353-5356.
[219] Wang, Y. F.; Zhao, D.; Ma, W. H.; Chen, C. C.; Zhao, J. C. *Environmental Science & Technology* 2008, *42*, 6173-6178.
[220] Marcus, R. A. *Science* 1992, *256*, 1523.
[221] Marcus, R. A.; Sutin, N. *Inorganic Chemistry* 1975, *14*, 213.
[222] Hubig, S. M.; Rathore, R.; Kochi, J. K. *Journal Of The American Chemical Society* 1999, *121*, 617-626.
[223] Mayer, J. M.; Rhile, I. J. *Biochim. Biophys. Acta* 2004, *1655*, 51-58.
[224] Eberson, L.; Shaik, S. S. *Journal Of The American Chemical Society* 1990, *112*, 4484-4489.
[225] Schwarz, H. A.; Dodson, W. R. *J. Phys. Chem.* 1989, *93*, 409.
[226] Cohen, H.; Meyerstein, D. *Journal Of The American Chemical Society* 1972, *94*, 6944-6948.
[227] Cohen, H.; Meyerstein, D. *Inorganic Chemistry* 1974, *13*, 2434-2443.
[228] Espenson, J. H.; Bakac, A.; Kim, J. H. *Inorganic Chemistry* 1991, *30*, 4830.

INDEX

A

absorption, 13, 42, 43, 44, 55, 65, 78, 80, 84, 88, 103, 189, 193
absorption spectra, 80
abstraction, 189, 195, 208
access, 11, 121
accessibility, 83
accuracy, 104
acetic acid, 76, 111
acetylcholine, 17, 19, 24, 26, 27, 29, 75, 81, 159, 160, 167
acetylcholinesterase, 19, 81, 159
achievement, 2
acid, 3, 7, 9, 10, 11, 18, 27, 31, 33, 34, 37, 45, 53, 54, 55, 61, 62, 63, 64, 65, 66, 67, 68, 73, 74, 75, 76, 77, 79, 80, 81, 82, 83, 88, 95, 96, 109, 111, 113, 114, 116, 117, 118, 120, 129, 131, 148, 150, 151, 153, 156, 159, 161, 162, 164, 168, 170, 171, 182, 199, 200, 201, 208
acidity, 16, 41, 44, 118, 122, 131
acidosis, 90
action potential, 15, 17
activated carbon, 108, 109, 120, 124, 146, 147, 153
activation parameters, 209
active centers, 80
active oxygen, 82
active site, 32, 33, 34, 35, 37, 38, 42, 43, 44, 45, 46, 49, 51, 53, 56, 57, 58, 61, 65, 66, 75, 83
adaptability, 20
adaptation, 16
adenine, 73, 77, 84, 86, 171
adenosine, 27, 73, 74, 76, 78, 80, 83
adenosine triphosphate, 73
adenovirus, 83
adhesion, 9, 64
adipose, 89
adipose tissue, 89
ADP, 74, 75, 77, 84
adrenal gland, 27
adsorption, viii, 95, 100, 104, 108, 109, 113, 118, 119, 122, 126, 127, 145, 146, 147, 149, 151, 152, 153, 204
adulthood, 9
advantages, 110, 114, 139
aetiology, 12, 182
age, 2, 20, 27, 156, 157, 158, 159, 162, 165, 171
ageing, 20, 23, 30, 180
agencies, 1, 102
aggregates, 4, 98, 161, 165, 166, 169, 170
aggregation, 4, 9, 21, 29, 102, 170, 172, 174, 175, 176, 177, 183, 185
aggression, 2
aging population, 155
aging process, 157, 159
albumin, 60
alcohol, 35, 75
alcohols, 131, 134, 191
alertness, 2
alkenes, 190, 199, 200
allergic reaction, 97
alopecia, 125
ALS, 26, 155, 156
alternatives, 109, 146
alters, 90
aluminium, vii, 1, 3, 10, 11, 12, 13, 14, 15, 16, 17, 18, 19, 20, 21, 22, 23, 24, 25, 26, 27, 28, 29, 30, 95, 96, 97, 98, 99, 100, 102, 103, 104, 105, 106, 107, 108, 109, 110, 111, 112, 113, 114, 115, 116, 117, 118, 119, 120, 121, 122
amines, 131, 132
amino acids, 23, 58, 69, 75, 76, 79, 81, 82, 162, 163, 164, 166, 167, 170, 171, 172
aminoglycosides, 69

ammonia, 83
ammonium, 90, 120, 130, 131, 153
ammonium salts, 130, 131
amorphous precipitate, 99
amygdala, 157, 158, 174
amyloid beta, 162, 181
amyloid deposits, 162, 165, 166, 175
amyloidosis, 180, 188
amyotrophic lateral sclerosis, 11, 13, 27, 155, 156
anatase, 204
anatomy, 2
anemia, 77, 84, 97, 125
anorexia, 103
anoxia, 82
antacids, 84
anticancer drug, 71
antigen, 11
antioxidant, vii, 12, 160, 161, 162, 175
apoptosis, 25, 26, 28, 29, 79
appetite, 160
applications, 31, 61, 69, 96, 108, 119, 132, 177, 186
aqueous solutions, 108, 109, 117, 120, 124, 132, 133, 135, 139, 146, 147, 150, 151, 152, 153, 188, 189, 194, 196, 198, 199, 201, 204, 206
arginine, 10
aromatics, 133, 189
arsenic, 117, 125, 153
arterial blood gas, 82
arteries, 74
aryl hydrocarbon receptor, 90
ascorbic acid, 88, 161
aspartate, 21, 34, 43, 83, 171
aspartic acid, 175
assessment, 2, 3
assessment models, 2
assignment, 47, 49
astrocytes, 11, 15, 19, 20, 29, 76
atherogenesis, 162
atherosclerosis, 89, 188
atomic force, 16
atoms, 35, 36, 45, 73, 78, 79, 80, 81, 84, 85, 112, 115, 124, 132, 149, 189, 199, 206, 208
ATP, vii, 45, 73, 74, 75, 76, 77, 78, 80, 81, 82, 83, 84, 86, 87, 90
atrophy, 157, 158, 175
attachment, 74, 81, 112
attacks, 32, 36, 170
autosomal recessive, 183
availability, 35, 37, 98, 119, 160
axons, 165

B

background, 51
bacteria, 84, 117, 172, 187
barriers, 13, 84, 147
base pair, 55
bauxite, 96
BBB, 177, 178
behavior, vii, 81, 87, 100, 115, 118, 120
bending, 164
beneficial effect, 82
benzene, 147
binding, 4, 8, 9, 16, 18, 19, 25, 35, 48, 51, 55, 57, 58, 61, 63, 65, 68, 69, 79, 81, 83, 85, 87, 115, 116, 174, 175, 176, 180, 186, 188, 204
bioavailability, 96, 177
biochemical action, 84
biochemistry, 51, 73, 95, 157, 175, 177, 187
biological activity, 71
biological processes, 73, 76, 80, 85, 87, 126, 161, 187, 190, 196
biological systems, viii, 12, 75, 80, 177, 187, 191, 194, 195, 208
biomass, 122, 147, 152
biomass materials, 147
biomaterials, 147
biopolymer, 114, 115, 117
bioremediation, 117
biosynthesis, 187
biotechnology, 51
bipolar disorder, 88
bladder, 74
blocks, 184
blood, 12, 13, 82, 84, 95, 160, 174, 177, 178, 188
blood flow, 82
blood pressure, 82, 188
blood stream, 95
blood-brain barrier, 12, 13, 177
bloodstream, 75
bonding, 59, 78, 80, 81, 208
bonds, 9, 31, 32, 51, 53, 56, 57, 58, 59, 60, 61, 69, 79, 80, 81, 87, 118, 132, 164, 171, 189, 195, 196, 198, 201, 202, 203, 204, 207
bone, 33, 62, 63, 73, 85, 87, 97, 103, 125, 126
bone form, 85
bone resorption, 33
bones, 14, 84, 95, 103
brain, 3, 9, 10, 11, 12, 13, 14, 17, 18, 19, 20, 21, 24, 25, 26, 27, 28, 29, 30, 75, 81, 82, 83, 84, 85, 95, 155, 156, 157, 158, 159, 160, 161, 162, 163, 164,

165, 166, 167, 168, 169, 170, 171, 174, 175, 176, 177, 178, 179, 183, 186
brain activity, 155
brain damage, 171
brain functions, 156
brainstem, 157
Brazil, 31, 70, 73, 91, 95, 96, 102, 109, 116, 155, 156, 178, 179
breakdown, 74, 148
breathing, 160
bridges, 41
Britain, 93
buffer, 55, 75, 82, 117
building blocks, 171
by-products, 102, 152, 160, 188

C

cadmium, 25, 27, 30, 120, 130, 151, 167
calcium, 4, 11, 13, 15, 16, 18, 19, 21, 24, 25, 26, 28, 29, 83, 84, 87, 89, 110, 117, 159, 178
calcium carbonate, 117
caloric restriction, 178
calyx, 152
cancer, 79, 125, 126, 188
cancer cells, 79
candidates, 20
capillary, 158
carbon, 80, 101, 108, 109, 120, 124, 131, 146, 147, 153, 177, 194, 195, 196, 197, 198, 201, 202, 203, 204, 205, 209
carbon atoms, 80
carbon nanotubes, 177
carbon-centered radicals, 194, 195, 205, 209
carboxylic acids, 130
carboxymethyl cellulose, 115
cardiomyopathy, 77, 89
carotene, 161
carrier, 13, 124, 136, 139, 140, 150, 151, 177
casein, 4, 83
catalysis, 33, 46, 49, 52, 53, 55, 57, 64, 170, 187, 188
catalyst, 54, 56, 57
catalytic activity, 41, 46, 54, 56, 57, 204
catalytic properties, 35, 37
catalytic system, 31, 45, 188, 194
catecholamines, 21, 159
cation, 14, 16, 25, 27, 33, 40, 43, 79, 83, 98, 99, 109, 129, 133, 137, 148, 187, 195, 196, 199, 208
cattle, 167
cell, 2, 3, 4, 9, 12, 14, 17, 18, 20, 21, 25, 28, 29, 63, 73, 74, 75, 80, 156, 158, 159, 160, 161, 162, 165, 167, 168, 169, 170, 171, 174, 175, 177, 180, 181, 183, 187, 191
cell body, 158
cell culture, 3
cell death, 14, 18, 21, 28, 158, 159, 162, 174, 183
cell line, 25, 29
cell lines, 29
cell membranes, 17, 160
cell metabolism, 74
cell surface, 75
cellular homeostasis, 162
cellulose, 113, 115, 117, 147, 153
cement, 11, 151
central nervous system (CNS), 20, 125, 156, 157, 165, 166, 175
ceramic, 138
cerebellum, 157, 158, 176
cerebral blood flow, 82
cerebral cortex, 11, 158, 159, 174
cerebrospinal fluid, 11, 12, 13, 159, 176, 181, 185, 186
ceruloplasmin, 161
channels, 14, 15, 16, 19, 27, 28, 77
character, 16, 106, 207
chemical bonds, 87, 207
chemical degradation, 148
chemical properties, 115, 204
chemical reactions, vii, 74, 131
chemical stability, 74
chemotherapeutic agent, 33
children, 27
China, 96
chitin, 114, 117, 118, 121, 122
cholesterol, 17, 21, 28, 176, 185
choroid, 175
chromium, 83, 117, 125, 149, 150, 151, 152, 153
chromosome, 167, 171, 181
chronic diseases, 160
chronic renal failure, 126
chymotrypsin, 172
circulation, 13
cirrhosis, 125, 184, 185
class, 32, 45, 61, 74, 79, 198, 208
classes, 191
classification, 67, 132
cleaning, 98
cleavage, 9, 36, 51, 52, 53, 54, 55, 56, 57, 58, 60, 61, 68, 69, 74, 166, 167, 184, 189

cleavages, 53
clinical trials, 176
CMC, 114, 115, 116, 117
CO2, 85, 189, 190, 200, 201, 206, 209
coagulation, 98, 99, 100, 101, 102, 104, 105, 106, 183
coagulation process, 104
coal, 118, 122, 146, 152
cobalt, 82, 83, 84, 117, 130, 150
coding, 2
coenzyme, 160
cognition, 176, 186
cognitive abilities, 20
cognitive ability, 166
cognitive deficit, 11, 76
cognitive deficits, 11, 76
cognitive dysfunction, 11
cognitive function, 159, 167
cognitive impairment, 76, 84, 176
color, iv, 33, 88, 99, 108
coma, 11
combustion, 187, 190
communication, 2, 15, 17, 20, 28, 85, 177
community, 95
competition, 20, 25, 84
competitive process, 118
compilation, 206
complaints, 103
complex interactions, 51
complexity, 149
complications, 11
components, 3, 25, 207
composition, 9, 28, 129, 145, 165, 189, 202, 204
compounds, vii, viii, 12, 13, 21, 28, 41, 45, 58, 59, 66, 67, 74, 76, 85, 87, 109, 124, 125, 126, 131, 132, 134, 147, 187, 189, 190, 191, 202, 206
comprehension, 104
concentration, 12, 15, 17, 18, 54, 56, 60, 74, 75, 77, 79, 81, 87, 88, 95, 98, 99, 102, 103, 104, 105, 106, 107, 108, 111, 112, 113, 119, 123, 124, 126, 127, 134, 135, 136, 145, 146, 150, 156, 159, 160, 161, 162, 163, 166, 167, 169, 175, 193
conditioning, 27
conduction, 158
conductors, 188, 201
configuration, 16, 138, 139, 140, 141, 164, 166
conjugation, 13
consensus, 4, 7
conservation, 3, 33, 75
construction, 37, 96
consumption, 12, 77, 103, 109
consumption rates, 77
contact time, 117
contamination, vii, 123, 135
control, 1, 23, 24, 33, 74, 77, 84, 88, 90, 102, 105, 142, 170, 183, 190
control group, 88
controversies, 13, 178
conversion, 73, 169, 202
coordination, 34, 37, 40, 42, 43, 46, 48, 59, 62, 65, 78, 79, 80, 81, 83, 84, 85, 86, 87, 103, 109, 113, 114, 115, 132, 149, 162, 166, 206
copolymers, 149
copper, 59, 60, 61, 64, 67, 68, 69, 76, 77, 79, 80, 81, 83, 117, 118, 125, 130, 150, 153, 160, 161, 162, 167, 175, 176, 177, 178, 180, 185, 186
corpus callosum, 157
correlation, 50, 102, 103, 207
correlations, 68, 88
cortex, 11, 13, 16, 25, 28, 29, 157, 158, 159, 174, 175
cortical neurons, 25
cosmetics, 96
cost, 1, 3, 96, 98, 108, 109, 113, 114, 119, 120, 123, 126, 152
costs, 1, 3, 109, 145
couples, 43, 45, 162, 206, 208, 209
coupling, 15, 34, 35, 41, 45, 208
covalent bond, 118
cranial nerve, 158
creatine, 75, 77, 78, 81, 83
Creutzfeldt-Jakob disease, 166, 168, 170
crown, 145
crude oil, 133
crystal structure, 34, 40, 43, 58, 60, 64, 83, 174
crystalline, 83, 152
crystallization, 41
crystals, 81
culture, 18, 27, 28
currency, 73
cyanide, 151
cycles, 105
cytoarchitecture, 158
cytochrome, 25, 26, 90, 161, 175, 188
cytokines, 160
cytoplasm, 12, 161, 167
cytosine, 171
cytoskeleton, 74, 167
cytotoxicity, 69

D

daily living, 186
death, 14, 21, 28, 84, 125, 156, 165, 166, 168, 170, 175
decay, 57
decomposition, 74, 161, 194, 195, 196, 197, 198, 201, 203, 204
decomposition reactions, 194, 198
deconvolution, 79
defects, 170
defence, vii, 13
defense, 161, 187
deficiency, 75, 77, 90, 125, 184
deficit, 2, 19
definition, 100
deformation, 164, 184
degradation, 9, 16, 76, 79, 81, 88, 148, 158, 170, 174, 202
degradation mechanism, 76
dementia, 10, 12, 13, 24, 84, 160, 165, 166, 167, 168, 174, 178, 180, 182, 185
denaturation, 164
dendrites, 158, 165
dendritic spines, 158
density, 16, 45, 48, 98, 125, 157, 165
deoxyribonucleic acid (DNA), vii, 13, 20, 32, 41, 45, 46, 48, 51, 52, 53, 54, 55, 56, 57, 59, 60, 61, 67, 68, 69, 70, 73, 79, 80, 82, 160, 163, 171, 179, 187
dephosphorylation, 33, 74, 81
depolarization, 17, 19
deposition, 3, 9, 13, 103, 162, 166, 169, 176, 182
deposits, 26, 162, 165, 166, 171, 173, 175, 180
depression, 88, 89
deregulation, 61
derivatives, 34, 38, 44, 49, 57, 64, 74, 129, 130, 132, 147
designers, 61
destruction, 156, 171
detection, 112, 117
developed countries, 166
developing countries, 2
deviation, 208
DFT, 79, 85, 86, 204
diabetes, 89
diagnosis, 75, 181, 182
dialysis, 11, 12, 28, 103, 109, 112
dialysis dementia, 12
diarrhea, 125
diet, 24, 77, 176
differentiation, 9, 25, 75
diffusion, 107, 113, 170, 194, 195, 206
diffusivity, 162
digestion, 170
digestive enzymes, 17, 172
diluent, 123, 124, 129, 132, 133, 134, 137, 145
dimerization, 190
dipeptides, 59, 70
direct adsorption, 100
disadvantages, 98, 104, 109
discourse, 22
discs, 11
diseases, 11, 14, 20, 61, 75, 76, 81, 97, 104, 125, 126, 156, 164, 166, 167, 168, 170, 174, 177, 178, 183, 187, 188
disinfection, 102, 103, 104
disorder, 88, 162, 166, 170, 174, 175
dispersion, 112, 124, 134, 140, 142, 143, 144
dissociation, 44, 46, 113
distillation, 130
distortion, 164
distortions, 43, 87
distribution, 4, 13, 27, 54, 64, 95, 99, 102, 103, 106, 116, 185
distribution function, 16
disturbances, 19, 29, 77, 97, 103, 157, 159, 162, 177
diversity, 2
division, 12, 75
DNA damage, 68, 179
DNA ligase, 53
DNA polymerase, 60
donors, 78
dopamine, 18, 20, 21, 27, 29, 30, 159, 160, 161, 170, 175
dopaminergic, 13, 21, 159
dosage, 12, 99, 105
double bonds, 189
Down syndrome, 181
drainage, 117
dreaming, 22
dreams, 22, 23
drinking water, 1, 24, 95, 97, 98, 99, 100, 103, 104, 105, 107, 108, 109, 111, 112, 119, 120
drug delivery, 177
drug design, 58, 61, 62, 70, 71
drug discovery, 70
drug interaction, 55
drugs, 52, 55, 58, 61, 70, 71, 76, 77, 88, 156, 176, 177
D-serine, 76

duality, 87
dumping, 128
duodenum, 172
duration, 2, 82
dyes, 158
dynamical properties, 16
dysarthria, 12

E

earth, 125
efferent nerve, 158
effluent, 102, 105, 106, 107, 118, 126, 128, 131, 136, 148, 149, 151
effluents, 110, 114, 116, 118, 122, 123, 126, 128, 135, 147, 149, 151
elasticity, 23, 138
elderly, 2, 158, 159, 165
electrical properties, 18
electrocatalyst, 201
electrochemistry, 188, 190
electrodes, 202
electrolysis, 43
electron, viii, 42, 44, 48, 67, 68, 78, 79, 87, 130, 131, 187, 189, 191, 195, 200, 202, 204, 205, 206, 207, 208, 209
electron paramagnetic resonance, 67, 79
electrons, 43, 159, 162, 191
elucidation, 198
emphysema, 184
emulsions, 133
encephalitis, 97, 156
encephalopathy, 11, 12, 27, 28, 82, 103, 156, 170, 174
encoding, 163, 171
endocrine, 15
endocrine system, 15
endorphins, 161
endothelium, 158
endothermic, 196
energy, 19, 45, 73, 74, 75, 77, 81, 90, 160, 177, 202, 205, 208, 209
engineering, 57, 103
enlargement, 16, 157
environment, 1, 12, 13, 34, 37, 40, 46, 59, 82, 96, 104, 125, 126, 136, 146
environmental conditions, 16
environmental factors, 13
Environmental Protection Agency (EPA), 102
environmental quality, 2
enzymatic activity, 73
enzymes, 12, 14, 17, 29, 31, 32, 33, 34, 35, 45, 51, 52, 53, 57, 58, 62, 65, 66, 69, 70, 74, 76, 87, 125, 159, 160, 161, 162, 170, 171, 172, 175, 176, 177, 191
epidemic, 12
epilepsy, 11
epithelial cells, 79
equilibrium, 28, 47, 99, 129, 150, 152, 189, 197, 200, 207
equipment, 108, 123, 135, 139, 145
ester, 33, 36, 57, 61, 66
ethanol, 25, 46, 47, 48, 89, 112
ethers, 131, 145
ethylene, 199, 200, 201
etiology, 13, 88, 182, 183
evolution, 3, 15, 57, 69, 175, 179, 182
EXAFS, 36, 48, 65
exchange rate, 196, 207, 208, 209
excitability, 15, 18, 19, 159, 185
exclusion, 136
excretion, 13
exercise, 81
exocytosis, 15, 18
exons, 4
experiences, 98
experimental condition, 54, 60, 200
expertise, vii
exposure, 2, 3, 4, 12, 13, 16, 19, 21, 24, 25, 28, 84, 95, 97, 103, 120, 178, 179
expulsion, 175
external environment, 15
extra help, 177
extracellular matrix, 20
extraction, 104, 109, 110, 111, 112, 113, 116, 120, 121, 123, 124, 126, 127, 128, 129, 130, 131, 132, 133, 134, 135, 139, 140, 141, 143, 150, 151, 152
extrapolation, 3

F

failure, 21, 27, 126, 164, 170
family, 14, 16, 75
family members, 75
fat, 12, 74, 160
fatigue, 11, 12
fatty acids, 79
ferritin, 81, 87
fiber, 17, 124, 139, 140, 141, 142, 143, 144, 150, 151
fibers, 9, 139, 140, 141, 144, 157

fibrogenesis, 162
fibrosis, 27
filters, 98, 102, 110
filtration, 95, 98, 99, 102, 103, 104, 105, 106, 107, 109, 110, 119, 120, 126
fine tuning, 16
fish, 95, 104, 120
fixed rate, 146
flammability, 132
flexibility, 61, 108
flight, 22
flocculation, 98, 99, 104, 105, 126
flooding, 123, 135
flotation, 95, 104, 108, 109, 119, 120
fluid, 11, 12, 13, 62, 64, 112, 157, 159, 176, 181, 185, 186
fluidized bed, 112, 113, 121
fluorescence, 55, 76
focusing, 75
folate, 156, 161
food, 96, 97, 126, 160
Ford, 212
forebrain, 27
formula, 98
fouling, 109
fragility, 2
fragments, 60, 79, 161, 166, 175
France, 11, 91
free energy, 207, 208
free radicals, 12, 87, 157, 159, 160, 161, 162, 163, 165, 166, 171
freshwater, 104
frontal cortex, 13, 16, 158
fruits, 22
fuel, 74
fungi, 124, 147
fusion, 15, 16, 25, 26

G

ganglion, 77
gases, 189
gel, 60, 107, 140
gene, 4, 21, 27, 59, 83, 90, 167, 168, 170, 171, 174, 183
gene expression, 21, 27
gene therapy, 59
generation, 10, 12, 15, 28, 33, 58, 60, 77, 87, 132, 162, 191
genes, 4, 13, 26, 61, 163, 170
genetic code, 33
genetic defect, 181
genetic information, 73, 79
genome, 75
Germany, 179
germination, 34
gestation, 2
gland, 27
glia, 24, 28
glial cells, 15, 18, 158
globus, 159
glucose, 26, 77, 81, 82, 87, 89, 90, 115, 158, 159
glutamate, 18, 20, 21, 24, 28, 30, 76, 159, 161, 174, 175, 186
glutathione, 12, 160, 161, 180
glycerol, 56, 77
glycine, 10, 18, 29, 81, 87
glycogen, 4, 26, 90
glycolysis, 159
glycosylation, 7
gold, 151, 153, 204
gout, 81
grants, 210
granules, 158
gravity, 105
gray matter, 84
groundwater, 96
groups, 16, 17, 40, 44, 53, 55, 59, 70, 73, 74, 77, 78, 79, 80, 82, 85, 88, 99, 109, 112, 113, 114, 115, 118, 131, 133, 156, 161, 188, 198, 200
growth, 30, 74, 75, 77, 79, 88, 125, 175, 182
growth factor, 29, 30, 88, 182
guanine, 171
guidelines, 2, 109
Guinea, 13

H

hafnium, 131
half-life, 13, 32, 33, 54, 58, 60
Hamiltonian, 35
harm, 160
health, 1, 11, 14, 95, 97, 102, 103, 104, 109, 119, 125, 156
health effects, 14
heart disease, 89
heart rate, 82
heat, 82, 190
heating, 146
heavy metals, 123, 126, 128

height, 20, 146
heme, 161
hemodialysis, 11, 84
hemoglobin, 188
hepatocytes, 23, 29, 173
hepatomegaly, 125
hexane, 199
hippocampus, 3, 16, 19, 25, 82, 158, 159, 174, 175, 185, 186
histamine, 69
histidine, 10, 34, 84, 171
HIV, 69
homeostasis, 13, 21, 26, 27, 159, 162, 175, 177, 185
homocysteine, 156, 161, 178, 180
homolytic, 188, 195, 197
homopolymers, 115
hormone, 79, 90
host, 33
House, 186
human brain, 4, 9, 10, 11, 12, 18, 20, 21, 27
human exposure, 12
human genome, 75
Hunter, 91
hybrid, 149, 150, 151
hydrocephalus, 13, 25
hydrogen, 59, 74, 80, 81, 82, 83, 130, 160, 189, 191, 200, 204, 208
hydrogen abstraction, 189
hydrogen atoms, 189
hydrogen bonds, 81
hydrogen peroxide, 160, 191
hydrogenation, 201
hydrolases, 38, 39, 45, 52, 65, 66
hydrolysis, 14, 28, 31, 32, 33, 41, 46, 48, 49, 50, 51, 52, 54, 55, 56, 57, 58, 59, 60, 61, 62, 66, 67, 68, 70, 74, 75, 83, 85, 88, 98, 99, 103, 106, 171
hydrophilicity, 118
hydrophobicity, 61
hydroxide, 11, 38, 50, 51, 65, 97, 99, 100, 101, 104, 117, 118
hydroxyapatite, 73
hydroxyl, 52, 56, 61, 81, 82, 87, 112, 114, 118, 189, 192, 193
hydroxyl groups, 52, 114, 118
hydroxylapatite, 87
hypercholesterolemia, 77
hyperglycemia, 89, 90
hypertension, 82, 188
hypertrophy, 158
hypothalamus, 29
hypothesis, 15, 16, 19, 76, 78, 170, 174, 176, 180, 185
hypothyroidism, 89
hypoxia, 81

I

ideal, 36
identification, 3, 24, 61, 115
identity, 33
idiopathic, 11
IFM, 122
image, 60
imagination, 22
imbalances, 160, 162, 167
immobilization, 116, 151
immune response, 11, 125, 188
immune system, 160
impacts, 1, 4
impairments, 76, 81, 84
impregnation, 139, 149
impulses, 167
impurities, 105, 108, 128
in utero, 42, 67
in vitro, vii, 3, 10, 13, 14, 17, 19, 20, 21, 26, 28, 33, 64, 88, 89, 169, 177
in vivo, vii, 4, 13, 19, 20, 24, 28, 33, 81, 83, 176, 182
incentives, 1
incidence, 12, 13
inclusion, 82, 140, 173, 174
India, 96
indication, 17, 45, 108, 115
individual development, 2
induction, 21, 25, 90, 179
industrial processing, 116
industry, 1, 96, 126, 135, 152, 153
infarction, 75
inflammation, 81, 160
inflammatory responses, 14
information processing, 15
ingestion, 13, 28, 95, 96, 103
inhibition, 14, 19, 21, 29, 30, 33, 65, 77, 83, 87, 88, 172, 183, 184
inhibitor, 83, 173, 183, 184
initiation, 12, 18
injury, 1, 2, 11, 12, 28, 167, 178
inositol, 88
insertion, 172, 173, 184, 198
insight, vii, 37, 45, 51
instability, 139, 164, 170

insulin, 30, 81, 87, 89, 176, 177
integration, 3, 15, 16
integrity, 19
interaction, vii, 18, 19, 26, 28, 52, 54, 55, 59, 67, 69, 80, 82, 85, 87, 98, 125, 176, 188, 207
interactions, 13, 17, 76, 79, 80, 81, 82, 83, 84, 175, 208
interface, 124, 144
intoxication, 1, 12, 25
intramuscular injection, 11
intravenously, 177
ion channels, 14, 18
ion-exchange, 113, 145, 146, 148, 152
ionic strength, 54, 60
ionization, 148
ionizing radiation, 188, 189
ions, vii, viii, 17, 18, 21, 23, 27, 35, 40, 42, 43, 49, 51, 53, 59, 70, 76, 80, 81, 83, 84, 104, 106, 107, 108, 110, 112, 113, 115, 116, 118, 121, 122, 124, 130, 149, 151, 152, 153, 156, 160, 161, 162, 163, 164, 166, 167, 168, 171, 175, 176, 177, 178, 185, 190, 194, 199, 201
IR spectra, 78
IR spectroscopy, 115
iron, 12, 13, 33, 34, 41, 43, 62, 63, 64, 65, 67, 68, 81, 84, 87, 117, 118, 150, 160, 161, 162, 163, 167, 175, 176, 178, 185, 188, 202
iron transport, 33
ischemia, 77, 90
isotope, 50
Israel, 92, 187, 210, 215
issues, 1

J

Jamaica, 96
Japan, 1, 13, 216
joints, 81
justification, 11

K

K^+, 19, 28, 29, 74, 89, 90
ketones, 131
kidney, 33, 34, 39, 40, 43, 44, 64, 65, 66, 77, 125, 126
kidneys, 81
killing, 29, 63, 187
kinase activity, 75
kinetic equations, 170
kinetic parameters, 48
kinetic studies, 118, 192
kinetics, 57, 113, 133, 151, 152, 169, 189, 193, 194, 196, 198, 199, 204
KOH, 47

L

lakes, 103
lanthanide, 54
latency, 11
lateral sclerosis, 11, 13, 27, 155, 156
LD, 91
leakage, 11
learning, 18, 20, 21, 23, 157, 158, 160, 185
lesions, 9, 11
leucine, 70
life cycle, 2
life expectancy, 156
life quality, 2
life span, 17
lifetime, 202
ligand, 15, 33, 34, 37, 38, 39, 40, 41, 42, 43, 46, 48, 54, 55, 58, 59, 60, 61, 65, 66, 67, 78, 79, 80, 81, 82, 85, 86, 109, 110, 111, 112, 113, 114, 121, 191, 192, 195, 196, 198, 209
line, 111, 121, 172, 173
linear dependence, 48
linkage, 61, 184
links, 10, 103, 164
lipid metabolism, 75
lipid peroxidation, 13, 14, 88, 163
lipids, vii, 13, 25, 74, 75, 160
liposomes, 88, 177
liquid phase, 126
liquids, 132, 133, 150
lithium, 88
liver, 75, 77, 89, 125, 126, 178, 184
liver cirrhosis, 125
localization, 64
locomotor, 159
locus, 175
lumen, 75

M

machinery, 15, 16, 19
macrophages, 191

macropores, 112, 113
mad cow disease, 156, 166, 168
magnesium, 13, 27, 83
magnetic properties, 191
magnetic resonance, 77
maintenance, 33, 84, 97, 115, 175
majority, 52, 77, 188, 193, 196, 202
mammal, 30
mammalian brain, 18, 19
management, 1, 178
manganese, 13, 27, 30, 69, 83, 117, 163, 175
manic-depressive illness, 88
manipulation, 59
mapping, 58
markers, 166
mass spectrometry, 56
mastery, 13
matrix, 20, 33, 119, 124, 145, 147, 148, 149, 161
maturation, 125
media, 78, 89, 105, 106, 108, 120, 129, 132, 147, 202, 209
medication, 160
Mediterranean, 96
melanin, 81
melting, 132
membranes, 15, 16, 17, 18, 20, 25, 73, 74, 75, 77, 107, 124, 126, 136, 137, 138, 139, 140, 145, 150, 151, 160, 168
memory, 2, 10, 12, 18, 20, 21, 103, 157, 158, 160, 166, 176, 177, 185
memory loss, 160
men, 22, 97, 156
meninges, 157
mental capacity, 165
mercury, 1, 21, 29, 30, 64, 125, 150, 152, 167
messages, 167
messengers, 159, 161
metabolic disturbances, 77
metabolism, 12, 19, 21, 24, 26, 73, 74, 75, 82, 83, 87, 88, 90, 158, 159, 160, 161, 179, 187
metabolites, 74, 82, 176
metal extraction, 130, 131
metal oxides, 204
metals, 24, 40, 46, 57, 58, 80, 104, 110, 113, 117, 123, 124, 125, 126, 128, 129, 131, 132, 133, 146, 147, 149, 152, 161, 162, 171, 176, 177, 178, 182, 183, 188, 190, 201, 202, 203, 204
methanol, 204
methylation, 21
Mg^{2+}, 13, 33, 83
mice, 3, 11, 25, 28, 63, 178, 180, 185
micelles, 177
microdialysis, 13, 19
microscope, 170
microscopy, 16
microspheres, 118, 122
migration, 9, 76, 198
military, 11
mining, 1, 116, 118, 122
mitochondria, 25, 88, 160, 188
mitogen, 4
mixing, 17, 112, 143
MMPs, 161
model, vii, 16, 24, 29, 31, 32, 38, 39, 40, 41, 42, 43, 45, 46, 48, 49, 52, 53, 54, 55, 56, 58, 60, 61, 65, 66, 67, 70, 82, 87, 111, 118, 120, 121, 163, 170, 171, 172, 191, 208
model system, 42, 49, 52, 58, 82, 191
modeling, 67
modelling, 2, 151
models, vii, 2, 3, 10, 11, 12, 13, 14, 17, 19, 20, 21, 22, 37, 38, 41, 58, 65, 67, 70, 115, 153, 170, 194
modern society, 126
modification, 21, 153, 169
modules, 139, 140, 141
mole, 208
molecular biology, 51
molecular dynamics, 24
molecular mobility, 183, 185
molecular oxygen, 53
molecular structure, 71
molecular weight, 33, 85, 111
molecules, 16, 18, 31, 35, 37, 43, 51, 58, 59, 70, 73, 74, 76, 83, 85, 86, 89, 102, 113, 131, 132, 159, 162, 164, 165, 169, 171, 173, 183, 188, 189, 190, 191, 209
molybdenum, 81
monitoring, 55, 87
monoamine oxidase inhibitors, 88
monomers, 73, 83, 111, 149
monosaccharide, 115
mood, 160
morphology, 2, 21
mortality, 1, 89
motif, 65
motion, 172
motor neurons, 13, 175
movement, 74, 160, 170, 175, 184
mRNA, 3, 79, 83, 158
multicellular organisms, 16

multiple sclerosis, 160
muscarinic receptor, 26
muscles, 168
mutant, 174, 183, 185
mutation, 170, 171, 173, 174, 184, 185
myelin, 158
myocardial infarction, 75
myocardium, 90
myosin, 83, 89

N

Na^+, 19, 28, 29, 74, 89, 90
NaCl, 54
NAD, 77, 88
NADH, 88, 89, 191
nanoparticles, 177
nanostructures, 16, 177
nanotechnology, 177
natural selection, 179
necrosis, 26
neocortex, 3, 165
nerve, 15, 16, 17, 19, 24, 25, 26, 27, 29, 30, 75, 155, 157, 158, 160, 167
nerve conduction velocity, 158
nerve fibers, 157
nerve growth factor, 29
nervous system, 1, 2, 15, 16, 20, 24, 27, 33, 64, 77, 125, 156, 157, 158, 160, 165, 166, 175, 188
Netherlands, 215
network, 3, 15, 61, 95
neural function, 156, 162
neural systems, 163
neuroblastoma, 21, 25, 178, 179
neurodegeneration, 2, 19, 20, 21, 23, 24, 28, 156, 157, 174, 175, 176, 180
neurodegenerative dementia, 165
neurodegenerative diseases, vii, viii, 13, 27, 126, 155, 156, 157, 160, 161, 162, 164, 165, 166, 175, 176, 177, 178, 181
neurodegenerative disorders, 12, 20, 30, 164, 167, 168, 177, 183, 185, 186
neurofibrillary tangles, 3, 164, 165, 167, 174, 176, 181
neurofilaments, 167
neurogenesis, 2, 120
neurological disease, viii, 11, 97, 176
neurons, 13, 15, 17, 18, 19, 20, 24, 25, 29, 77, 120, 155, 158, 159, 161, 167, 168, 170, 175, 176, 177
neuropharmacology, 182
neuroscience, 177
neurotoxicity, 2, 3, 10, 12, 25, 26, 29, 126, 174, 175, 176, 185
neurotransmission, 15, 19, 25, 76, 174
neurotransmitter, 18, 19, 28, 74, 75, 88, 157, 158, 159, 161, 167
neutropenia, 125
neutrophils, 172, 191
next generation, 61
nickel, 117, 130, 150, 153, 198, 199
nicotinamide, 77
niobium, 131
NIR, 42
nitrate, 110, 188
nitric oxide, 24, 161, 188
nitrogen, 55, 74, 78, 79, 80, 85, 112, 117, 130, 188
nitrogen dioxide, 188
N-methyl-D-aspartic acid, 175
nodes, 15
norepinephrine, 30, 161
nuclear magnetic resonance (NMR), 35, 65, 77, 78, 85, 86, 90, 93, 162, 180, 183
nuclei, 158, 174
nucleic acid, 31, 54, 56, 73, 79, 80, 168
nucleophiles, 68
nucleotides, 77, 79, 80, 81, 83, 84, 85, 171
nucleus, 158, 159, 165
nutrients, 77

O

observations, 27, 77, 81
oil, 124, 133
oligomerization, 10
oligomers, 9, 21, 176, 177, 186
operon, 83
order, 20, 32, 34, 37, 38, 42, 44, 49, 51, 54, 57, 76, 78, 80, 81, 85, 87, 95, 98, 99, 105, 108, 109, 111, 112, 113, 114, 116, 118, 131, 139, 156, 160, 175, 177, 207
organ, 2, 15, 61, 160, 202
organelles, 162
organic compounds, 109, 134, 147, 202
organic materials, 100
organic matter, 100, 101, 107, 116
organic solvents, 189
organism, 23, 33, 126
oscillations, 15
ossification, 63
osteoarthritis, 188

osteomalacia, 84
osteoporosis, 33
overload, 12, 97
overproduction, 81
oxidation, 13, 23, 30, 33, 43, 56, 57, 79, 87, 88, 89, 161, 162, 176, 190, 192, 196, 198, 201, 205
oxidative damage, 13, 29, 175
oxidative stress, 10, 12, 14, 21, 84, 156, 160, 161, 162, 163, 165, 166, 171, 177, 178, 179, 180
oxides, 11, 96, 104, 150, 188, 201
oxygen, 12, 16, 28, 33, 35, 36, 45, 54, 56, 61, 63, 73, 77, 78, 79, 80, 81, 82, 83, 84, 85, 87, 112, 115, 125, 131, 132, 147, 160, 161, 162, 175, 188, 191, 198
oxygen consumption, 77
oxygen consumption rate, 77
ozone, 193

P

p53, 26
Pacific, 13, 26
packaging, 96
pain, 11, 103
pancreas, 17, 172
paradigm, 1, 3, 58, 62, 70
parallel, 116
parameter, 80, 111
parameters, 25, 41, 45, 48, 50, 54, 56, 57, 87, 108
parenchyma, 11
parkinsonism, 13, 170, 183
particles, 30, 96, 98, 99, 100, 101, 102, 103, 106, 107, 108, 112, 122, 188, 202, 203, 204
passive, 18, 117
pathogenesis, 3, 12, 176, 186, 187
pathogens, 33, 163
pathologist, 170
pathology, 61, 77, 168, 174, 175, 182
pathophysiology, 174
pathways, 13, 14, 19, 21, 26, 61, 69, 73, 82, 192, 193, 197
peptidase, 59, 70, 171
peptides, 9, 23, 26, 32, 58, 59, 64, 156, 161, 169, 174, 176, 177, 183, 186
perchlorate, 40
percolation, 111, 112
perforated synapses, 24
performance, 1, 2, 90, 102, 105, 107, 123, 130, 132, 134, 135, 136, 149, 151, 156, 170
pericardium, 158
permeability, 14, 136, 178
permeation, 139, 150
permission, iv
peroxidation, vii, 13, 14, 24, 88, 163
peroxide, 28, 160, 191, 192
peroxynitrite, 188
personality, 76, 84
pesticides, 2, 76
PET, 182
petroleum, 130
phagocytosis, 158
phenothiazines, 88
phenotype, 61
phenylalanine, 10
phosphates, vii, 65, 74, 76, 78, 80, 81, 83, 84, 85, 86, 87, 88, 90, 132
phosphatidylcholine, 88
phosphatidylserine, 14
phosphocreatine, 76, 77, 78, 87, 90
phospholipids, vii, 17, 45, 79, 81, 84, 85
phosphorous, 129, 130, 131
phosphorus, 33, 35, 48, 73, 74, 103
phosphorylation, 3, 7, 9, 14, 19, 26, 75, 76, 77, 81, 83, 85, 159, 165, 167
photocatalysis, 202
photosynthesis, 74, 187
phylogenetic tree, 3, 5
physical properties, 16, 115
physicochemical properties, 34, 37, 39, 40, 45
physiology, 157, 176, 187
pigments, 158
pigs, 33
placebo, 186
plants, 33, 34, 63, 76, 96, 98, 105, 117, 119, 146
plaque, 162, 166, 170, 176, 181
plasma, 3, 15, 18, 75, 76, 80, 88, 160, 187
plasma membrane, 3, 15, 18, 75, 76, 160
plasmid, 53, 54, 55, 56, 61
plasticity, 16, 20, 21, 23, 29, 159
plexus, 175
pneumonia, 125
poison, 15
polarization, 26
pollutants, 105, 107, 108, 118, 123, 126, 190, 198
pollution, 1, 190
poly(2-hydroxyethyl methacrylate), 121
polymer, 111, 112, 118, 124, 138, 140, 149, 150, 151, 173, 174
polymer chains, 118, 173, 174
polymer matrix, 149

polymerase, 60
polymerization, 112, 173, 187, 198
polymers, 124, 138, 149, 153, 173, 174, 184
polyunsaturated fat, 12
polyunsaturated fatty acids, 12
pools, 15
population, 2, 11, 95, 125, 155, 156
porosity, 146
posture, 170
potassium, 19, 27, 74
potato, 33, 34, 35, 36, 39, 40, 45, 64, 65
precedent, 1
precipitation, 81, 100, 102, 104, 109, 116, 117, 118, 120, 126, 161, 183
preference, 160
prefrontal cortex, 157
pressure, 112, 113, 144
prevention, 2, 11, 104, 156, 185
prions, 168, 170, 171, 175
probability, 7
probe, 44, 49
production, 1, 11, 26, 63, 84, 89, 90, 96, 97, 123, 159, 160, 161, 162, 163, 180, 188, 189, 191
profit, 23
pro-inflammatory, 27
proliferation, 75
promoter, 21, 30
propagation, 13, 170
properties, vii, 2, 17, 18, 20, 23, 31, 34, 36, 37, 39, 40, 41, 42, 43, 44, 45, 60, 65, 66, 68, 69, 80, 98, 115, 119, 124, 125, 126, 133, 134, 164, 167, 183, 185, 188, 205, 209
prophylactic, 13
proposition, 80
prosperity, 22
prostate, 125
protease inhibitors, 171, 172
proteases, 62, 70, 161, 171, 172, 173
protein folding, 176, 184, 185
protein kinase C, 4, 26, 28
protein kinases, 4, 75
protein sequence, 34
protein structure, 180
protein synthesis, 74, 158, 163
proteinase, 166, 170, 171, 172, 183, 184
proteins, vii, 3, 9, 10, 13, 14, 15, 16, 20, 25, 28, 29, 31, 33, 45, 52, 57, 58, 59, 61, 63, 74, 75, 79, 83, 84, 122, 156, 158, 160, 161, 162, 163, 164, 165, 166, 167, 168, 170, 171, 172, 174, 175, 179, 181, 182
proteolysis, 65, 167
proteolytic enzyme, 58
proteomics, 58
protocols, 15
protons, 59, 77, 136
Pseudomonas aeruginosa, 57
psychosis, 84, 88
psychotropic drugs, 88
public health, 1, 12
pulse, 198
pumps, 74, 143
purification, 112, 181
pyramidal cells, 82, 158

Q

quality of life, 1
quanta, 17
quaternary ammonium, 120, 131, 153
quinone, 208

R

race, 20
radial distribution, 16
radiation, 188, 189
Radiation, 210, 211, 217
radical formation, 63, 156, 208
radical mechanism, 199
radical pairs, 208
radical reactions, 187, 188, 191, 195, 197, 209
radicals, 12, 56, 87, 157, 159, 160, 161, 162, 163, 165, 166, 171, 187, 188, 189, 190, 191, 192, 193, 194, 195, 196, 197, 198, 199, 202, 203, 204, 205, 206, 207, 208, 209
radius, 109
rain, 95, 96
Raman spectroscopy, 67, 79
range, 13, 18, 31, 34, 40, 42, 46, 47, 48, 49, 63, 74, 79, 116, 124, 127, 145, 148, 198
reactants, 195, 206, 207
reaction mechanism, 53, 54, 68, 201
reaction rate, 76, 89, 113, 207
reaction time, 111
reactions, vii, viii, 16, 31, 38, 43, 45, 48, 57, 68, 74, 75, 79, 97, 99, 100, 131, 132, 171, 184, 187, 188, 190, 191, 192, 194, 195, 196, 197, 198, 199, 200, 201, 202, 203, 204, 205, 206, 207, 208, 209
reactive oxygen, 33, 160, 162, 175, 188

reactivity, 24, 48, 49, 50, 68, 108, 205
reagents, 124, 129, 130, 131, 132, 133, 189, 194, 205
reality, 134
reason, 2, 75, 76, 85, 87, 109, 111, 114, 116, 156, 165
reasoning, 2, 166
receptors, 18, 19, 21, 26, 29, 74, 159
recognition, 4, 9, 18, 21, 30, 59, 70
recommendations, iv, 182
recovery, 88, 90, 108, 111, 112, 122, 123, 127, 128, 143, 151, 152, 153, 167
recycling, 128
red shift, 55
redistribution, 17
regenerate, 75
regeneration, 139, 153
regenerative capacity, 2
region, 1, 4, 10, 42, 77, 156, 158, 160, 167, 170, 171, 207
regioselectivity, 41
regulation, 4, 25, 64, 185, 188
regulations, 1, 104
rejection, 15
relationship, 53, 126, 174, 175, 176
relevance, 21, 67, 76
remediation, 113, 114, 116, 117, 118, 121, 152, 153
remodelling, 17
renal failure, 126
repair, 20, 179
repetitions, 171
replacement, 34, 47, 67
replication, 80, 170
repression, 64
reproduction, 125
research and development, 128
reserves, 96
residuals, 102
residues, 7, 9, 10, 33, 34, 74, 75, 84, 115, 119, 166, 167, 168, 171, 172, 173, 179, 188
resins, viii, 111, 113, 124, 147, 148, 149, 152, 153
resistance, 71, 90
resolution, 16, 64, 183
resources, 97
respect, 2, 9, 16, 54, 57, 103, 159, 160, 170
respiration, 82, 175, 191
respiratory, 77, 103, 191
responsiveness, 17
retardation, 75, 77, 125
retention, 30
reticulum, 25, 158, 173
ribonucleic acid, 73
ribose, 73, 77
rice, 108, 147
rice husk, 108, 147
risk, 2, 3, 12, 84, 113, 119, 120, 125, 126, 156, 178, 180
risk assessment, 2, 3
RNA, vii, 45, 51, 59, 70, 73, 79, 83, 158
rocks, 95, 96
rodents, 89
rolling, 146
ROOH, 192
room temperature, 58, 99, 109, 132
ruthenium, 68, 69

S

safety, 2, 11, 23, 104, 107, 182
salt, 40, 120, 131, 133, 153, 202
salts, 11, 95, 96, 98, 102, 103, 105, 107, 119, 130, 131, 132, 192, 201, 202
sawdust, 147, 151
scaling, 20
scavengers, 197
schizophrenia, 76
sclerosis, 13, 160
screening, 3, 18
search, 52
secretion, 15, 17, 18, 19
sediment, 121
sedimentation, 98, 99, 105
sediments, 113, 126
seizure, 11
selectivity, 52, 111, 124, 126, 129, 133, 136, 147, 148, 149
self-assembly, 24
semiconductors, 204
senescence, 2, 20
sensing, 81
sensitivity, 17, 104, 126
separation, viii, 40, 43, 102, 104, 108, 120, 121, 126, 129, 131, 132, 136, 143, 150
sequencing, 58
serine, 75, 76, 81, 171, 172, 173
serotonin, 24, 27, 159, 160, 161
serum, 12, 60
serum albumin, 60
severity, 2, 174, 185
sex, 2

shape, 164
shares, 13
sharing, 33
sheep, 168
shock, 148
side effects, 178
signal transduction, 14, 19, 21, 25, 26, 28, 74, 75, 90, 159
signalling, 19
signals, 17, 20, 21, 23, 161
signs, 12, 13
silica, 105, 147, 153
silicon, 125
silkworm, 118, 122
silver, 150
simulation, 24
single-electron transfer reaction, 187
SiO2, 204
skeletal muscle, 75, 81
skeleton, 198
skin, 102, 125
sludge, 104, 109, 126, 147, 151
smooth muscle, 188
sodium, 18, 27, 38, 74, 81, 88, 90, 110, 114, 115, 118, 120, 130
sodium hydroxide, 118
soil, 34, 96, 105, 107, 111, 113, 121
solid matrix, 124, 145, 149
solid phase, 109, 110, 111, 112, 113
solid state, 43, 78, 79, 80, 114, 115
solid waste, 126, 153
solubility, 99, 102, 104, 108, 109, 115, 116, 131, 133, 148, 162, 164
solvation, 26, 129, 131, 132, 136, 137, 195, 209
solvent molecules, 43, 83
solvents, 132, 150, 189, 209
sorption, 108, 109, 113, 117, 120, 145, 151, 152
soybean, 33
space, 14, 18, 75, 157
Spain, 123
speciation, 114, 115, 121
species, 3, 9, 11, 13, 18, 33, 36, 41, 42, 43, 44, 46, 48, 49, 51, 53, 54, 56, 59, 61, 78, 80, 82, 85, 86, 98, 99, 100, 101, 102, 104, 105, 106, 107, 108, 111, 113, 116, 121, 124, 126, 127, 131, 132, 136, 137, 138, 145, 146, 147, 159, 160, 162, 168, 175, 176, 188, 193, 196, 198, 202, 207, 208, 209
spectra analysis, 71
spectroscopy, 35, 43, 48, 67, 79, 90, 115, 193
spectrum, 2, 43, 55, 61, 78, 79, 85, 86
speculation, 12
speech, 84
speed, 156, 158
spin, 41, 45, 48, 63
spinal cord, 11, 14
spleen, 34, 36, 62, 65, 68
stability, 3, 33, 79, 80, 81, 82, 85, 86, 114, 115, 133, 138, 139, 140, 158, 184
stabilization, 53, 167, 180
starch, 31
starvation, 81
steel, 150, 152, 153
stimulus, 15, 17
stock, 96
stoichiometry, 78, 113
stomach, 74
storage, 75
strain, 171
strategies, 11, 53, 61, 75
strategy, vii, 37, 45, 177
streams, 15, 123, 126, 149
strength, 20, 50, 200, 208
stress, 14, 17, 20, 21, 89, 156, 162, 180, 183
stretching, 79
striatum, 27
stroke, 82
strong interaction, 55
structural changes, 79, 164
structural protein, 83
students, vii
styrene, 121, 147, 149
subacute, 11
subcortical nuclei, 174
substitution, 47, 51, 118
substitutions, 7, 9, 10, 49
substrates, vii, 31, 32, 33, 45, 53, 64, 75, 79, 83, 161, 187, 188, 192, 193, 196, 197, 200
sugar, 73, 83, 88
sugarcane, 147
sulfur, 81, 162
sulfuric acid, 118
sulphur, 131
Sun, 89, 94, 185, 186, 216
supply, 96, 97, 98, 99, 102, 103, 109, 175
supported liquid membrane, 124, 137, 138, 139, 140, 150
surface area, 12, 108, 139, 146
surfactant, 109, 119, 120, 124, 137, 150, 152
surveillance, 178
survey, 12, 67

survival, 20, 21, 74, 75, 125
susceptibility, 14, 26, 45, 65
sustainability, 2
swelling, 148
switching, 15
symmetry, 43
symptoms, 13, 26, 61, 76, 81, 84, 156, 165, 166, 177
synapse, 14, 15, 16, 17, 18, 19, 20, 24, 75
synaptic plasticity, 16, 20, 21
synaptic strength, 20
synaptic transmission, 15, 18, 27, 76
synaptic vesicles, 17
syndrome, 181
synthesis, 19, 31, 37, 61, 67, 73, 74, 76, 78, 79, 81, 158, 159, 161, 163, 175
synthetic polymers, 124

T

tandem repeats, 7
tangles, 3, 164, 165, 167, 174, 176, 181, 182
tanks, 98, 104
tantalum, 131
targets, 51, 71, 133, 160, 188
tartrate-resistant acid phosphatase, 62, 63, 64
tau, 3, 5, 7, 9, 10, 25, 26, 29, 158, 182
TBP, 132, 134
TCC, 173
temperature, 55, 56, 58, 60, 99, 102, 108, 109, 118, 132, 138, 174, 185, 189, 203, 204
temporal lobe, 157
terminals, 16, 17, 24, 25, 29
thalamus, 159
therapeutic agents, 177
therapeutic targets, 76
therapeutics, 59
therapy, 59, 61, 90, 177, 178, 185
thermal analysis, 115
thermal stability, 132
thermodynamics, 152
thermolysis, 188, 189
thimerosal, 30
threonine, 75, 81, 171
threshold, 90
thymus, 55
thyroid, 81
tissue, 9, 20, 75, 77, 82, 90, 161, 162, 167, 168, 169, 171, 174
titanium, 152
TNF, 185
topology, 37, 75
toxic effect, 97, 104, 176, 178
toxic metals, 123
toxic products, 123
toxicity, 1, 2, 3, 21, 23, 24, 26, 28, 29, 30, 95, 96, 97, 103, 104, 111, 119, 125, 126, 161, 165, 170, 175, 190
toxicology, 30
TPA, 93
trace elements, 167
trade, 11
transcription, 79, 80
transduction, 14, 19, 21, 25, 26, 28, 74, 75, 90
transferrin, 81, 161
transformation, 2, 23, 57
transgene, 10
transition, 23, 33, 38, 42, 43, 44, 45, 50, 53, 57, 62, 87, 161, 162, 173, 178, 187, 188, 190, 191, 192, 193, 194, 195, 198, 199, 205, 207, 208, 209
transition metal, 23, 87, 161, 162, 187, 188, 190, 191, 194, 195, 198, 199, 205, 209
transition metal ions, 194, 199
transitions, 173, 183
translocation, 19, 184
transmission, 15, 18, 19, 27, 29, 76, 82, 158, 161, 167
transport, 12, 13, 18, 25, 30, 33, 74, 77, 81, 87, 90, 120, 124, 136, 150, 161, 185
transportation, 96
trends, 83, 176
trial, 182
triggers, 18, 20
trypsin, 172
tumors, 61
turnover, 17, 29, 52, 53, 63, 65, 83, 159
tyrosine, 4, 10, 21, 33, 34, 44, 45, 63, 75, 81
tyrosine hydroxylase, 21

U

UK, 25, 91, 179
ulcers, 102
ultrasound, 188, 189
ultrastructure, 16, 26
underlying mechanisms, 19
uniform, 100, 139
United States, 102, 212
uric acid, 81, 161
urinary bladder, 125
UV, 42, 43, 50, 60, 61, 115

V

valence, 38, 40, 41, 42, 43, 45, 46, 47, 55, 66, 67, 107, 208
valine, 168
vanadium, 75, 87, 88, 89, 90
variability, 12, 183
variables, 2, 25, 111
variations, 3, 15, 88, 105, 156, 168
vasodilator, 188
vegetables, 97
vein, 13
velocity, 19, 158
Venezuela, 96
ventricle, 13
ventricles, 157
versatility, 69, 129
vesicle, 15, 16
vessels, 20, 135
viruses, 73
viscosity, 134
vision, 103
vitamin B1, 82
vitamin B12, 82
vitamin E, 156, 161, 178, 179
vitamins, 156
vulnerability, 27

W

waste, vii, 74, 120, 126, 128, 147, 149, 151, 152, 153
waste water, vii
wastewater, 116, 118, 122, 151
water quality, 104, 106, 107
wave number, 79, 86
weak interaction, 55
weakness, 103
wealth, 206
weight loss, 157
welding, 13
wetting, 139
white matter, 157, 158
wood, 146
workers, 13, 38, 41, 49, 54, 55, 56, 57, 59, 61, 210
World Health Organisation, 2
writing, 22

X

X-ray, 36, 37, 38, 40, 44, 48, 65, 70, 180, 193

Z

zeolites, 147
zinc, 21, 32, 64, 65, 83, 84, 117, 120, 130, 151, 161, 163, 167, 175, 176, 178, 185
zirconium, 131